Anjan Kundu

Tsunami and Nonlinear Waves

Anjan Kundu
(Editor)

Tsunami and Nonlinear Waves

With 170 Figures

Springer

PROF. DR. ANJAN KUNDU

Theory Group & Centre
for Applied Mathematics
and Computational Science
Saha Institute of Nuclear Physics
Sector 1, Block AF, Bidhan Nagar
Calcutta 700064
India

e-mail: anjan.kundu@saha.ac.in

Library of Congress Control Number: 2007921989

ISBN-13 978-3-540-71255-8 Springer Berlin Heidelberg New York

Springer is a part of Springer Science+Business Media
springer.com

Cover design: deblik, Berlin
Typesetting: camera-ready by the editor
Production: Christine Adolph
Printing: Krips bv, Meppel
Binding: Stürtz GmbH, Würzburg

Printed on acid-free paper 30/2133/ca 5 4 3 2 1 0

In memory of those died on December 26, 2004
in the Indian Ocean Tsunami

Preface

Unimaginable catastrophe struck the coasts of Indian Ocean in the morning of January 26, 2004, wiping out more than 275,000 human life at a stroke from the face of the earth. It was the killer Tsunami, that originated its journey at the epicenter of the earthquake (of intensity 9.2) near Banda Aceh in Indonesia and traveled as long as to Port Elizabeth in South Africa, covering a distance of more than 8,000 km and bringing unprecedented devastation to the countries like Indonesia, Thailand, Sri Lanka, India and others.

All of us were shocked saddened and felt helpless, wanted to do something in accordance to our own ability. I as a scientist working in India and interested in nonlinear dynamics, soliton and related phenomena, decided to contribute by organizing a dedicated effort by world experts to study different aspects of the Tsunami and other oceanic waves with special emphasis on the nonlinear connection of this problem. Our Centre for Appl. Math. & Comp. Sc. (CAMCS) of our Institute, specially my colleague Prof Bikas Chakrabarti enthusiastically supported the idea and came along with the support of a generous fund.

In contrast to the conventional linear theory of Tsunami, our emphasis on nonlinearity is in part related to my own conviction for its need, especially for describing the near-shore evolution of the waves with varying depth. The other motivation was the realization that, though a large mass of literature is already devoted to Tsunami and related topics, no consolidated collective study has been dedicated to nonlinear aspects of Tsunami and other oceanic waves. This was in spite of the fact that the results obtained through conventional studies are not all convincing and conclusive and in spite of a group of internationally well known experts, as evident from the present volume, have long been emphasizing on the importance of nonlinearity in this regard.

Therefore as a first step we organized an international meeting on the same topic: *Tsunami & Nonlinear Waves* in Saha Institute of Nuclear Physics, Calcutta (March 6-10, 2006). That helped us not only to identify and contact the leading experts in this field, but also to spend a highly beneficial and stimulating week in interacting and exchanging thoughts and experiences with some of them. I am also thankful to the Springer-Verlag for offering to publish this edited volume with interest in their Geo-Science series. This volume is based not only on selected lectures presented in the conference (Caputo (France), Dias (France), Fujima (Japan), Lakshmanan (India), Rao (India), Segur (USA), Shankar (India)), but also on the contributions from other experts well known in the field: Grimshaw (UK), Kharif (France), Madsen (Denmark), Weiss (USA), Yalciner (Turkey), Zakharov (USA) and their collaborators, who could not participate in the conference.

This volume has 14 chapters which I have divided loosely into 2 parts: *Propagation* and *Source & Run up*, for convenience, though many chapters in fact are overlapping. I have also tried to arrange the chapters from more theoretical to more application oriented, though again not in a strict sense. The overall emphasis is on theoretical and mathematical aspects of the oceanic waves, though the authors have given ample introduction to their subjects, starting the material from the beginning before taking the readers to the applicable research level with needed scientific rigor.

Hope this volume will be equally interesting and fruitful to the experts actively working or planning to work in this field, as well as to the common people who got interested in the subject just after 2004 and even to the Government bureaucrats, who are forced now to take interest in such events.

Calcutta, December 2006 *Anjan Kundu*

Contents

Part I Propagation

Part II Source & Run up

List of Contributors

Jean-Guy Caputo
Laboratoire de Mathématiques,
INSA de Rouen,
B.P. 8, 76131 Mont-Saint-Aignan
cedex, France.
&
Laboratoire de Physique théorique
et modelisation,
Université de Cergy-Pontoise and
C.N.R.S.
caputo@insa-rouen.fr

David R. Fuhrman
Technical University of Denmark,
Mechanical Engineering
Department, Nils Koppels Allé,
Building 403, DK-2800
Kgs. Lyngby, Denmark
drf@mek.dtu.dk

Denys Dutykh
Centre de Mathématiques
et de Leurs Applications, Ecole
Normale Supérieure de Cachan,
61 avenue du Président Wilson,
94235 Cachan cedex, France
dutykh@cmla.ens-cachan.fr

Frédéric Dias
Centre de Mathématiques
et de Leurs Applications, Ecole
Normale Supérieure de Cachan,
61 avenue du Président Wilson,
94235 Cachan cedex, France
dias@cmla.ens-cachan.fr

Irina Didenkulova
Institute of Applied Physics, Nizhny
Novgorod, Russia
dii@hydro.appl.sci-nnov.ru

Koji Fujima
Dept. of Civil and Environmental
Eng., National Defense Academy.
1-10-20 Hashirimizu, Yokosuka,
239-8686 Japan.
fujima@nda.ac.jp

Roger Grimshaw
Loughborough University, Loughborough, LE11 3TU, UK
R.H.J.Grimshaw@lboro.ac.uk

H. Karakus, C. Ozer & G. Ozyurt
Department of Civil Engineering,
Middle East Technical University,
Ocean Engineering Research Center,
06531 Ankara, Turkey
khulya@metu.edu.tr,
cozer@metu.edu.tr,
gulizar@metu.edu.tr

Christian Kharif
Institut de Recherche sur les phénomènes Hors Equilibre, Marseille, France
kharif@irphe.univ-mrs.fr

Alexander O. Korotkevich
Landau Institute for Theoretical Physics RAS 2, Kosygin Str., Moscow 119334, Russian Federation
kao@landau.ac.ru

A. Kurkin & A. Zaitsev
Department of Applied Mathematics, Nizhny Novgorod State Technical University, 24 Minin Street, 603950 Nizhny Novgorod, Russia
kurkin@kis.ru,
aizaytsev@mail.ru

M. Lakshmanan
Centre for Nonlinear Dynamics, School of Physics, Bharathidasan University, Tiruchirapalli - 620 024
lakshman@cnld.bdu.ac.in

Per A. Madsen
Technical University of Denmark, Mechanical Engineering Department, Nils Koppels Allé, Building 403, DK-2800 Kgs. Lyngby, Denmark
prm@mek.dtu.dk

Efim Pelinovsky
Institute of Applied Physics, Nizhny Novgorod, Russia
pelinovsky@hydro.appl.sci-nnov.ru

Andrei N. Pushkarev
Lebedev Physical Institute RAS,53, Leninsky Prosp., GSP-1 Moscow, 119991, Russian Federation
Waves and Solitons LLC, 918 W. Windsong Dr., Phoenix, AZ 85045, USA
andrei@cox.net

N. Purnachandra Rao
National Geophysical Research Institute, Hyderabad 500 007, India
raonpc@ngri.res.in

Don Resio
Coastal and Hydraulics Laboratory, U.S. Army Engineer Research and Development Center, Halls Ferry Rd., Vicksburg, MS 39180, USA

Harvey Segur
Department of Applied Mathematics, University of Colorado, Boulder, Colorado, USA
Segur@colorado.edu

R. Shankar
The Institute of Mathematical Sciences, C.I.T Campus, Chennai 600113, INDIA
shankar@imsc.res.in

Tarmo Soomere
Institute of Cybernetics, Tallinn, Estonia
soomere@cs.ioc.ee

Y. A. Stepanyants
Reactor Operations, ANSTO, PMB 1, Menai (Sydney), NSW, 2234, Australia.
Yury.Stepanyants@ansto.gov.au

Robert Weiss
Joint Institute for the Study of the Atmosphere and Ocean, University of Washington–NOAA Center for Tsunami Research, 7600 Sand Point Way NE, Seattle WA 98115, USA
weiszr@u.washington.edu

Kai Wünnemann
Institut für Mineralogie, Museum für Naturkunde,
Humboldt-Universität zu Berlin,
Invalidenstrae 43,
10115 Berlin, Germany
kai.wuennemann@museum.hu-berlin.de

Ahmet C. Yalciner,
Department of Civil Engineering,
Middle East Technical University,
Ocean Engineering Research Center,
06531 Ankara Turkey,
yalciner@metu.edu.tr

Narcisse Zahibo
University of Antilles and Guyane,
Guadeloupe, France
narcisse.zahibo@univ-ag.fr

Vladimir E. Zakharov
Department of Mathematics, University of Arizona, Tucson, AZ 85721, USA
& Lebedev Physical Institute RAS,53, Leninsky Prosp.,
GSP-1 Moscow, 119991, Russian Federation
& Landau Institute for Theoretical Physics RAS 2,
Kosygin Str., Moscow 119334,
Russian Federation
& Waves and Solitons LLC, 918 W. Windsong Dr.,
Phoenix, AZ 85045, USA
zakharov@math.arizona.edu

Part I

Propagation

Waves in shallow water, with emphasis on the tsunami of 2004

Harvey Segur

Department of Applied Mathematics,
University of Colorado, Boulder, Colorado, USA
Segur@colorado.edu

1 Introduction

This conference was organized in response to the 2004 tsunami, which killed nearly 300,000 people in coastal communities around the Indian Ocean. We can expect more tsunamis in the future, so now is a good time to think carefully about how to prepare for the next tsunami. With that objective, this paper addresses three broad questions about tsunamis.
1) *How do tsunamis work?* Is there a simple explanation of the dynamics of tsunamis? What makes them so much more destructive than other ocean waves?
2) Our understanding of the theory of nonlinear waves has advanced significantly in the last forty years because of the development of "soliton theory", which began with the Korteweg-de Vries (KdV) equation, to be discussed below. But Korteweg & de Vries derived their now-famous equation in 1895 to describe approximately the evolution of long waves of moderate amplitude in shallow water of uniform depth. *What does KdV theory tell us about tsunamis in general, and about the 2004 tsunami in particular?*
3) In response to the tsunami of 2004, India and other affected countries have begun plans to implement an early warning system for tsunamis in the Indian Ocean. On logical grounds, it seems that the requirements for such a system should be:

- Reliability – the system must not fail when it is needed;
- Accuracy – people will lose confidence in a system that fails to predict an important tsunami, or that predicts tsunamis that do not materialize;
- Speed – an accurate tsunami-alert issued after the tsunami hits is useless.

What kind of warning system is feasible with today's technology and meets these requirements?

The sections that follow address each of these questions in turn. Section 2, on the basic dynamics of tsunamis, is intentionally written in nontechnical language to make it accesible to as broad an audience as possible. The sections that follow it are more technical, but the entire paper has been written to minimize the technical expertise required by the reader.

2 Basic dynamics of tsunamis and other water waves

Water waves have broad appeal as a scientific topic, because we all have personal experience with water waves – at the beach or in the kitchen sink. In this paper, "water waves" refers to the waves that occur on the free surface of a body of water, under the force of gravity. These include the waves that one commonly sees at the beach, those in the kitchen sink, and tsunamis. Except for very short waves (with wavelengths less than a few millimeters), waves on the ocean's free surface are due to the restoring force of gravity. Other kinds of waves in the ocean, including internal waves, inertial waves and sound waves, are not considered here.

Almost everyone has personal experience with water waves and sound waves. But even for waves of small amplitude, these two kinds of wave systems behave differently. Sound waves have an important property: *All sound waves travel with the same speed*, independent of the frequency or wavelength of the wave. We define "the speed of sound" to be this common speed at which all sound waves travel. If sound waves at different frequencies traveled at different speeds, then human communication by speech would be difficult or impossible.

Unlike sound waves, *water waves with different wavelengths travel with different speeds.* For gravity-induced water waves, longer waves have lower frequencies, and they travel faster. Figure 1 shows a series of snapshots that illustrate this effect, but anyone can carry out a similar experiment. Drop a rock into a quiet pond, and observe the waves patterns created. Longer waves travel faster, so in each snapshot in Figure 1, the waves with longer wavelengths are further away from the center of the pattern (i.e., from the source of the disturbance), while waves with shorter wavelengths are closer to it. As time goes on, more and more waves propagate away from the center, but in each snapshot the longest waves in that snapshot are farthest out, and the shortest waves are closest to the center. This property of water waves is called "wave dispersion".

Long water waves travel faster than short waves, but there is an upper limit. For gravity-induced water waves of small amplitude, the maximum speed of propagation is

$$c = \sqrt{gh} \tag{1}$$

where g represents the acceleration due to gravity (about 9.8 m/sec^2 at sea level), and h is the local water depth, measured from the bottom of the water (at the floor of the ocean, or the bottom of the water tank) up to the quiescent free surface. All surface waves with wavelengths much longer than the local water depth (and with small amplitudes) travel with an approximate speed of $\sqrt{gh}$. Thus these very long waves have a common speed, $\sqrt{gh}$, which acts

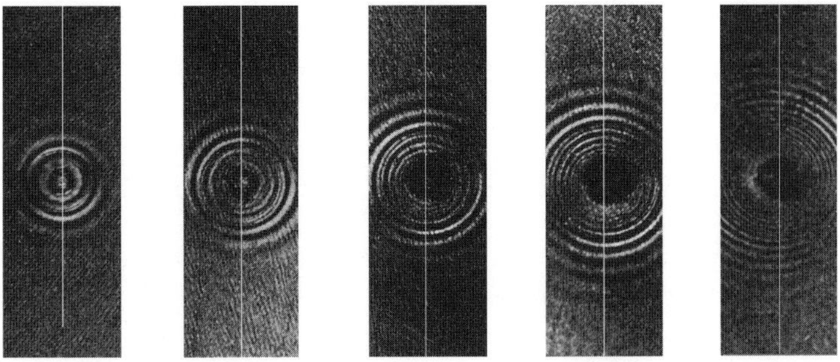

Fig. 1. Concentric water waves, propagating outward from a concentrated disturbance at the center. The longest waves in each snapshot are the furthest from the disturbance, showing that long waves travel faster than short waves for gravity-induced waves on the water's surface. These photos are a subset of the series shown in (15) pp. 172,173, originally taken by J.W. Johnson.

like an approximate "sound speed" for the long waves (and only for them). Because of this special property, we call a water wave a "long wave" if its wavelength is much longer than the local water depth; equivalently, we call a body of water "shallow water" if its depth is much less than the wavelength of the waves in question. Both phrases indicate that the relevant waves all travel with an approximate speed of $\sqrt{gh}$. This criterion is especially pertinent for tsunamis, which have very long wavelengths.

The information in (1), plus a few measurements, is enough to provide some understanding of the basic dynamics of the tsunami of 2004. The earthquake that generated that tsunami changed the shape of the ocean floor, by *raising* the ocean floor to the west of the epicenter, and *lowering* it to the east. The scale of this motion is impressive. Measured wave records indicate that horizontal scale of the piece of seabed raised was about 100 km in the east-west direction, and maybe 900 km in the north-south direction. The piece of lowered seabed had similar scales. In each case, the vertical motion was a few meters. (All of these lengths are crude estimates. The qualitative results are unchanged if one changes any of these estimates by a factor of 2. The numbers quoted here were given by S. Ward, at http://www.es.ucsc.edu/~ward.) Figure 2 shows the initial shape of the 2004 tsunami, according to a computer simulation by K. Satake, of Japan. In Figure 2, the region in red is where the water surface was raised up by the earthquake, while the region in blue is where it was lowered [1] (Go to http://staff.aist.go.jp/kenji.satake/animation.html to see the entire computer simulation of the tsunami that evolved from these initial data. A comparable simulation by S. Ward can be found at http://www.es.ucsc.edu/~ward.)

[1] Editor's note: Due to conversion to b/w the color code is not visible. See the website for colored figure.

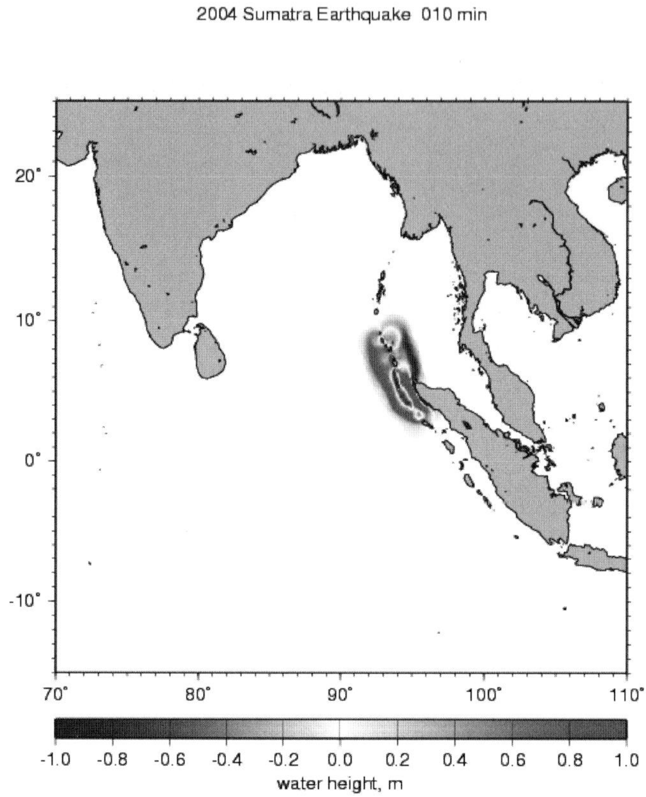

Fig. 2. The shape and intensity of the initial water wave, 10 minutes after the beginning of the earthquake. These initial conditions generated the tsunami simulated by K. Satake, at http://staff.aist.go.jp/kenji.satake/animation.html. As the scale below the figure shows, red indicates a locally elevated water surface, while blue indicates a depressed water surface. (Figure courtesy of K. Satake.)

We also need an estimate of the average ocean depth. According to http://www.infoplease.com/ce6/world/A0825114.html, the average depth of the Indian Ocean is about 3400 m. The average depth in the Bay of Bengal is slightly less, and the earthquake that generated the 2004 tsunami occurred near a sharp change in the ocean depth. Let us take the ocean depth *west* of the epicenter of the quake (i.e., in the red region in Figure 2, or in the Bay of Bengal) to be about 3 km, and the ocean depth east of it (in the blue region of Figure 2, or in the Andaman Sea) to be about 1 km. (See http://www.ngdc.noaa.gov/mgg/image/2minrelief.html.)

Now we can compare scales. The ratio of ocean depth to wavelength was

$$\frac{h}{\lambda} = \frac{3}{100} = 0.03 \ll 1 \tag{2}$$

for the waves traveling west, and even smaller for waves traveling east. So the Indian Ocean was indeed "shallow water" for the 2004 tsunami. Moreover the initial height of the of the wave relative to the ocean depth was

$$\frac{a}{h} = \frac{1}{3000} = 3.3 \cdot 10^{-4} \ll 1 \ , \tag{3}$$

so the tsunami was initially very small, and (1) is applicable. According to (1), the wave traveling west (i.e., towards India) had a speed of about 620 km/hr, while the wave traveling east (towards Thailand) moved slower, at about 350 km/hr. (For comparison, recall that the cruising speed of a commercial jet is about 800 km/hr.)

As we discuss in the next section, an approximate governing equation for such a wave pattern is the linear, one-dimensional wave equation, with a propagation speed of $\sqrt{gh}$. A feature of that equation is that if the initial shape of the wave is given by an east-west slice through the wave pattern shown in Figure 2, with no initial vertical velocity (for simplicity), then this initial shape splits into two – a wave with this spatial pattern and half the amplitude travels east and an identical half travels west. As long as the water depth remains (approximately) constant, these waves travel with almost no change of form. Thus, the coastal regions of India and Sri Lanka should have experienced a positive wave (with water levels higher than normal) followed by a negative wave (with water levels lower than normal), while the coastal regions of Thailand should have experienced the opposite: a negative wave, followed by a positive wave. This is indeed what was reported, and this is what the computer simulation at http://staff.aist.go.jp/kenji.satake/animation.html shows.

How would an observer experience this wave, as it traveled across the ocean? Based on the scales quoted above, the positive wave that traveled to the west (say) is 100 km long, and 1 m. high. That's a lot of water, and the wave is traveling at 620 km/hr. (A delicate point: What travels at 620 km/hr is the rise in water height. The horizontal velocity of the water in the wave is much smaller.) If you were sitting in a boat in the middle of the Indian Ocean, what would you experience? The wave is moving towards you at 620 km/hr, but it's 100 km long, so it takes almost 10 minutes to move past your boat. In the course of 10 minutes, therefore, your boat would move up by about 1 m, and then back down by 1 m. Unless you were extremely sensitive, you probably would not even notice that a wave had gone by. So because of their very long wavelengths, tsunamis are barely noticeable in the open ocean.

As the wave approaches shore, everything changes. The speed of propagation is still $\sqrt{gh}$, but near shore the water depth (h) decreases, and the wave must slow down. More precisely, the front of the wave must slow down. The back of the wave is still 100 km out at sea, so it does not slow down. The consequence is that the back of the wave starts to catch up with the front, and the wave compresses (horizontally) as it moves into shallower water. But water is nearly incompressible, so if the wave compresses horizontally, then it must grow vertically to accommodate the extra water that is piling up. And the volume of water involved is enormous: about 10^5 m^3 of water per meter of shoreline. The deadly result is that a wave that was barely noticeable in the open ocean can become very large and destructive near shore.

Summary: A tsunami is a very long ocean wave, usually generated by a submarine earthquake or landslide. The wave propagates across the ocean with a speed given approximately by (1). From these two facts, it follows that the tsunami is barely noticeable in the open ocean, and the same tsunami can become large and destructive near shore.

3 Theoretical models of long waves in shallow water

The mathematical theory of water waves goes back at least to Stokes (16), who first wrote down the equations for the motion of an incompressible, inviscid fluid, subject to a constant gravitational force, where the fluid was bounded below by a rigid bottom and above by a free surface. (See Figure 3.)

If the motion is irrotational, then the fluid velocity can be written in terms of a velocity potential,

$$\mathbf{u} = \nabla\phi \; ,$$

and the velocity potential satisfies

$$\begin{array}{lr} \nabla^2\phi = 0 & \text{for } -h(x,y) < z < \zeta(x,y,t) \; , \\ \nabla\phi\cdot\nabla(z+h(x,y)) = 0 & \text{on } z = -h(x,y) \; , \\ \partial_t\zeta + \partial_x\zeta\cdot\partial_x\phi + \partial_y\zeta\cdot\partial_y\phi = \partial_z\phi & \text{on } z = \zeta(x,y,t) \; , \\ \partial_t\phi + \frac{1}{2}|\nabla\phi|^2 + g\zeta = 0 & \text{on } z = \zeta(x,y,t) \; . \end{array} \tag{4}$$

The first equation, the Laplace equation, applies where there is water (i.e., between the fixed bottom boundary and the moving free surface). This equation says that the motion is irrotational and that water is incompressible. The second equation says that no water can flow through the bottom boundary. The third and fourth equations both apply on the free surface, at $z = \zeta(x,y,t)$. The third equation says that the surface at $z = \zeta(x,y,t)$ is a free surface, while while fourth equation says that the fluid pressure vanishes there.

These equations have been known for more than 150 years, and they are still too hard to solve in any general sense. The difficulty is not due to the Laplace equation, which can be solved by a variety of methods. The complication arises because an essential part of the problem is to locate the boundary of the domain, at $z = \zeta(x,y,t)$. Until we know where the boundary is, we cannot solve Laplace equation easily. But the boundary moves with time, and its (changing) location is determined by solving two coupled, nonlinear, partial differential equations. So we need the solution of Laplace equation in

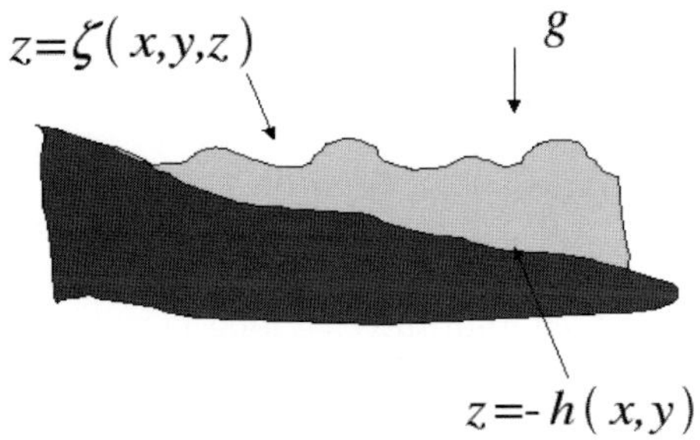

Fig. 3. The equations of water waves apply where there is water, shown in gray. No water passes through the solid lower boundary at $z = -h(x, y)$. The (moving) upper surface is at $z = \zeta(x, y, t)$. Here x, y are horizontal coordinates, z is vertical, $\nabla = (\partial_x, \partial_y, \partial_z)$, and gravity g acts downward. Other possible effects (surface tension, viscosity, wind, fish, *etc.*) are ignored in this formulation.

order to provide the information needed for these coupled equations, and we need information from these two coupled equations in order to solve Laplace equation. That is the basic difficulty.

Because of this intrinsic difficulty, most advances in the theory of water waves have come through approximations. In this approach, we abandon hope for solving the equations in (4) in any general sense, and concentrate instead on solving the equations approximately in some limiting situation, where the motion simplifies. The limit relevant to tsunamis is that of long waves of small or moderate amplitude, propagating in nearly one direction in wave of uniform (and shallow) depth. In 1895, D. J. Korteweg & G. de Vries derived what we now call the Kortewg-deVries (or KdV) equation,

$$\partial_\tau f + f\partial_\xi f + \partial_\xi^3 f = 0 \ , \tag{5}$$

to describe wave motion in this limit. This limit has attracted a lot of attention, so there are several nearby equations, all of which are aimed at approximately the same limiting situation: (i) long waves, (ii) of small or moderate amplitude, (iii) traveling in one direction or nearly so, (iv) in water of uniform, shallow depth, and (v) neglecting dissipation. Alternatives to (5) that have been studied extensively in recent years are the equation of Kadomtsev & Petviashvili (KP, 1970),

$$\partial_\xi\{\partial_\tau f + f\partial_\xi f + \partial_\xi^3 f\} + \partial_\eta^2 f = 0 \ , \tag{6}$$

an equation due to Boussinesq (1871),

$$\partial_\tau^2 f = c^2\{\partial_\xi^2 f + \partial_\xi^2(f^2) + \partial_\xi^4 f\} \ , \tag{7}$$

and the equation of Camassa & Holm (1993),

$$\partial_\tau m + c\partial_\xi f + f\partial_\xi m + 2m\partial_\xi f + \gamma\partial_\xi^3 f = 0 \ , \tag{8}$$

where

$$m = f - \alpha^2 \partial_\xi^2 f \ ,$$

and $\{c, \alpha, \gamma\}$ are constants. These four equations share two unrelated properties: (i) each can be derived as an approximate model of the evolution of long waves of moderate amplitude, propagating in nearly one direction in shallow water of uniform depth; and (ii) each has a rich mathematical structure, called complete integrability, which guarantees a long list of other properties including exact N-soliton solutions. (For details see (1), or any decent reference on soliton theory.)

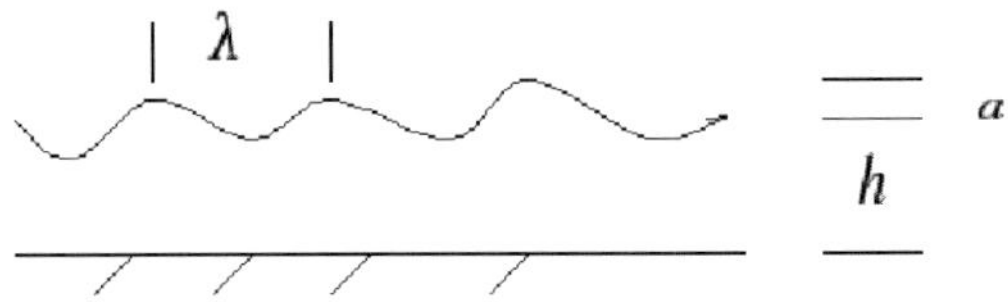

Fig. 4. Relevant length scales, needed to derive either the KdV or KP equation: h is the time-averaged water height, a is a typical wave amplitude, and λ is a typical wavelength in the direction of propagation of the waves. The KdV equation allows for no variation normal to the direction of propagation of the waves. The KP equation requires that the scale of variations in this normal direction (i.e., coming out of the page in this figure) be much longer than λ.

The limit in which any of these equations apply can be stated in terms of length scales, which must be arranged in a certain order. Three of the four relevant lengths are shown in Figure 4. The derivation of either the KdV or KP equation from (4) is based on four assumptions:

- Long waves (or shallow water) $h \ll \lambda$
- Small amplitude $a \ll h$
- The waves move primarily in one direction
  If this is exactly true, it leads to the KdV equation, (4)
  If it is approximately true, it can lead to the KP equation, (5)
- All these small effects are comparable in size. For KdV, this means

$$\varepsilon = \frac{a}{h} = O\left(\left(\frac{h}{\lambda}\right)^2\right) \tag{9}$$

One imposes these assumptions self-consistently on both the velocity potential, $\phi(x, y, z, t)$, and the location of the free surface, $\zeta(x, y, t)$, in (4). (See §4.1 of (1), for details.) At leading order ($\varepsilon = 0$ in a formal expansion), the waves in question are infinitely long, infinitesimally small, and the motion at the free surface is exactly one-dimensional. The result (if we also take $h =$ constant) is the one-dimensional wave equation :

$$\partial_t^2 \zeta = c^2 \partial_x^2 \zeta, \text{ with } c^2 = gh \ . \tag{10}$$

The general solution of (10) is known. Inserting it back into the expansion for $\zeta(x,y,t;\varepsilon)$ yields

$$\zeta(x,y,t;\varepsilon) = \varepsilon h[F(x-ct;y,\varepsilon t) + G(x+ct;y,\varepsilon t)] + O(\varepsilon^2) \ , \qquad (11)$$

where F and G are arbitrary functions, determined from given initial data. In words, (11) says that a signal (F) propagates to the right, and another signal (G) propagates to the left, both with speed $\sqrt{gh}$, as predicted by (1). Neither signal changes shape as it propagates, at this order.

This partial result already provides useful information about the tsunami of 2004. Let x represent east-west distance from the epicenter of the earthquake, with x increasing to the east. Then F represents the wave that propagated towards Thailand, while G represents the wave that propagated towards India. The initial shape of F and G were determined by the initial data from the earthquake, shown approximately in Figure 2. The G-wave, which propagated towards India, traveled approximately 1500 km cross the Bay of Bengal in slightly over 2 hours. The west-going G-wave wave had a positive region (i.e., extra water) in front, with a negative region (a deficit of water) behind. As it propagated across the Bay of Bengal, this shape remained approximately constant according to (11), and as shown in the animation at http://staff.aist.go.jp/kenji.satake/animation.html. As discussed in Section 2, the wave changed its shape entirely when it entered the shallow coastal regions where h changes, and where the assumptions in (9) break down. Even so, the wave that entered India's coastal region had a positive wave (i.e., with extra water) in front, with a negative wave (with a deficit of water) behind. This is what inundated regions of India experienced. The F-wave, which propagated towards Thailand, was moving slower, but the distance across the Andaman Sea was also smaller. It took 1-2 hours to reach land in Thailand. As it traveled, it had a negative wave in front, followed by a positive wave behind. This is consistent with what inundated regions of Thailand experienced.

Now return to the derivation of (5), the KdV equation, from (4). We may follow (for example) the F-wave by changing to a coordinate system that moves with the F-wave, at speed $\sqrt{gh}$. Set

$$\xi = \frac{\sqrt{\varepsilon}}{h}(x - t\sqrt{gh}) \ . \qquad (12)$$

At leading order, according to (10), F does not change in this coordinate system, so we may proceed to the next order, $O(\varepsilon^2)$. Now the small effects that were ignored at leading order (i.e., that the wave amplitude is small but not infinitesimal, that the wave length is long but not infinitely long, and that slow transverse variations are allowed) can be observed. Each effect is small, but over a long distance these small effects can build up, to have a significant cumulative effect on F. To capture this slow evolution of F, we introduce a slow time-scale,

$$\tau = \varepsilon t \sqrt{\frac{\varepsilon g}{h}} \ , \qquad (13)$$

and find that F satisfies approximately the KdV equation,

$$2\partial_\tau F + 3F\partial_\xi F + \frac{1}{3}\partial_\xi^3 F = 0 \ , \tag{14}$$

if the surface waves are strictly one-dimensional, or the KP equation,

$$\partial_\xi(2\partial_\tau F + 3F\partial_\xi F + \frac{1}{3}\partial_\xi^3 F) + \partial_\eta^2 F = 0 \tag{15}$$

if the surface patterns are weakly two-dimensional. After rescaling the variables in (14) or (15) to absorb constants, (14) becomes (5), and (15) becomes (6).

In words, (10) & (11) say that on a short time-scale, the right-going wave does not change (so $\partial_\tau F = 0$) in the coordinate system given by (12,13). On a longer time-scale, the KdV equation (14) describes how F changes slowly, due to weak nonlinearity ($F\partial_\xi F$) and weak dispersion ($\partial_\xi^3 F$). Alternatively, the KP equation (15) allows F to change because of these two weak effects and also because of weak two-dimensionality ($\partial_\eta^2 F$).

The KdV and KP equations have been derived in many physical con texts, and they always have the same physical meaning: on a short time-scale, the leading-order equation is the one-dimensional, linear wave equation; on a longer time scale, each of the two free waves that make up the solution of the 1-D wave equation satisfies its own KdV (or KP) equation, so each of the two waves changes slowly because of the cumulative effect of weak nonlinearity, weak dispersion and (for KP) weak two-dimensionality.

How does this theory apply to the tsunami of 2004? For the wave that propagated towards India and Sri Lanka, the two parameters required to be small are (from (2))

$$\frac{a}{h} = \frac{1}{3000} = 3 \cdot 10^{-4}, \qquad \left(\frac{h}{\lambda}\right)^2 = \left(\frac{3}{100}\right)^2 = 9 \cdot 10^{-4} \ .$$

Both numbers are much smaller than 1, and they are comparable to each other. In addition, the length scale of the affected seabed in the north-south direction (900 km) was significantly longer than the wavelength in the east-west direction (100 km), so the initial wave propagation was approximately one-dimensional. At leading order, therefore, the 2004 tsunami is a good candidate for KdV theory.

But a problem arises at the next order. The KdV equation describes approximately the dynamics of the propagating wave on a slow time-scale. One can see from (12,13) that the time required to see KdV dynamics is longer by a factor of about $\left(\frac{1}{\varepsilon}\right)$ than a typical time scale for (10). Equivalently, the propagation distance required to see KdV dynamics is approximately $\left(\frac{1}{\varepsilon}\right)$ longer than a typical length scale of the problem. The scaling above uses the water depth, h, as the fundamental length-scale, so the distance required for the westward propagating wave (which struck India and Sri Lanka) to show KdV dynamics was about

$$D \sim \frac{h}{\varepsilon} = 3 \cdot 3000 \sim 10^4 \ km \ .$$

But the distance across the Bay of Bengal is nowhere much more than 1500 km, much too short for KdV dynamics to develop. For the eastward propagating wave (which struck Thailand) the wave speed is slower, but the maximal distances are also smaller, and the conclusion is the same. *For the 2004 tsunami, the propagation distances from the epicenter of the earthquake to India, Sri Lanka, or Thailand were much too short for KdV dynamics to develop.*

This conclusion applies to the 2004 tsunami, and probably to any future tsunami generated in same geological fault region (near Sumatra, and where the tectonic plate that contains India is subducting beneath the plate that contains Burma). Even so, during this conference Prof. M. Lakshmanan observed correctly that the conclusion does not apply to all tsunamis. He pointed out that the 1960 Chilean earthquake, the largest earthquake ever recorded (magnitude 9.6 on a Richter scale), produced a tsunami that propagated across the Pacific Ocean. It reached Hawaii after 15 hours, Japan after 22 hours, and it caused massive destruction in both places. This tsunami propagated over a long enough distance that KdV dynamics were probably relevant. For more information about this earthquake and its tsunami, see (13), (14), or http://neic.usgs.gov/neis/eq_depot/world/1960_05_22_tsunami.html.

Why does it matter whether KdV dynamics apply to tsunamis? One appealing feature of integrable equations, like those in (5)-(8), is that they are nonlinear partial differential equations that can be solved exactly, as initial-value problems. (See (1), for details.) For the KdV equation, (14), starting with arbitrary initial data that are smooth and sufficiently localized in space, the solution that evolves from these data evolves into a finite number of discrete, localized, positive waves (called solitons), plus an oscillatory tail. Each soliton retains its localized identity forever, while the oscillatory tail disperses and spreads out in space. All solitons travel slightly faster than $\sqrt{gh}$, and taller solitons travel faster than shorter ones. The oscillatory tail travels slightly slower than $\sqrt{gh}$, so after a long time the solution evolves into an ordered set of solitons, with the tallest in front, followed by an oscillatory tail. The details of this general picture can be predicted fairly easily from detailed knowledge of the initial data.

In the early 1970s, Joe Hammack carried out a series of laboratory experiments on the dynamics of long waves in shallow water, and Hammack & Segur (1974,1978a,b) used his data to test the predictions of KdV theory. The motivation for their work was closely related to the motivation for this conference: *Can KdV theory be used effectively to predict tsunamis?* One of their conclusions was that KdV dynamics do not occur unless the propagation distance is long enough, as discussed above. A second major conclusion was the importance of the *wave volume* in the initial data, which we discuss next.

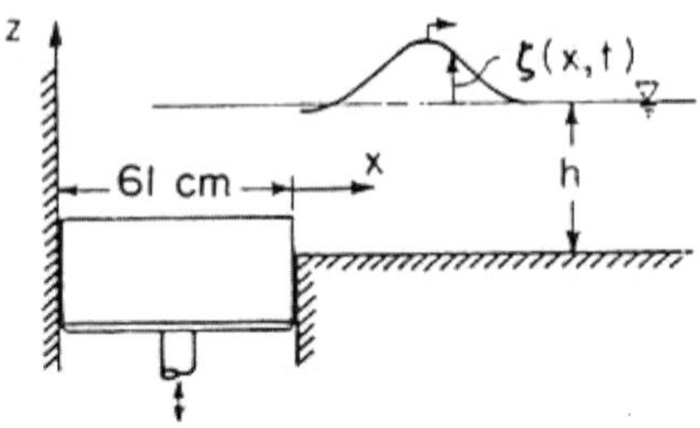

Fig. 5. Schematic diagram of wave maker, used by J.L. Hammack to create waves in shallow water. See (6) or (7) for more details.

Hammack's experiments were carried out in a long wave tank. At one end of the tank was a piston that spanned the width of the tank, as shown in Figure 5 and as discussed in detail in (6). The piston was programmed to move up or down in a controlled way, and its vertical motion was intended to approximate the motion of the ocean floor during a submarine earthquake. If the piston moved up (or down) quickly enough, then the water surface above the piston moved up (or down) with it, after which this positive (or negative) surface wave propagated from one end of the tank to the other. Measuring probes, positioned at either four or five separate locations along the tank, measured the shape of the wave as it propagated the length of the tank.

The results of one set of experiments are shown in Figure 6. Each column in this figure provides information about one of the three experiments in this set. The top picture in each column shows the time-history of the paddle motion for that experiment – in the first experiment the paddle was raised smoothly from one elevation to another. The piston motion was fast enough that the water above it simply rose along with it, so the shape of the wave observed at the first measuring station (at $x/h = 0$ in the first column) is closely related to the shape of the paddle – approximately a rectangular box. Moving down the first column, the next picture shows the wave observed at $x/h = 20$, in a coordinate system moving with speed $\sqrt{gh}$. Equation (10) predicts that the wave does not change, provided we travel with speed $\sqrt{gh}$, and little or no change is observed over this short distance. Over long distances, KdV theory predicts that this initially positive wave should evolve into four solitons, ordered in size, and four solitons are observed at $x/h = 400$. [Within each wave record the wave is propagating to the *left*, so at $x/h = 180$ or at $x/h = 400$ the tallest soliton is out in front, as KdV theory predicts.] In this experiment, the oscillatory tail is very small and barely visible.

Anyone who has experienced a serious earthquake knows that the monotonically rising piston motion shown in the first column is too simple to describe ground motion during an actual earthquake. So the experiments in the second and third columns had the same mean piston motion as that in the

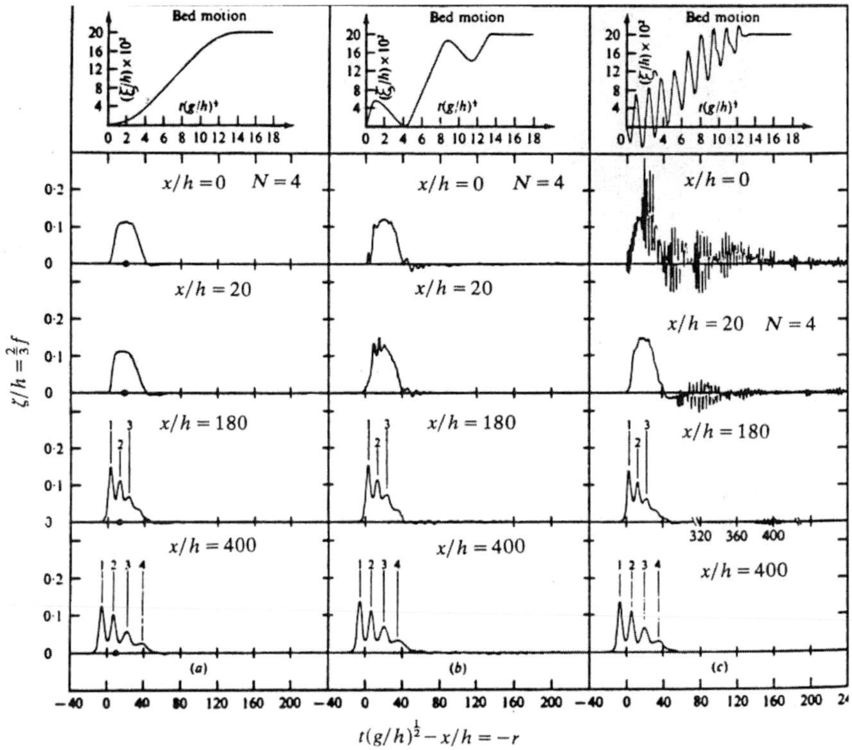

Fig. 6. A set of three experiments, each with net-upward piston motion. The top figure in each column shows the piston height as a function of time. The four wave records beneath it show the measured height of the water wave generated by this piston motion, as it passed four measuring locations along the tank. In the coordinates used here, the wave in each record should be interpreted as a wave moving to the left. From (7).

first column, but with extra complications. The experiment summarized in the second column had a somewhat more complicated piston motion, and the wave observed at $x/h = 0$ is slightly more complicated than that observed in the first experiment. At $x/h = 20$, this slightly more complicated initial wave had started to change its shape, more than that in the first experiment. By $x/h = 400$, however, the leading wave is almost identical to that in the first experiment.

Looking closely at the regions behind (i.e., to the right of) the lead waves in these two experiments, one sees more trailing small oscillations in the second experiment. These extra waves carry the energy from the extra complications in the piston motion of the second experiment. This effect is even more pronounced in the third experiment.

The piston motion in the third experiment had the same mean motion as that in the first experiment, but its overall piston motion was much more complicated. As a result, the wave measured at $x/h = 0$ was quite a mess, and Joe Hammack reported that water was splashing completely out of the tank at the beginning of this experiment. But the extra complications in the piston motion can be viewed as higher frequency motion superimposed on the basic piston motion. The higher frequency piston motion generated higher frequency water waves, with shorter wavelengths, and these travel slower than $\sqrt{gh}$. Already by $x/h = 20$ in the third experiment, these additional high-frequency waves have started to drop behind (i.e., to the right of) the leading wave. By $x/h = 400$, the leading wave in each of the three experiments look nearly identical.

In these three experiments, it is clear that the mean piston motion determined the details of the leading wave. If we view these waves as possible models of tsunamis, it is clear that the leading wave would cause the most damage. (9) posed the question: What parameter (or set of parameters) from the initial wave record (at $x/h = 0$) provides the crucial information about the leading wave at $x/h = 400$? Their analysis identified the *wave volume* (i.e., the area under the curve at $x/h = 0$) as an important quantity for predicting the final state of the wave train, and an estimate of the time scale on which KdV dynamics become important.

All of the experiments in Figure 6 were initiated by upward (mean) piston motion, which produced initial wave shapes that were mostly positive, and led to (positive) solitons. But an earthquake can raise, lower, or leave alone the elevation of the ocean floor Figure 2 shows that the earthquake that generated the 2004 tsunami raised the ocean floor west of the epicenter of the quake and lowered it in the east. Joe Hammack did other experiments to see what kinds of waves evolved from other kinds of seabed motion. Figure 7 shows the waves generated by quickly lowering the piston. This experiment can be viewed as comparable to the first experiment in Figure 6, but turned upside down, and the wave measured at $x/h = 0$ here is approximately the shape of the wave at $x/h = 0$ in the first column of Figure 6, but turned upside down. If the wave evolution were linear, then all of the waves measured in Figure 7 should look like the wave measured at the same location in the first column of Figure 6, but upside down. But the wave evolution in Figure 7 is quite different from that in Figure 6, showing the importance of nonlinearity in the waves dynamics. KdV theory predicts that a purely negative initial wave, like that in Figure 7(a), generates no solitons, so all of the wave energy must go into the oscillatory tail. Figure 7 shows a typical oscillatory tail. [As in Figure 6, these wave records were taken in a coordinate system moving with speed $\sqrt{gh}$, and the wave in each record is moving to the *left*.] Figure 7 shows clearly wave dispersion – the longest waves are in front (i.e., to the left), while waves with shorter wavelength fall further behind (to the right). Waves of each wavelength travel with their own group velocity (as shown by the arrows), so

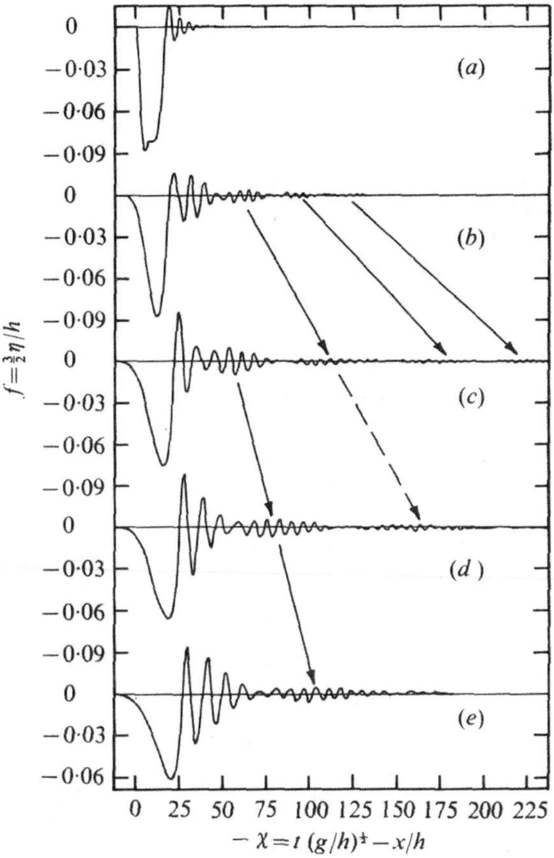

Fig. 7. An experiment with downward piston motion. The five wave records show the measured height of the water wave, as it passed five measuring locations along the tank: (a) $x/h = 0$; (b) $x/h = 50$; (c) $x/h = 100$; (d) $x/h = 150$; (e) $x/h = 200$. From (8).

the entire wave train spreads out in space. Then energy conservation forces the wave amplitudes to decrease as time goes on.

The largest waves, which would be the most destructive if this were an actual tsunami, lie at the front (i.e., to the left) of the wave train, so we should focus our attention on that region of the wavetrain. The first (or leftmost) wave is negative (i.e., with a lowered water level) because the initial wave is negative, and this first wave carries the entire wave volume. As the wave train evolves, the first wave becomes more nearly triangular in shape, and it keeps the same volume. Immediately behind the first wave, a sequence of steep oscillatory waves form, and Figure 7 shows that the number of large, steep oscillatory waves increases slowly as the wave propagates over longer and longer distances.

It is important to keep in mind that this wave evolution is well described by the KdV equation, and it is not necessarily like what occurred the Indian Ocean in December, 2004. Even so, some qualitative features of these waves are quite similar to what was reported at various locations in 2004.

Figure 2 shows that the wave that propagated *east* from the epicenter of the earthquake in 2004 had the shape of a negative wave followed by a positive wave. So the leading wave was negative, perhaps like that shown in Figure 7. Imagine how a small fishing village might have experienced a wave pattern like this as the wave came ashore. The first thing that would have happened in this village was that the water level began to drop. Then it continued to drop, to levels lower than anyone in this village had ever seen. Parts of the coast that had been underwater as long as anyone could remember were exposed for the first time, and people rushed to the beach to see this marvel. Then at some point the water level stopped dropping, and an enormous, very steep wave rushed in and killed almost all of the people on the beach at that time. At various locations, people described being inundated by one, or two, or three of these large, very steep waves after the initial negative wave. Figure 7 suggests that these apparently conflicting stories might all be correct. At location (c) in Figure 7, there is one large, steep wave following the initial negative wave. At location (e) there are at least two. The number of very steep waves following the initial negative wave grows, as the wave propagates over longer and longer distances.

The waves in Figure 7 might also be relevant for the coastal regions of India and Sri Lanka. For those places, the first wave that reached shore was positive, and it flooded entire coastal areas near shore. Then the wave equation in (10) and the initial data in Figure 2 predict that the positive wave should have been followed by a large negative wave, which might have evolved in a way similar to that shown in Figure 7.

The experiments in Figures 6 and 7 demonstrate the importance of the *wave volume* in predicting the evolution of long waves in the KdV regime. They show that the wave volume is especially important in determining the nature of the leading waves, which are often the largest waves, and the most damaging. The experiments suggest but do not prove that the wave volume might also be important in predicting the evolution of tsunamis, even outside the KdV regime. But one should interpret “wave volume” appropriately. For initial data like those in Figure 2, relevant for the 2004 tsunami, the water wave generated by the earthquake was initially positive (i.e., red) to the west of the epicenter of the quake, and negative (blue) to the east of the epicenter. In this situation, one should not add the positive volume from the western portion to the negative volume from the eastern portion and conclude zero volume overall. Instead, the positive wave to the west and the negative wave to the east were far enough apart that they never interacted during this tsunami, so one should consider them as two separate waves, and measure the (positive or negative) volume of each. In 2004, the two waves were each destructive, with no mitigating cancellation.

Finally, let us summarize this section. The Korteweg-de Vries equation, (5), does **not** apply to the 2004 tsunami, because the distance across either the Bay of Bengal (for the westward-propagating wave) or the Andaman Sea (for the eastward-propagating wave) is not long enough for the small effects that control KdV dynamics to build up. This conclusion applies as well to the models in (6), (7) or (8). A realistic description of the evolution the 2004 tsunami as it it propagated across the Indian Ocean is the following.

- Tsunamis are caused by submarine earthquakes or landslides, which provide the initial disturbance in the height of the oceans surface. For short times after that initial disturbance, the motion of the tsunami is governed by a linear wave equation, like that in (10). (A more realistic model would be a linear wave equation with variable water depth, as discussed below in Section 4.) After the earthquake creates the disturbance, the linear wave equation splits that disturbance into two sets of waves, propagating in different directions.
- A linear wave-equation model like (10) can break down for two quite different reasons.
  - The wave equation applies over short distances, but over long distances small effects can build up and create cumulative effects. The KdV equation, (14), describes this kind of cumulative effect.
  - The wave equation applies as long as the assumptions in (9) are satisfied. When one or both of these assumptions breaks down, then the linear wave equation is no longer the correct equation.
- For the tsunami of 2004, the propagation distance across the Indian Ocean was too short for KdV dynamics to become relevant. A linear wave-equation model described the propagation of the tsunami clear across the Indian Ocean, until it entered shallower coastal regions. As discussed in Section 2, as h decreases in coastal regions, wavelengths become shorter, wave amplitudes become larger, so (9) fails. How the wave changes its shape in the region near shore is important, but it is not described by either (10) or (14).
- Laboratory experiments in the KdV regime show the importance of the wave volume in determining how the wave evolves. The wave volume determines the time-scale over which the linear wave-equation model applies, and it also determines the shape of the leading wave as it evolves. The leading waves of a wave train are often the most destructive, so measuring the wave volume of the initial wave provides crucial information for wave evolution, in the KdV regime. This argument suggests but does not prove that the wave volume would also be an important quantity for tsunamis like that in 2004, which lay outside the KdV regime.

4 How well can we predict tsunamis?

After the destructive tsunami of 2004, several governments around the Indian Ocean began planning an early warning system for tsunamis in the Indian Ocean. (See http://ioc3.unesco.org/indotsunami/) The system would be comparable to the Tsunami Warning System that has operated in the Pacific Ocean for more than forty years. This final section considers some of the questions that might be important in designing such a system. *Where in the Indian Ocean are tsunamis likely to originate? What information about the origin of a tsunami is essential for predicting its propagation and evolution? What kind of theoretical model is effective in predicting the propagation of a tsunami in the open ocean? As discussed above, tsunamis behave one way away from shore, and quite differently near shore. What kind of theoretical model is effective for predicting the dynamics of a tsunami near shore? How should the information be disseminated?*

In designing an early warning system, it is important to realize that the warnings issued must be reliable and accurate, and also that the warnings must be issued early enough to be effective. These two objectives (reliable accuracy *vs.* speed) can compete with each other. An important part of the design of a warning system is to decide when to sacrifice speed for the sake of greater accuracy, or to accept a less accurate prediction that can be obtained sooner.

Next we consider separately three pieces of this warning system, dealing with the source of the tsunami, its propagation in the open ocean, and its propagation near shore.

4.1 The source of the tsunami

Consider first the source of the disturbance that generates a tsunami. As discussed above, a tsunami is a very long wavelength wave, generated by an underwater earthquake or landslide. Tsunamis should be distinguished from other very long waves, including tides, and storm surges that often accompany hurricanes or tropical cyclones. The storm surge that struck New Orleans (in the US) in August 2005 demonstrated how destructive a storm surge can be, but both storm surges and tides can be predicted by other means. We concentrate here on underwater earthquakes, including the one that generated the tsunami of 2004.

Almost all large earthquakes occur at the boundary of tectonic plates, where one plate is sliding over, under, or past another. Seismologists classify earthquakes into three types of faults, depending on how the relative motion of adjacent plates affects the shape of the solid earth. (See http://www.abag.ca.gov/bayarea/eqmaps/fixit/ch2/sld001.htm for more information about many of the assertions made in this subsection. And see articles in a special issue of *Science*, **308**, pp. 1125-1146, 2005 for detailed information about the earthquake that generated the tsunami of 2004.)

- In a *thrust fault*, one tectonic plate moves up and over an adjacent plate.
- In a *normal fault*, one plate moves down relative to an adjacent plate.
- In a *strike-slip fault*, two plates slide past each other horizontally, with neither plate being raised or lowered significantly.
- *Reverse normal faults* and *thrust faults* are closely related. We do not distinguish between them here.

In terms of tsunami generation, a thrust fault raises the floor of the ocean, which in turn raises the water above it, and creates a positive water wave (with extra water in the region above the fault). A normal fault lowers the floor of the ocean, so it creates a negative water wave above it (with a "hole" in the water surface above the fault). Strike-slip faults do not change the shape of the ocean floor, and they do not generate tsunamis. Thus, the magnitude of an earthquake (in terms of a reading on a Richter scale) by itself does not determine whether that earthquake will generate a significant tsunami – the kind of fault is also important. The earthquake of December 2004 was a combination of thrust fault and normal fault, and both parts contributed to the tsunami.

Knowing that large earthquakes occur at the boundaries of the earth's tectonic plates simplifies immensely the problem of locating possible sources of tsunamis, because the approximate shape and location of the earth's tectonic plates are known. (For a map, see http://www.abag.ca.gov/bayarea/eqmaps/fixit/ch2/sld005.htm.) In fact, the problem is even simpler, because the history of any particular earthquake zone shows what kind of faults occur there. For example, the most famous earthquake fault line in the US is the San Andreas fault line. Its earthquakes are invariably strike-slip faults, so even if this fault line were under water, it would not generate tsunamis. In contrast, the earthquake of December 2004 occurred where the tectonic plate that contains India is sliding under the plate that contains Burma (Myanmar). Earthquakes along this fault line often occur as thrust faults, as normal faults or as combinations of these, so we can expect more tsunamis from future earthquakes along this boundary.

Once a submarine earthquake occurs, essential information for a tsunami warning system includes: *When did the earthquake occur? Where did it occur? What information about the details of the fault are available?* (As discussed in Section 3, the *wave volume* is an important piece of this detailed information.) The earthquake generates a variety of seismic waves that travel through the solid earth. Among these seismic waves, different kinds of waves travel at different speeds. By identifying when each kind of seismic wave reached various measuring stations around the world, seismologists can deduce when an earthquake occurred, where its epicenter was (on the surface of the earth), and how deep below the surface was its hypocenter. This information can be obtained and processed within a few minutes of the earthquake, and it provides accurate information about *when* and *where* the earthquake occurred.

In this discussion, please keep in mind the relevant times for the 2004 tsunami. It reached Bandeh Aceh a few minutes after the earthquake began, it reached parts of Thailand after about an hour, and it reached most of the eastern coast of India in about two hours. An effective warning system must be able to provide sensible warnings faster than these times to be useful.

For tsunami prediction, the most important questions about an earthquake are: { *When? Where? What wave volume?*}. The time required to answer these questions can be reduced by installing measuring devices near known fault lines in the Indian Ocean. The earthquakes of most interest are strong, and a very strong earthquake will probably destroy some of these instruments. Even so, if enough instruments have been installed and if they are spaced appropriately, then some will survive, and the remaining instruments can provide information about the details of the quake sooner than other instruments further away. Spending extra money here, to install good instrumentation closer to the epicenter of the quake, could save many lives.

4.2 Tsunami propagation in the open ocean

As discussed in Section 3, the propagation of a tsunami far from shore is described approximately by a linear wave equation. If we relax two assumptions: (i) that the ocean depth, $h(x, y)$, is constant and (ii) that the surface motion is one-dimensional (so the fluid motion is two-dimensional), then (10) is replaced by

$$\begin{aligned} &\partial_t \zeta + \partial_x (u \cdot h) + \partial_y (v \cdot h) = 0 \ , \\ &\partial_t (u \cdot h) + c^2 \partial_x \zeta = 0 \ , \\ &\partial_t (v \cdot h) + c^2 \partial_y \zeta = 0 \ , \end{aligned} \qquad (16)$$

where

$$c^2 = g \cdot h(x, y) \ ,$$

x, y are orthogonal horizontal coordinates, $u(x, y, t), v(x, y, t)$ are the components of horizontal velocity in the x- and y-directions, g represents gravity, and $\zeta(x, y, t)$ is the height of the free surface above the still-water level. [If the surface motion happens to be one-dimensional, then set $v = 0$ in (16a), and ignore (16c).] Equations (16) can be combined into a single, equivalent equation:

$$\partial_t^2 \zeta = \partial_x (g \cdot h(x, y) \partial_x \zeta) + \partial_y (g \cdot h(x, y) \partial_y \zeta) \qquad (17)$$

This is the two-dimensional, linear wave equation, with a spatially variable speed of propagation. [It describes fluid motion in three dimensions, but there is no appreciable vertical motion for these very long waves.] Once $h(x, y)$ is known, then (17) determines $\zeta(x, y, t)$ in terms of given initial conditions, $\zeta(x, y, 0), \partial_t \zeta(x, y, 0)$ and boundary conditions. [For an isolated earthquake, sensible boundary conditions would be $u = 0, v = 0, h = 0$ far from the location of the earthquake.] Equation (17) can be solved numerically by a variety of means, including Clawpack (http://www.amath.washington.edu/~claw/).

But independent of any numerical procedure, some conclusions can be drawn directly from the structure of these equations.

One important conclusion is that (16) conserves the total wave volume of the initial conditions. This can be seen by integrating (16a) over space, and using the boundary information that u, v, h all vanish at large distances:

$$\frac{d}{dt} \int \int \zeta(x, y, t) \cdot dx dy = 0 . \tag{18}$$

Thus, the total wave volume is a constant of the motion, so it is important to obtain a good estimate of it from initial seismic data. [This point deserves some emphasis. Equation (18) shows that the total wave volume is conserved by (16). Independently, the measured wave records in Figures 6 and 7 show that the wave volume determines the shape of the leading part of the wave, regardless of other details of the initial data, at least for waves that propagate in the KdV regime. These two facts, taken together, emphasize the importance of determining the wave volume from the initial seismic readings, even if other details of the initial shape of the wave are not known as well.]

A second important conclusion is that any solution of (17) propagates with (variable) local speed

$$c = \sqrt{g \cdot h(x, y)} . \tag{19}$$

Without even solving (15), one can determine the time required for a wave that starts at the epicenter of the earthquake to reach any particular coastal region. Simply draw a curve from the given starting point to the identified coastal region. Let s represent arclength along this curve. Then the total time for a wave to propagate the length of this chosen curve is

$$T = \int_0^S \frac{ds}{\sqrt{gh(x(s), y(s))}} . \tag{20}$$

To obtain the shortest time for *any* wave that started at the epicenter of the quake to reach the identified coastal region, find the minimum T over all possible paths. [Because $h(x, y)$ is not constant, the shortest *time* may or may not correspond to the shortest *distance*.] To obtain the shortest time for a wave, starting anywhere in the earthquake region, to reach this coastal region, repeat this calculation, starting at all places where the earthquake created an initial disturbance. For the 2004 tsunami, the earthquake itself spread northward over about 900 km from its original epicenter, so there were many starting points for this calculation.

For a tsunami warning system, the time at which waves will first arrive at a given coastal region is one of the most important pieces of information needed. These times can be obtained from (20), as described above. But they can also be obtained by solving either (16) or (17) numerically. It is not necessary to do both.

A third conclusion that follows from (16) is that waves can diffract around objects, and do serious damage even to coastlines that face away from the epicenter of the earthquake. See http://www.agu.org/eos_elec.000929e2.html for observations of Chris Chapman, a geologist who experienced the diffracted tsunami on the western coast of Sri Lanka. Again, the time required for a wave to propagate from the source region of the earthquake to a location (like the western coast of Sri Lanka) that does not face the source region can be obtained from (20), using paths that curve around Sri Lanka. Or, one can obtain the same information by solving (16) or (17) numerically.

Finally, it is important not to ask too much of (16) or (17). Recall the two assumptions underlying both (16) and (17): (i) *long waves* – wavelengths are much longer than the fluid depth; and (ii) *small amplitude* – wave amplitudes are much smaller than the fluid depth. But as argued in Section 2, as a long wave of small amplitude approaches the shoreline, the fluid depth decreases (to $h = 0$ at the shoreline), which makes wavelengths shorten, which makes wave amplitudes grow, which eventually guarantees that the assumptions underlying the model break down. As a result, *predictions from* (16) *or* (17) *of wave evolution near shore are almost certainly wrong.* We discuss next what happens near shore, where (16) and (17) fail.

4.3 Tsunami evolution near shore

To see exactly how the assumptions underlying (16) and (17) fail near shore, we can consider a special case of bottom topography. For simplicity, assume that there are no variations in y, so $v = 0$, $\partial_y = 0$ in (16) and (17). In addition, assume a simple form for the bottom topography:

$$\begin{aligned} h(x) &= H(\text{const.}) \quad L < x \,, \\ h(x) &= sx \qquad\qquad 0 < x < L \,, \end{aligned}$$

where $s = H/L$, as shown in Figure 8. As a result, (17) takes the form

$$\partial_t^2 \zeta = (gH)\partial_x^2 \zeta, \qquad \text{for } L < x \,, \tag{21}$$

$$\partial_t^2 \zeta = \partial_x (gs \cdot x \cdot \partial_x \zeta), \qquad \text{for } 0 < x < L \,. \tag{22}$$

Matching conditions at $x = L$ are that both ζ and $\partial_x \zeta$ are continuous there.

Fig. 8. Cartoon of a model ocean, with uniform depth H in the open ocean ($x > L$), and a uniformly sloping bottom in the coastal region ($0 < x < L$).

The detailed calculation for this model is carried out in the Appendix. [Or see §2.2 of (10).] One sees from the calculation that as an oscillatory wave approaches the shoreline, its wavelength necessarily shortens, and its amplitude grows. Both features support the assertions made in Section 2. In fact, the linear model in (22) becomes quite unphysical near $x = 0$, because wave amplitudes becomes infinite there, showing that nonlinear terms necessarily become important near shore.

Many researchers have used a nonlinear, long wave model ("the shallow water equations") to describe wave dynamics near shore. (See (4), (5), and the references cited therein.) Tsunami evolution is vitally important near shore, where the waves are dangerous and destructive. And details of this wave evolution matter – the tsunami of 2004 completely destroyed some locations, while it spared other locations a few kilometers away. Such large variations in wave inundation must be due to variations in the bottom topography offshore, which steered the tsunami away from one location and towards another. Thus, several effects that are negligible in the open ocean become crucial near shore, including details of the bottom topography, nonlinear wave interactions, wave breaking, energy dissipation, and return flow.

Accurate computations of this flow near shore will necessarily be complicated, and costly in terms of computing time. But time is precious for a tsunami warning system, which cannot afford to wait for these necessarily detailed calculations of wave dynamics near shore. Here the conflicting objectives of reliable accuracy and speed clash.

There is a possible way around this clash of objectives. Split the process of predicting tsunamis into two parts: one part must be done quickly, in response to a particular earthquake that might (or might not) generate a tsunami; the other part can be done slowly and carefully, to obtain answers that are very reliable, but are not carried out in response to a specific tsunami.

a) The tsunami warning system should be designed to predict accurately when a tsunami will approach a coastal region. It should **not** be responsible for predicting how the tsunami behaves in shallow coastal waters. If the system is only responsible for predicting when the tsunami will approach a given coastal region, then the necessary calculations for that prediction are fairly simple, so they can be carried out quickly and broadcast in time to move people to safe locations.

b) Separately, much more complicated computer simulations can be carried out to analyze wave dynamics in each particular coastal region. The appropriate numerical models for these simulations need to include the specific bottom topography for that region, nonlinear wave interactions, wave breaking, energy dissipation, and any other effects that might be important to that region. The incoming waves for these simulations can be parameterized, so the analysis for each coastal region will demonstrate how that region responds to a variety of possible incoming tsunamis – coming from different directions, carrying different wave volumes, with different combinations of wavelength and wave amplitude.

Once the detailed study of a specific coastal region has been completed, then the information from that study can be used to improve local zoning laws and building codes. Does the shape of the bottom topography offshore in this region always funnel tsunamis into specific places and away from other places? If so, then buildings can be approved in some regions and not in others. Perhaps mangroves could be planted in the most at-risk places, to absorb some of the incoming tsunami. Or, the buildings in the at-risk areas could be built far enough back from the shoreline that they would not be inundated by a tsunami. Or, if buildings must be built in at-risk areas, then building codes could require that buildings in these at-risk locations be able to withstand tsunamis. Many options can be pursued here, depending on details for each coastal region. The main point is to realize that these studies of local coastal regions can be carried out carefully and deliberately, independent of specific tsunami warnings.

Finally, a local analysis like this is especially important for a place like Bandeh Aceh, which was inundated almost immediately after the earthquake. Perhaps a tsunami warning system will never be effective for a coastal region very close to the epicenter of an earthquake, because the time between the earthquake and the resulting tsunami is too short. But these places can be identified now, when there is no immediate danger, by comparing a map of earthquake fault lines with a map of the shoreline. Warning systems might not help these places, because there is not enough time to generate a reliable warning, but strict zoning laws and building codes might do wonders.

Acknowledgements

I am grateful to Prof. Anjan Kundu for organizing this conference, and for inviting me to participate in it. I also thank the Saha Institute of Nuclear Physics for its hospitality during my stay. Kenji Satake kindly provided me with the content of Figure 2. Finally, I owe thanks to Joe Hammack, who introduced me to the fascinating subject of tsunamis more than 30 years ago. Joe died quite unexpectedly in September 2004.

References

[1] Ablowitz MJ, Segur H (1981) Solitons and the Inverse Scattering Transform, SIAM, Philadelphia, PA

[2] Boussinesq J (1871) Theorie de l'intumescence liquide appele onde solitaire ... . Comptes Rendus 72:755–759

[3] Camassa R, Holm DD (1993) An integrable shallow water equation with peaked solitons. Phys. Rev. Lett. 71:1661–1664

[4] Carrier, GF, Greenspan HP (1958) Water waves of finite amplitude on a sloping beach. J. Fluid Mech. 4:97–109

[5] Carrier GF, Wu TT, Yeh, H (2003) Tsunami runup and drawdown on a plane beach. J. Fluid Mech. 475:79–99
[6] Hammack JL (1973) A note on tsunamis: their generation and propagation in an ocean of uniform depth. J. Fluid Mech. 60:769–800
[7] Hammack JL, Segur H (1974) The Korteweg-deVries equation and water waves, part 2. Comparison with experiments. J. Fluid Mech. 65:289–314
[8] Hammack JL, Segur H (1978a) The Korteweg-deVries equation and water waves, part 3. Oscillatory waves. J. Fluid Mech. 84:337–358
[9] Hammack JL, Segur H (1978b) Modelling criteria for long water waves. J. Fluid Mech., 84:359–373
[10] Johnson RS (1997) An Introduction to the Mathematical Theory of Water Waves, Cambridge Univ. Press
[11] Kadomtsev BB, Petviashvili VI (1970) On the stability of solitary waves in weakly dispersive media. Sov. Phys. Doklady 15:539–541
[12] Korteweg DJ, de Vries G (1895) On the change of form of long waves advancing in a rectangular canal. Philos. Mag. Ser. 5, 39:422–443
[13] Lakshmanan M, Rajasekar R (2003) Nonlinear Dynamics — Integrability, Chaos and Patterns, Springer, NY
[14] Scott AC (1999) Nonlinear Science: Emergence and Dynamics of Coherent Structures, Oxford Univ. Press, NY
[15] Stoker JJ (1957) Water Waves. Wiley Interscience, NY
[16] Stokes GG (1847) On the theory of oscillatory waves. Trans. Camb. Phil. Soc. 8:441-455

Appendix: A linear model for waves on a sloping beach

Equation (21) is a special case of (10), so its general solution is given by (11),

$$\zeta(x,t) = \varepsilon h[F(x - t\sqrt{gH}) + G(x + t\sqrt{gH})],\ L < x\ ,$$

where F and G are arbitrary functions. If $x = 0$ represents the shoreline in India, then G represents the (known) incoming tsunami, and F represents the (unknown) reflected wave. Because G represents the tsunami in the open ocean, then necessarily G vanishes outside a finite region, and G consists of only long waves (i.e., long compared to H) inside this finite region.

Equation (22) has constant coefficients in t, so we may represent its solution in terms of a Fourier transform

$$\zeta(x,t) = \int_{-\infty}^{\infty} e^{i\omega t}[\hat{\zeta}(x,\omega)]d\omega\ . \tag{23}$$

Then (22) reduces to an ordinary differential equation in x for $\hat{\zeta}(x,\omega)$. Change variables: define

$$\chi = 2|\omega|\sqrt{\frac{x}{gs}}, \quad \text{and set } \hat{\zeta}(x,\omega) = \bar{\zeta}(\chi) . \tag{24}$$

Then the differential equation for $\bar{\zeta}(\chi)$ is

$$\frac{d^2\bar{\zeta}}{d\chi^2} + \frac{1}{\chi}\frac{d\bar{\zeta}}{d\chi} + \bar{\zeta} = 0 . \tag{25}$$

This is the equation for a Bessel function of order 0, so its general solution is a linear combination of two Bessel functions,

$$\bar{\zeta}(\chi; a, b) = a \cdot J_0(\chi) + b \cdot Y_0(\chi) .$$

Substituting this back into (23) yields the form of the general solution of (19b):

$$\zeta(x,t) = \int_{-\infty}^{\infty} e^{i\omega t}[a(\omega)J_0(\chi) + b(\omega)Y_0(\chi)]d\omega , \tag{26}$$

with χ given in (21). The coefficients, $a(\omega), b(\omega)$, must be found by matching at $x = L$.

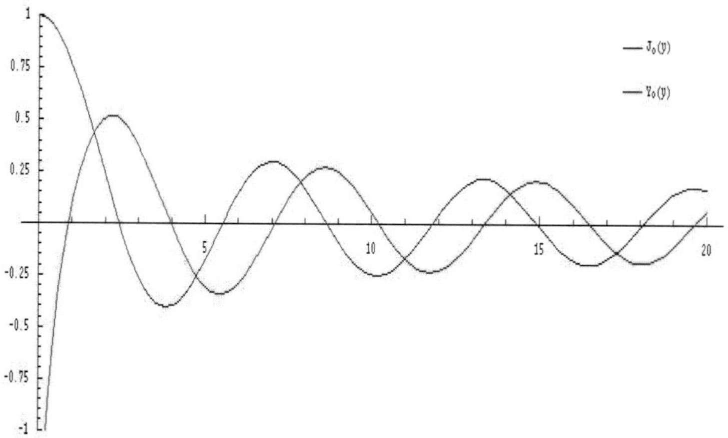

Fig. 9. Two linearly independent solutions of (22): $J_0(\chi)$ and $Y_0(\chi)$.

Figure 9 shows a plot of the two Bessel functions in (26). Both oscillate for large values of χ; one can show that as $\chi \to \infty$, the period of oscillation approaches a constant, for either Bessel function. But χ , defined in (24), is not the physical length. When Figure 9 is replotted in terms of x, we obtain Figure 10. This plot shows clearly that as each wave (with fixed temporal frequency, ω) approaches shore (at $x = 0$), the distance between successive zeroes decreases. In other words, *wavelengths get shorter as a wave approaches*

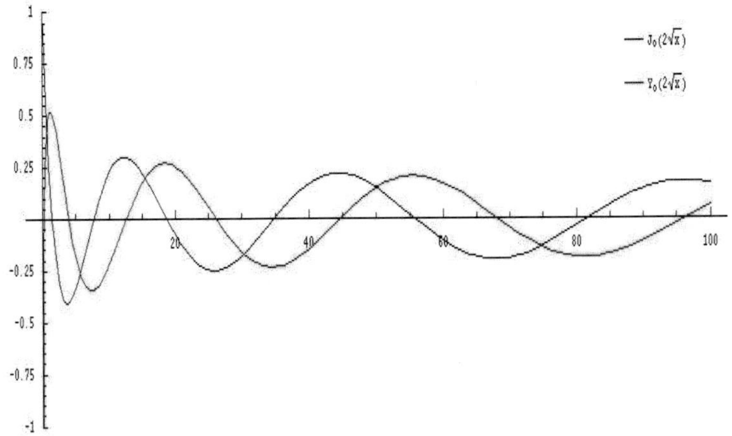

Fig. 10. The same two solutions of (22), but plotted in terms of horizontal distance, x, with g, ω, s all fixed.

shore. This confirms what was asserted in Section 2, but from a different argument.

Either Figure 9 or 10 also shows that as $\chi \to 0$, $J_0(\chi) \to 1$, while $Y_0(\chi)$ becomes infinite. Thus, unless the matching condition at $x = L$ miraculously assures that $b(\omega) = 0$ for all ω, then (26) virtually guarantees that $\zeta(x,t)$ blows up as $\chi \to 0$. This does not imply that the water surface becomes infinitely high there, but only that the linear model necessarily breaks down near shore, where nonlinear terms become important.

Integrable Nonlinear Wave Equations and Possible Connections to Tsunami Dynamics

M. Lakshmanan

Centre for Nonlinear Dynamics, School of Physics, Bharathidasan University, Tiruchirapalli - 620 024,
lakshman@cnld.bdu.ac.in

Summary. In this article we present a brief overview of the nature of localized solitary wave structures/solutions underlying integrable nonlinear dispersive wave equations with specific reference to shallow water wave propagation and explore their possible connections to tsunami waves. In particular, we will discuss the derivation of Korteweg-de Vries family of soliton equations in unidirectional wave propagation in shallow waters and their integrability properties and the nature of soliton collisions.

1 Introduction

The term 'tsunami' which was perhaps an unknown word even for scientists in countries such as India, Srilanka, Thailand, etc. till recently has become a house-hold word since that fateful morning of December 26, 2004. When a powerful earthquake of magnitude 9.1-9.3 on the Richter scale, epicentered off the coast of Sumatra, Indonesia, struck at 07:58:53 local time described as the 2004 Indian Ocean earthquake or Sumatra-Andaman earthquake it triggered a series of devastating tsunamis as high as 30 meteres that spread throughout the Indian Ocean killing about 2,75,000 people and inundating coastal communities across South and Southeat Asia, including parts of Indonesia, Srilanka, India and Thailand and even reaching as far as the east coast of Africa. The catastrophe is considered to be one of the deadliest diasters in modern history (see Figs. 1 and 2 for some details [1]) (1; 2).

Since this earthquake and consequent tsunamis, several other earthquakes of smaller and larger magnitudes keep occurring off the coast of Indonesia. Even as late as July 17, 2006 an earthquake of magnitude 7.7 on the Richter scale struck off the town of Pangadaran at 15.19 local time and set off a tsunami of 2m high which had killed more than 300 people.

These tsunamis, which can become monstrous tidal waves when they approach coastline, are essentially triggered due to the sudden vertical rise of the seabed by several meters (when earthquake occurs) which displaces massive volume of water.

[1] Fig. 1 http:// www.blogaid.org.uk
Fig. 2a) http:// www.hinduonnet.com/gallery/0071/007108.htm
Fig. 2b) http:// www.bhoomikaindia.org

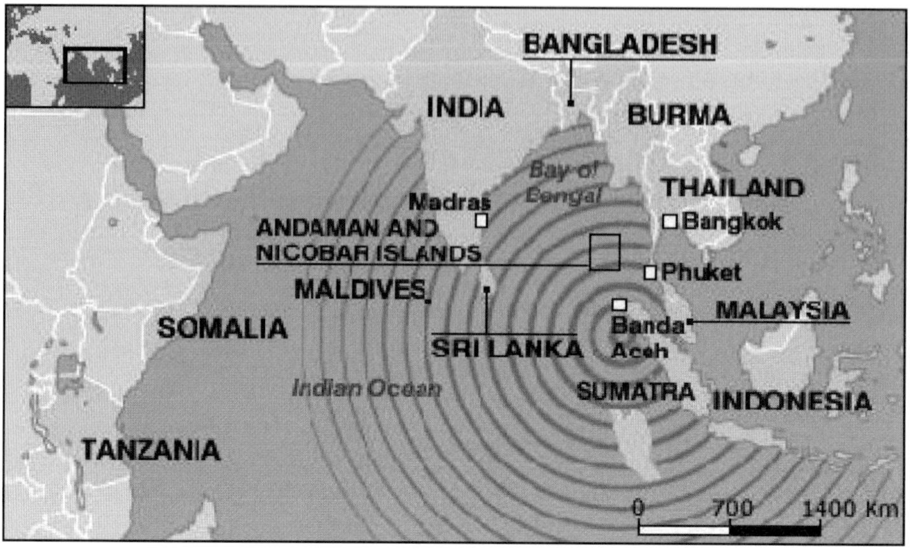

Fig. 1. December 2004 Tsunami in Indian Ocean

Fig. 2. (a) Thiruvalluvar statue in Kanyakumari (in the southern tip of peninsular India), height 133ft (b) Tsunami waves rising near the statue on December 26, 2004 (to almost its top)

The tsunamis behave very differently in deep water than in shallow water as pointed out below. By no means the tsunami of 2004 and later ones are exceptional; More than two hundred tsunamis have been recorded in scientific literature since ancient times. The most notable earlier one is the tsunami triggered by the powerful earthquake (9.3 magnitude) off southern Chile on May 22, 1960 (3); Fifteen hours after the devastating earthquake, the tsunami hit Hawai (namely 10,000 kms away from the epicenter of earthquake), killing 61 people. Seven hours later, the Japanese islands of Honshu and Hokkaido were struck by a wall of water 21-feet high and 197 people drowned.

It is clear from the above events that the tsunami waves are fairly permanent and powerful waves, having the capacity to travel extraordinary distances without practically diminishing in size or speed. In this sense they seem to have considerable resemblance to shallow water nonlinear dispersive waves, particularly solitary waves and solitons. In particular, the Kortweg-de Vries family of nonlinear dispersive wave equations admit such solitary waves and solitons and describe unidirectional wave propagation in shallow waters, and it is appropriate to critically review the derivation of KdV and related equations. We will also briefly mention the nature of the solitary wave and soliton solutions and other integrability properties associated with KdV family of equations, including variable KdV and recently derived Camassa-Holm equation. Possible two dimensional generalizations will be also briefly touched.

The plan of the article is as follows. In sec.2, we will summarize the characteristic properties of tsunami waves. In sec.3, we will critically analyse how the Korteweg-de Vries equation was originally derived to describe the Scott-Russel phenomenon to describe shallow water wave propagation. In sec.4, we will discuss the properties of solitary waves and soliton solutions of the KdV equation. We will also briefly touch upon the complete integrability properties of the KdV equation. In sec.5, other interesting KdV type dispersive wave equations will be discussed. Finally, we summarize the discussion in sec.6.

2 Basics of Tsunami Waves

As noted above tsunami (tsu: harbour, nami: wave) waves of the type described earlier are essentially triggered by massive earthquakes which lead to vertical displacement of a large volume of water. Other possible reasons also exist for the formation and propagation of tsunami waves: underwater nuclear explosion, larger meteorites falling into the sea, volcano explosions, rockslides, etc. But the most predominanat cause of tsunamis appear to be large earthquakes as in the case of the Sumatra-Andaman earthquake of 2004. Then there are three major aspects associated with the tsunami dynamics:

1. Generation of tsunamis
2. Propagation of tsunamis
3. Tsunami run up and inundation

There exist rather successful models to approach the generation aspects of tsunamis when they occur due to the earthquakes (4). Using the available seismic data it is possible to reconstruct the permanent deformation of the sea bottom due to earthquakes and simple models have been developed (see for example, the article of F. Dias in this volume). Similarly the tsunami run up and unundation problems are extremely complex and they require detailed critical study from a practical point of view in order to save structures and lives when a tsunami strikes.

However, in this article we will be more concerned with the propagation of tsunami waves and their possible relation to wave propagation associated with nonlinear dispersive waves in shallow waters. In order to appreciate such a possible connection, we first look at the typical characteristic properties of tsunami waves as in the case of 2004 Indian Ocean tsunami waves or 1960 Chilean tsunamis.

Considering the Indian Ocean 2004 tsunami, satellite observations after a couple of hours after the earthquake establish an amplitude of approximately 60 cms in the open ocean for the waves. The estimated typical wavelength is about 200 kms (5). The maximum water depth h is between 1 and 4 kms.

Consequently, one can identify the following small parameters (ϵ and δ^2) of roughly equal magnitude:

$$\epsilon = \frac{a}{h} \approx 10^{-4} << 1, \tag{1}$$

$$\delta = \frac{h}{l} \approx 10^{-2} << 1 \tag{2}$$

As a consequence, it is possible that a nonlinear shallow water wave theory where dispersion also plays an important role has some relevance. However, we also wish to point out here that there are other points of view: Constantin and Johnson (6) estimate $\epsilon \approx 0.002$ and $\delta \approx 0.04$ and conclude that for both nonlinearity and dispersion to become significant the quantity $\delta\epsilon^{-\frac{3}{2}}\times$wavelength estimated as 90,000 kms is too large and shallow water equations with variable depth (without dispersion) should be used. However, it appears that these estimates can vary over a rather wide range and with suitable estimates it is possible that the range of 10,000 20,000 kms could be also possible ranges and hence taking into account the fact that both the Indian Ocean 2004 and Chilean 1960 tsunamis have travelled over 10 hours or more before encountering land mass appears to allow for the possibility of nonlinear dispersive waves as relevant features for the phenomena.

From this point of view in the next section we discuss the shallow water wave theory to deduce KdV equation.

3 Scott Russel Phenomenon and KdV Equation

It is a folklore that the first scientificaly recorded observation of a solitary wave was made by the Scottish naval engineer John Scott Russel in the year 1837 (7) when he identified a large solitary heap of water travelling with undiminished speed or shape over a distance in the Union Canal connecting the cities of Edinburg and Glasgow in Scotland. He went on to repeat the phenomenon at the laboratory in a rectangular channel of water by dropping wieghts at one end. By measuring the velocity and height of the wave he also established a phenomenological relation connecting these quantities which has stood the test of time.

In 1895, Kortweg and de Vries (8) considered the wave phenomenon underlying the observations of Scott Russel, from first principles of fluid dynamics. The basic features of their analysis can be summarized as follows (9; 10).

Consider the one-dimensional (x-direction) wave motion of an incompressible and inviscid fluid (water) in a shallow channel of height h, and of sufficient width with uniform cross-section leading to the formation of a solitary wave propagating under gravity. Let the length of the wave be l and the maximum value of its amplitude, $\eta(x,t)$, above the horizontal surface be a (see Fig.3). Then we can introduce two natural small parameters into the problem ϵ and δ as defined in Eqs. (1) and (2). Then we can proceed with the analysis as follows.

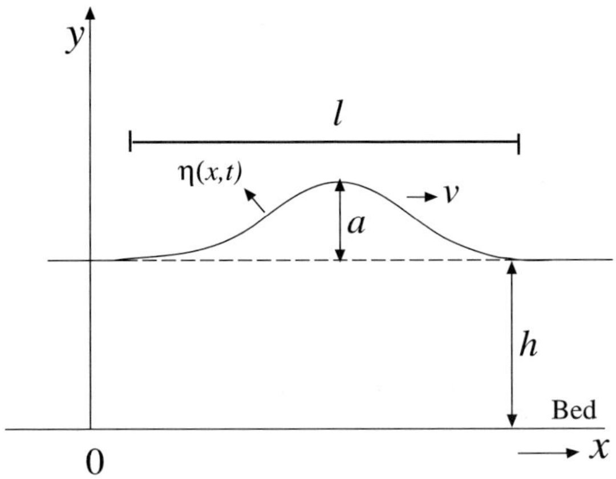

Fig. 3. One-dimensional wave motion in a shallow channel

A. *Equation of Motion*

The fluid motion can be described by the velocity vector

$$\mathbf{V}(x,y,t) = u(x,y,t)\mathbf{i} + v(x,y,t)\mathbf{j}\ , \tag{3}$$

where $\mathbf{i}$ and $\mathbf{j}$ are the unit vectors along the horizontal and vertical directions, respectively. As the motion is irrotational, we have

$$\nabla \times \mathbf{V} = 0\ . \tag{4}$$

Consequently, we can introduce the velocity potential $\phi(x,y,t)$ by the relation

$$\mathbf{V} = \nabla\phi\ . \tag{5}$$

(i) *Conservation of Density*

The system obviously admits the following conservation law for the mass density $\rho(x,\ y,\ t)$ of the fluid,

$$\frac{\mathrm{d}\rho}{\mathrm{d}t} = \rho_t + \nabla\cdot(\rho\mathbf{V}) = 0\ , \tag{6}$$

where $\mathbf{V}(x,y,t)$ is the velocity vector of the fluid. As ρ is a constant, from (6) we have

$$\nabla\cdot\mathbf{V} = 0\ . \tag{7}$$

Then using (5) in (7), we find that ϕ obeys the Laplace equation

$$\nabla^2\phi(x,y,t) = 0\ . \tag{8}$$

(ii) *Euler's Equation*

As the density of the fluid $\rho = \rho_0$ =constant, using Newton's law for the rate of change of momentum, we can write

$$\frac{\mathrm{d}\mathbf{V}}{\mathrm{d}t} = \frac{\partial \mathbf{V}}{\partial t} + (\mathbf{V} \cdot \boldsymbol{\nabla})\,\mathbf{V} = -\frac{1}{\rho_0}\boldsymbol{\nabla} p - g\mathbf{j}\,, \tag{9}$$

where $p = p(x, y, t)$ is the pressure at the point (x, y) and g is the acceleration due to gravity, which is acting vertically downwards (here $\mathbf{j}$ is the unit vector along the vertical direction). Making use of (5) in (9), we obtain (after one integration)

$$\phi_t + \frac{1}{2}(\boldsymbol{\nabla}\phi)^2 + \frac{p}{\rho_0} + gy = 0\,. \tag{10}$$

(iii) *Boundary Conditions*

The above two equations (8) and (9) or (10) for the velocity potential $\phi(x, y, t)$ of the fluid have to be supplemented by appropriate boundary conditions, by taking into account the fact (see Fig.2) that

(a) the horizontal bed at $y = 0$ is hard and
(b) the upper boundary $y = y(x, t)$ is a free surface .

As a result
(a) the vertical velocity at $y = 0$ vanishes,

$$v(x, 0, t) = 0\,, \tag{11}$$

which implies (using (3) and (5))

$$\phi_y(x, 0, t) = 0\,. \tag{12}$$

(b) As the upper boundary is free, let us specify it by $y = h + \eta(x, t)$ (see Fig.2). Then at the point $x = x_1$, $y = y_1 \equiv y(x, t)$, we can write

$$\frac{\mathrm{d}y_1}{\mathrm{d}t} = \frac{\partial \eta}{\partial t} + \frac{\partial \eta}{\partial x} \cdot \frac{\mathrm{d}x_1}{\mathrm{d}t} = \eta_t + \eta_x u_1 = v_1\,. \tag{13}$$

Since $v_1 = \phi_{1y}$, $u_1 = \phi_{1x}$, the last two parts of (13) can be rewritten as

$$\phi_{1y} = \eta_t + \eta_x \phi_{1x}\,. \tag{14}$$

(c) Similarly at $y = y_1$, the pressure $p_1 = 0$. Then from (10), it follows that

$$u_{1t} + u_1 u_{1x} + v_1 v_{1x} + g\eta_x = 0\,. \tag{15}$$

Thus the motion of the surface of water wave is essentially specified by the Laplace equation (8) and (10) along with one fixed boundary condition (12) and two *variable nonlinear* boundary conditions (14) and (15). One has to then solve the Laplace equation subject to these boundary conditions.

(iv) *Taylor Expansion of $\phi(x, y, t)$ in y*

Making use of the fact $\delta = h/l \ll 1$, $h \ll l$, we assume $y(= h + \eta(x,t))$ to be small to introduce the Taylor expansion

$$\phi(x,y,t) = \sum_{n=0}^{\infty} y^n \phi_n(x,t) . \tag{16}$$

Substituting the above series for ϕ into the Laplace equation (8), solving recursively for $\phi_n(x,t)$'s and making use of the boundary condition (14), $\phi_y(x,0,t) = 0$, one can show that

$$u_1 = \phi_{1x} = f - \frac{1}{2} y_1^2 f_{xx} + \text{ higher order in } y_1 , \tag{17}$$

$$v_1 = \phi_{1y} = -y_1 f_x + \frac{1}{6} y_1^3 f_{xxx} + \text{ higher order in } y_1 , \tag{18}$$

where $f = \partial \phi_0 / \partial x$. We can then substitute these expressions into the nonlinear boundary conditions (13),(14) and (15) to obtain equations for f and η.

(v) *Introduction of Small Parameters ϵ and δ*

So far the analysis has not taken into account fully the shallow nature of the channel ($a/h = \epsilon \ll 1$) and the solitary nature of the wave ($a/l = a/h \cdot h/l = \epsilon\delta \ll 1$, $\epsilon \ll 1$, $\delta \ll 1$), which are essential to realize the Scott Russel phenomenon. For this purpose we stretch the independent and dependent variables in the defining (13)–(15), (17) and (18) through appropriate scale changes, but retaining the overall form of the equations. To realize this we can introduce the natural scale changes

$$x = l x' , \quad \eta = a \eta' \tag{19}$$

along with

$$t = \frac{l}{c_0} t' , \tag{20}$$

where c_0 is a parameter to be determined. Then in order to retain the form of (17), (18) we require

$$u_1 = \epsilon c_0 u_1' , \quad v_1 = \epsilon \delta c_0 v_1' , \quad f = \epsilon c_0 f' . \tag{21}$$

We also have

$$y_1 = h + \eta(x,t) = h\left(1 + \epsilon \eta'\left(x', t'\right)\right) . \tag{22}$$

Substituting the transformations (19)–(22) into (17), we obtain

$$\begin{aligned} u_1' &= f' - \frac{1}{2}\delta^2 \left(1 + \epsilon\eta'\right)^2 f'_{x'x'} \\ &= f' - \frac{1}{2}\delta^2 f'_{x'x'} , \end{aligned} \tag{23}$$

where we have omitted terms proportional to $\delta^2 \epsilon$ as small compared to terms of the order δ^2. Similarly from (18), we obtain

$$v_1' = -\left(1 + \epsilon\eta'\right) f'_{x'} + \frac{1}{6}\delta^2 f'_{x'x'x'} . \tag{24}$$

Now considering the nonlinear boundary condition (14) in the form

$$v_1 = \eta_t + \eta_x u_1 \ , \tag{25}$$

it can be rewritten, after making use of the transformations (19)–(22) and neglecting terms involving $\epsilon\delta^2$, as

$$\eta'_{t'} + f'_{x'} + \epsilon\eta' f'_{x'} + \epsilon f' \eta'_{x'} - \frac{1}{6}\delta^2 f'_{x'x'x'} = 0 \ . \tag{26}$$

Similarly considering the other boundary condition (16) and making use of the above transformations, it can be rewritten, after neglecting terms of the order $\epsilon^2\delta^2$, as

$$f'_{t'} + \epsilon f' f'_{x'} + \frac{ga}{\epsilon c_0^2}\eta'_{x'} - \frac{1}{2}\delta^2 f'_{x'x't'} = 0 \ . \tag{27}$$

Now choosing the arbitrary parameter c_0 as

$$c_0^2 = gh \tag{28}$$

so that η'_x term is of order unity, (27) becomes

$$f'_{t'} + \eta'_{x'} + \epsilon f' f'_{x'} - \frac{1}{2}\delta^2 f'_{x'x't'} = 0 \ . \tag{29}$$

For notational convenience we will hereafter omit the prime symbol in all the variables, however remembering that all the variables hereafter correspond to rescaled quantities. Then the evolution equation for the amplitude of the wave and the function related to the velocity potential reads

$$\eta_t + f_x + \epsilon\eta f_x + \epsilon f \eta_x - \frac{1}{6}\delta^2 f_{xxx} = 0 \ , \tag{30}$$

$$f_t + \eta_x + \epsilon f f_x - \frac{1}{2}\delta^2 f_{xxt} = 0 \ . \tag{31}$$

Note that the small parameters ϵ and δ^2 have occurred in a natural way in (30), (31).

(vi) *Perturbation Analysis*

Since the parameters ϵ and δ^2 are small in (30), (31), we can make a perturbation expansion of f in these parameters:

$$f = f^{(0)} + \epsilon f^{(1)} + \delta^2 f^{(2)} + \text{ higher order terms} \ , \tag{32}$$

where $f^{(i)}$, $i = 0, 1, 2, ...$ are functions of η and its spatial derivatives. Substituting this into (30), (31) and regrouping, we obtain

$$\eta_t + f_x^{(0)} + \epsilon\left[f_x^{(1)} + \eta f_x^{(0)} + \eta_x f^{(0)}\right] + \delta^2\left[f_x^{(2)} - \frac{1}{6}f_{xxx}^{(0)}\right] + \text{ higher order terms in } (\epsilon, \delta^2) = 0 \ , \tag{33}$$

$$\eta_x + f_t^{(0)} + \epsilon\left[f_t^{(1)} + f^{(0)} f_x^{(0)}\right] + \delta^2\left[f_t^{(2)} - \frac{1}{2}f_{xxt}^{(0)}\right] + \text{ higher order terms in } (\epsilon, \delta^2) = 0 \ . \tag{34}$$

In order that (33) and (34) are self consistent as evolution equations for an one-dimensional wave propagating to the right, we can choose

$$f^{(0)} = \eta + O\left(\epsilon, \delta^2\right) , \tag{35}$$

where $O(\epsilon, \delta^2)$ stands for terms proportional to ϵ and δ^2. Then (33), (34) become

$$\eta_t + \eta_x + \epsilon\left[f_x^{(1)} + 2\eta\eta_x\right] + \delta^2\left[f_x^{(2)} - \frac{1}{6}\eta_{xxx}\right] = 0 , \tag{36}$$

$$\eta_t + \eta_x + \epsilon\left[f_t^{(1)} + \eta\eta_x\right] + \delta^2\left[f_t^{(2)} - \frac{1}{2}\eta_{xxt}\right] = 0 , \tag{37}$$

where higher order terms in ϵ and δ^2 are neglected. Since $f^{(1)}$ and $f^{(2)}$ are functions of η (and its spatial derivatives)

$$f_x^{(1)} = f_\eta^{(1)}\eta_x, \quad f_t^{(1)} = f_\eta^{(1)}\eta_t = -f_\eta^{(1)}\eta_x + O\left(\epsilon, \delta^2\right) = -f_x^{(1)} , \tag{38}$$

where in the last relation, (33), (34) have been used for η_t and η_x. Similarly, we can argue that

$$f_x^{(2)} = f_\eta^{(2)}\eta_x, \quad f_t^{(2)} = -f_\eta^{(2)}\eta_x + O\left(\epsilon, \delta^2\right) = -f_x^{(2)} , \tag{39}$$

Substituting (38), (39) into (36), (37) , we obtain

$$\eta_t + \eta_x + \epsilon\left[f_x^{(1)} + 2\eta\eta_x\right] + \delta^2\left[f_x^{(2)} - \frac{1}{6}\eta_{xxx}\right] = 0 , \tag{40}$$

$$\eta_t + \eta_x + \epsilon[-f_x^{(1)} + \eta\eta_x] + \delta^2\left[-f_x^{(2)} + \frac{1}{2}\eta_{xxx}\right] = 0 . \tag{41}$$

Compatibility of these two equations require that

$$f_x^{(1)} = -\frac{1}{2}\eta\eta_x, \quad f_x^{(2)} = \frac{1}{3}\eta_{xxx} . \tag{42}$$

Integrating, we find

$$f^{(1)} = -\frac{1}{4}\eta^2, \quad f^{(2)} = \frac{1}{3}\eta_{xx} . \tag{43}$$

Substituting $f^{(1)}$ and $f^{(2)}$ into (40), (41) , we ultimately obtain the KdV equation in the form

$$\eta_t + \eta_x + \frac{3}{2}\epsilon\eta\eta_x + \frac{\delta^2}{6}\eta_{xxx} = 0 , \tag{44}$$

describing the unidirectional propagation of shallow water waves.

(vii) *The Standard (Contemporary) Form of KdV Equation*

Finally, changing to a moving frame of reference,

$$\xi = x - t, \quad \tau = t \tag{45}$$

so that

$$\frac{\partial}{\partial x} = \frac{\partial}{\partial \xi} , \quad \frac{\partial}{\partial t} = -\frac{\partial}{\partial \xi} + \frac{\partial}{\partial \tau} , \tag{46}$$

(44) can be rewritten as

$$\eta_\tau + \frac{3}{2}\epsilon\eta\eta_\xi + \frac{1}{6}\delta^2\eta_{\xi\xi\xi} = 0 . \tag{47}$$

Then introducing the new variables

$$u = \frac{3\epsilon}{2\delta^2}\eta, \quad \tau' = \frac{6}{\delta^2}\tau , \tag{48}$$

(47) can be expressed as

$$u_{\tau'} + 6uu_\xi + u_{\xi\xi\xi} = 0 . \tag{49}$$

Redefining the variables τ' as t and ξ as x, again for notational convenience, we finally arrive at the ubiquitous form of the KdV equation as

$$u_t + 6uu_x + u_{xxx} = 0 . \tag{50}$$

4 Solitary Wave, Solitons and Complete Integrability of KdV equation

The Korteweg-de Vries equation (50) admits cnoidal wave solution and in the limiting case solitary wave solution as well. More importantly, the KdV solitary wave is a soliton : it retains its shape and speed upon collision with another solitary wave of different amplitude, except for a phase shift (11). In fact for an arbitrary initial condition, the solution of the Cauchy initial value problem consists of N-number of solitons of different amplitudes in the background of small amplitude dispersive waves. All these results ultimately lead to the result that the KdV equation is a completely integrable, infinite dimensional, nonlinear Hamiltonian system. It possesses

(i) a Lax pair of linear differential operators and is solvable through the so called inverse scattering transform (IST) method (12),
(ii) infinite number of conservation laws and associated infinite number of involutive integrals of motion
(iii) N-soliton solution
(iv) Hirota bilinear form
(v) Hamiltonian structure

and a host of other interesting properties (see for example (9; 10)). We will very briefly consider some of these properties.

4.1 Korteweg–de Vries Equation and the Solitary Waves and Cnoidal Waves

Let us look for elementary wave solutions of (50) in the form

$$u = 2f(x - ct) \tag{51}$$

$$= 2f(\xi) , \quad \xi = x - ct . \tag{52}$$

Then the KdV equation reduces to an ordinary differential equation of the form

$$-cf_\xi + 6\left(f^2\right)_\xi + f_{\xi\xi\xi} = 0\,. \tag{53}$$

Integrating twice the above equation, the solution of can be expressed in terms of Jacobian elliptic function as

$$f(\xi) = f(x-ct) = \alpha_3 - (\alpha_3 - \alpha_2)\mathrm{sn}^2\left[\sqrt{\alpha_3-\alpha_1}\,(x-ct)\,, m\right], \tag{54}$$

where

$$(\alpha_1+\alpha_2+\alpha_3) = \frac{c}{4}\,, \quad m^2 = \frac{\alpha_3-\alpha_2}{\alpha_3-\alpha_1}\,. \tag{55}$$

Here α_1, α_2 and α_3 are related to the three integration constants. Equations (54), (55) represent in fact the so-called *cnoidal wave* for obvious reasons.

Special Cases:

(i) $m \approx 0$: *Harmonic wave*

When $m \approx 0$, (54) leads to elementary progressing harmonic wave solutions. This can be verified by taking the limit $m \to 0$ in (54), (55).

(ii) $m = 1$: *Solitary wave*

When $m = 1$, we have

$$f = \alpha_3 - (\alpha_3 - \alpha_2)\tanh^2\left[\sqrt{\alpha_3-\alpha_1}\,(x-ct)\right]\,, \tag{56}$$

that is,

$$f = \alpha_2 + (\alpha_3 - \alpha_2)\ \mathrm{sech}^2\left[\ \sqrt{\alpha_3-\alpha_1}\,(x-ct)\right]\,. \tag{57}$$

Choosing now $\alpha_2 = 0$, $\alpha_1 = 0$, we have

$$f = \frac{\alpha_3}{4}\,\mathrm{sech}^2\left[\sqrt{\alpha_3}\,(x-ct)\right]\,. \tag{58}$$

Using (55), (58) can be written as

$$f = \frac{c}{4}\,\mathrm{sech}^2\left[\frac{\sqrt{c}}{2}\,(x-ct)\right]\,. \tag{59}$$

Substituting (59) into (51), the solution can be written as

$$u(x,t) = 2f = \frac{c}{2}\,\mathrm{sech}^2\left[\frac{\sqrt{c}\,(x-ct)}{2}\right]\,. \tag{60}$$

This is of course the Scott Russel solitary wave (Fig.4),

The characteristic feature of the above solitary wave is that the velocity of the wave ($v = c$) is directly proportional to the amplitude ($a = c/2$): the larger the wave, the faster it moves. Unlike the progressing wave, it is fully localized, decaying exponentially fast as $x \to \pm\infty$ (see Fig.4).

4.2 Lax pair and linearization

The KdV equation is well known to posses the Lax pair (12)

$$L = -\frac{\partial^2}{\partial x^2} + u(x,t) \tag{61}$$

and

$$B = -4\frac{\partial^3}{\partial x^3} + 3\left(u\frac{\partial}{\partial x} + \frac{\partial}{\partial x}u\right) \tag{62}$$

so that the Lax equation

$$L_t = [B, L] \tag{63}$$

is equivalent to the KdV equation. Or in other words the KdV equation is linearizable in the sense that it is the compatibility condition corresponding to a linear eigenvalue problem (the Schrödinger spectral problem) and a linear time evolution equation for the eigen function

$$L\psi = \lambda\psi \ , \tag{64}$$

$$\psi_t = B\psi \ . \tag{65}$$

Consequently a nonlinear generalization of the Fourier transform method, namely the inverse scattering transform (IST) technique can be formulated to solve the Cauchy initial value problem of the KdV equation. Schematically it is shown in Fig.5, which is self explanatory. For more details, see for example refs. [9,10]. The final result is that the general solution of the KdV equation can be written as

$$u(x,t) = -2\frac{\mathrm{d}}{\mathrm{d}x}K(x, x+0, t) \ . \tag{66}$$

where $K(x,y,t)$ is the solution of the Gelfand-Levitan-Marchenko linear integral equation

$$\begin{aligned} K(x,y,t) &+ F(x+y,t) \\ &+ \int_x^\infty F(y+z,t)K(x,z,t)\mathrm{d}z = 0 \ , \quad y > x \end{aligned} \tag{67}$$

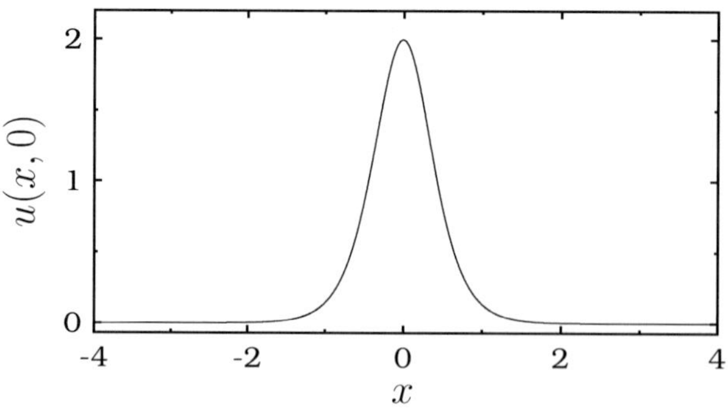

Fig. 4. Solitary wave solution (60) of the KdV equation.

and

$$F(x+y,t)=\sum_{n=1}^{N} C_n^2(0)\mathrm{e}^{8\kappa_n^3 t}\mathrm{e}^{-\kappa_n(x+y)}+\frac{1}{2\pi}\int_{-\infty}^{\infty} R(k,0)\mathrm{e}^{-8\mathrm{i}k^3 t}\mathrm{e}^{\mathrm{i}k(x+y)}\mathrm{d}k\,. \quad (68)$$

Here C_n, κ_n and $R(k,0)$ are the spectral data associated with the Schrödinger spectral problem (64) for $t=0$. Then the discrete states in eq.(68) essentially lead to the soliton picture: N discrete states correspond to N-soliton solutions.

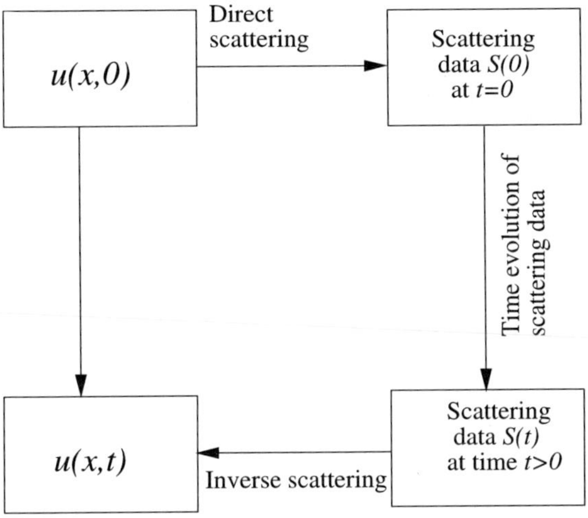

Fig. 5. Schematic diagram of the inverse scattering transform method

For example, the two soliton solution can be written as

$$u(x,t)=-2\left(\kappa_2^2-\kappa_1^2\right)\frac{\kappa_2^2\ \mathrm{cosech}^2\gamma_2+\kappa_1^2\ \mathrm{sech}^2\gamma_1}{(\kappa_2\ \coth\gamma_2-\kappa_1\ \tanh\gamma_1)^2}\,, \quad (69)$$

where $\gamma_i=\kappa_i x-4\kappa_i^3 t-\delta_i$, $\delta_i=\frac{1}{2}\log\left(\frac{C_{i0}^2}{2\kappa_i}\frac{(\kappa_2-\kappa_1)}{(\kappa_2+\kappa_1)}\right)$, $i=1,2$. When the solution (69) is plotted as in Fig.6, it clearly demonstrates the basic soliton property of elastic collision.

4.3 KdV as a Hamiltonian system

Defining the Hamiltonian density

$$\mathcal{H}=\frac{1}{2}u_x^2+u^3 \quad (70)$$

so that the Hamiltonian becomes

$$\widehat{\mathcal{H}}=\int\left(\frac{1}{2}u_x^2+u^3\right)\mathrm{d}x\,. \quad (71)$$

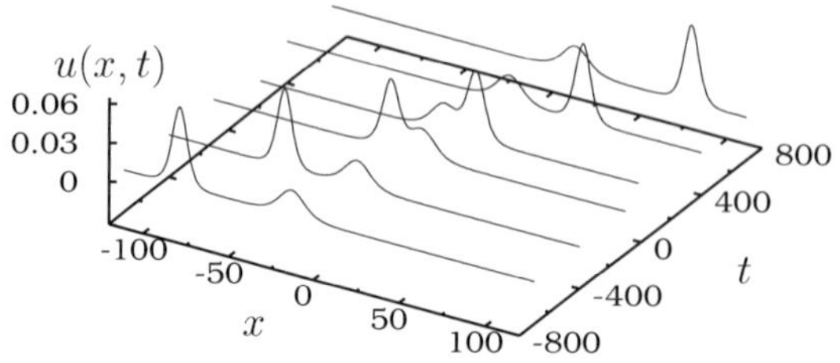

Fig. 6. Two soliton interaction of the KdV equation

KdV equation can be written as

$$u_t = \frac{\partial}{\partial x}\frac{\delta \widehat{\mathcal{H}}}{\delta u} \,. \tag{72}$$

Thus, KdV equation has a Hamiltonian structure.

Now, defining the Poisson bracket between two functionals U and V as

$$\{U, V\} = \int \mathrm{d}x \frac{\delta U}{\delta u(x)} \frac{\partial}{\partial x} \frac{\delta V}{\delta u(x)} \,. \tag{73}$$

it has been shown (13) that the inverse scattering transform discussed earlier allows one to identify appropriate set of (infinite number of) action-angle variables. Further one can show that in terms of these new variables, the Hamiltonian (71) can be expressed purely in terms of action variables alone. Consequently the corresponding canonical equations can be trivially integrated. In this sense KdV equation has been proved to be a complete integrable infinite dimensional Hamiltonian system.

4.4 Bilinearization of KdV

The KdV equation is not only linearizable but also can be bilinearized (14) under the transformation

$$u = 2\frac{\partial^2}{\partial x^2} \log F \,. \tag{74}$$

Eq. (52) can be rewritten in the bilinear form as

$$F_{xt}F - F_xF_t + F_{xxxx}F - 4F_{xxx}F_x + 3F_{xx}^2 = 0 \,. \tag{75}$$

Then it is fairly straightforward to obtain the soliton solutions by expanding F formally as a power series in terms of a small parameter ϵ so that eq. (75) is written as a system of linear partial differential equations. Restricting to a finite number of terms in the power series and solving the resultant system of linear partial differential equation recursively, one can obtain the N-soliton solution explicitly and the soliton property can be analysed.

Besides the above properties, KdV equation possesses many other characteristic features of integrable systems: (i) Existence of infinite number of conservation laws and constants of motion (ii) Bäcklund transformations (iii) Lie-Bäcklund symmetries, (iv) Painlevé property and so on. Again for details see refs. (9; 10)

5 KdV related Integrable and Nonintegrable Equations

Depending on the actual physical situation, the derivation of the shallow water wave equation can be suitably modified to obtain other forms of nonlinear dispersive wave equations in (1+1) dimensions as well as in (2+1) dimensions without going into the actual derivations. Some of the important equations are listed below (11).

- Boussinesq equation

$$u_t + uu_x + g\eta_x - \frac{1}{3}h^2 u_{txx} = 0 \tag{76}$$

$$\eta_t + [u(h+\eta)]_x = 0 \tag{77}$$

- Benjamin-Bona-Mahoney (BBM) equation

$$u_t + u_x + uu_x - u_{xxt} = 0 \tag{78}$$

- Camassa-Holm equation (15)

$$u_t + 2\kappa u_x + 3uu_x - u_{xxt} = 3u_x u_{xx} + uu_{xxx} \tag{79}$$

- Kadomtsev-Petviashville (KP) equation

$$(u_t + 6uu_x + u_{xxx})_x + 3\sigma^2 u_{yy} = 0 \tag{80}$$

σ^2=-1: KP-I, σ^2=+1: KP-II.

There also exist some interesting nonlinar dispersive wave equations to describe deep water wave propagation. These include the following.

- Nonlinear Schrödinger (NLS) equation

$$iq_t + q_{xx} + |q|^2 q = 0, \tag{81}$$

- Davey-Stewartson equation

$$\begin{aligned} iq_t + q_{xx} + q_{yy} + 2|q|^2 q + qu &= 0 \, , \\ u_{xx} - u_{yy} &= 4\left(|q|^2\right)_{xx} \, . \end{aligned} \tag{82}$$

In the derivation of the above equations, generally the bottom of water column or fluid bed is assumed to be flat. However in realistic situations the water depth varies as a function of the horizontal coordinates. In this situation, one often encounters inhomogeneous forms of the above wave equations. Typical example is the variable coefficient KdV equation (16):

$$u_t + f(x,t)uu_x + g(x,t)u_{xxx} = 0, \tag{83}$$

where f and g are functions of x, t. More general forms can also be deduced depending upon the actual situations, see for example ref. (17).

All these equations can be helpful to deal with tsunami wave propagation at different situations. Which one will suit which situation requires detailed analysis depending upon the experimental observations. Many of the above equations are integrable such as the Boussinesq, Camassa-Holm, KP, nonlinear Schrödinger and Davey-Stewartson equations and certain forms of inhomogeneous KdV equations, while BBM equation and general forms of inhomogeneous KdV equations are non-integrable but may possess solitary wave solutions and are amenable to perturbation analysis. Integrable equations in the above list admit interesting new types of solutions. For example, the Camassa-Holm equation admits peakon solution (see Fig. 7), while the KP equation and Davey-Stewartson equation can admit lump (algebraically decaying) solutions and line soliton solutions. The latter one also admits dromion (exponentially localized) solutions (see Figs. 8–10). These solutions can also be used for possible description of tsunami wave propagation in the appropriate situations.

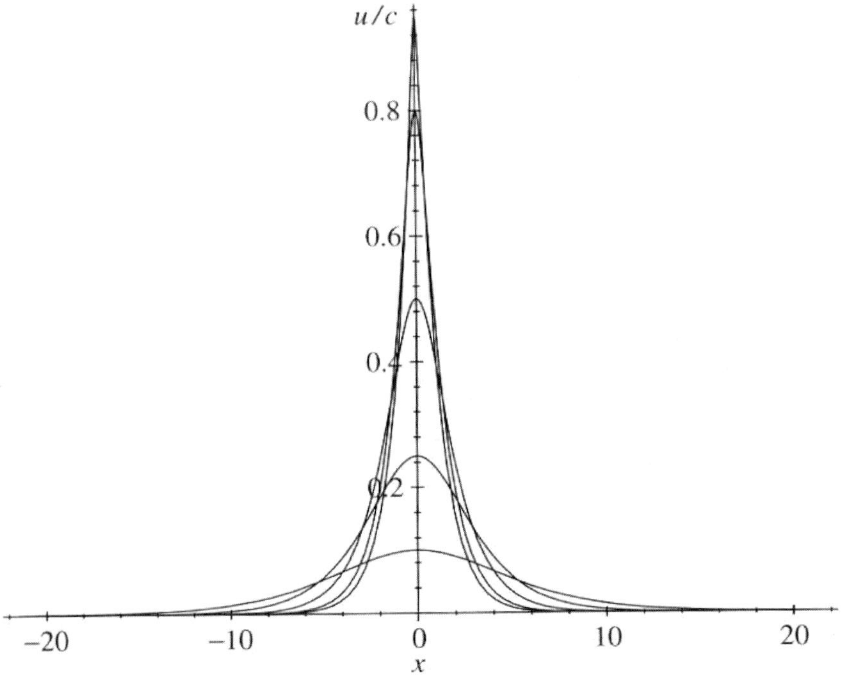

Fig. 7. Solitary wave (including peakon) solution of Camassa-Holm equation

6 Summary and Conclusions

In this article, we have discussed briefly the possibility of describing tsunami waves of the type which occurred in the Indian Ocean 2004 earthquake in terms of nonlinear shallow water wave equation of dispersive type like the Korteweg-de Vries equation. In particular, we have pointed how the KdV wave equation can be derived to describe the Scott-Russel phenomenon of unidirectional shallow water wave propagation. Existence of solitary waves, solitons and complete integrability properties of the KdV

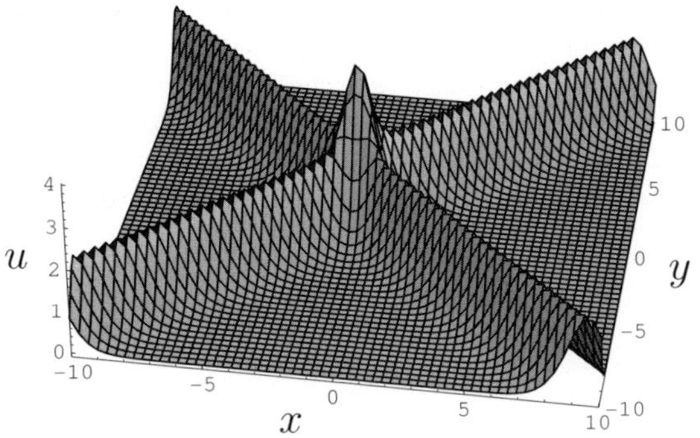

Fig. 8. Two line soliton solution of KP-II equation

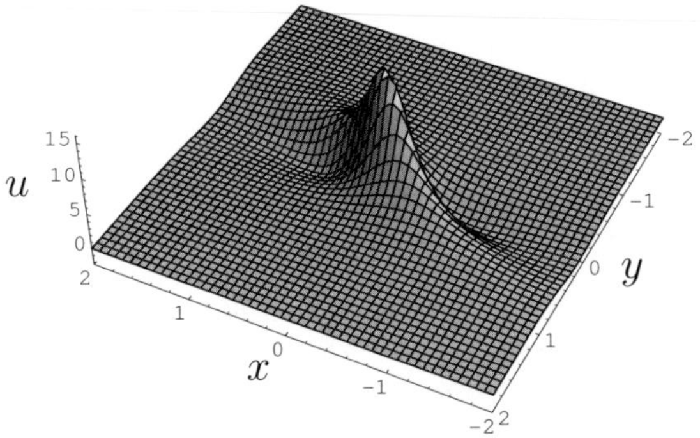

Fig. 9. Lump soliton solution of KP-I equation

equation was briefly explained. Other related equations which can be of some use in tsunami dynamics were also briefly touched upon. The generation and propagation of tsunami waves is an extremely complex process. Yet nonlinear dispersive waves of shallow water may be of considerable importance in describing the tsunami dynamics and much work remains to be done in this direction.

Acknowledgements

This work forms a part of a Department of Science and Technology, Government of India research project. The author has been supported by a DAE-BRNS Raja Ramanna Fellowship during the course of this work. The assistance of Mr. C. Senthil Kumar in the preparation of this article is thankfully acknowledged.

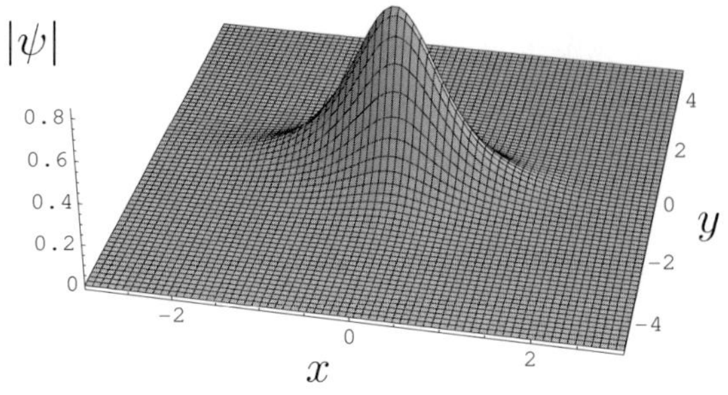

Fig. 10. Dromion solution of Davey-Stewartson equation

References

[1] Lay T etal (2005) The Great Sumatra-Andaman earthquake of 26 December 2004. Science 308:1127–1133
[2] Ram Mohan V etal (2006) Impact of South Asian Tsunami on 26 December 2004 on south east coast of India—A field report (preprint)
[3] Dudley WC and Min L (1988) Tsunami!. Honolulu, Hawaii: Univeristy of Hawaii press
[4] Okada Y (1985) Surface deformation due to sheer and tensile fault in a half space. Bull Seism Soc Am 82:1135–54
[5] Gower J (2005) Jason 1 detects the December 2004 tsunami. EOS 86:37-38
[6] Constantin A and Johnson RS (2006) Modelling tsunami. J Phys A: Math Gen 39:L215-L217
[7] See for example, Bullough RK (1988) The wave par excellence: the solitary progressive great wave of equilibrium of the fluid-an early history of solitary wave. In: Lakshmanan M (Ed.) Solitons: Introduction and applications. Springer, Berlin
[8] Kortweg DJ and de Vries D (1895) On the change of form of long waves advancing in a rectangular canal, and on a new type of long stationary waves. Phil Mag 39:422-443
[9] Lakshmanan M and Rajasekar S (2003) Nonlinear Dynamics: Integrability, Chaos and Patterns. Springer, Berlin
[10] Ablowitz MJ and Clarkson PA (1991) Solitons, Nonlinear evolution equations and inverse scattering. Cambridge University Press, Cambridge
[11] Zabusky NJ and Kruskal MD (1965) Interaction of "solitons" in a collisionless plasma and the recurrence of initial states. Phys Rev Lett 15:240-43
[12] Gardner CS, Greene JM, Kruskal MD and Miura RM (1967) Method for solving the Kortweg-de Vries equation, Phys Rev Lett 19:1095-97
[13] Zakharov VE and Faddeev LD (1971) The Kortweg-de Vries equation is a fully integrable Hamiltonian system. Funct Anal Appl 5:280-87
[14] Hirota R (1971) Exact solution of the Kortweg-de Vries equation for multiple collisions of solitons. Phys Rev Lett 27:1192-94

[15] Camassa R and Holm D (1993) An integrable shallow water equation with peaked solitons. Phys Rev Lett 71:1661-64
[16] Calogero F and Degasperis A (1982) Spectral transform and solitons. North-Holland, Amsterdam
[17] Caputo JG and Stepanyants YA (2003) Bore formation, evolution and disintegration into solitons in shallow inhomogeneous channels. Nonlinear Processes in Geophysics 10:407-24

Solitary waves propagating over variable topography

Roger Grimshaw

Loughborough University, Loughborough, LE11 3TU, UK
R.H.J.Grimshaw@lboro.ac.uk

1 Introduction

Solitary water waves are long nonlinear waves that can propagate steadily over long distances. They were first observed by Russell in 1837 in a now famous report (27) on his observations of a solitary wave propagating along a Scottish canal, and on his subsequent experiments. Some forty years later theoretical work by Boussinesq (8) and Rayleigh (26) established an analytical model. Then in 1895 Korteweg and de Vries (22) derived the well-known equation which now bears their names. Significant further developments had to wait until the second half of the twentieth century, when there were two parallel developments. On the one hand it became realised that the Korteweg-de Vries equation was a valid model for solitary waves in a wide variety of physical contexts. On the other hand came the discovery of the *soliton* by Kruskal and Zabusky (28), with the subsequent rapid development of the modern theory of solitons and integrable systems.

In this chapter, we are mainly concerned with the behaviour of solitary waves as they propagate through a variable medium, with a particular emphasis on water waves over variable topography. But, first, we consider the well-known situation when the background medium is uniform. Solitary waves owe their existence to a balance between nonlinearity and wave dispersion. When both these effects are weak, the leading order model evolution equation is the Korteweg-de Vries (KdV) equation,

$$A_t + cA_x + \mu AA_x + \delta A_{xxx} = 0\,. \tag{1}$$

Here $A(x,t)$ is the amplitude of the relevant wave mode, which is assumed to be propagating in the x-direction, c is the speed of a linear long wave, μ, δ are the coefficients of the quadratic nonlinear term, and the third-order linear dispersive term, respectively, and, like c depend on the particular physical system being considered. The leading terms here are the first two, which describe a linear long wave propagating with speed c; relative to these dominant terms, the remaining two terms describe a balance between weak nonlinear steepening and weak linear dispersion. It is precisely this balance which describes solitary waves. For water waves propagating over a constant depth h it is well known that (22)

$$c = \sqrt{gh}\,, \quad \mu = \frac{3c}{2h}\,, \quad \delta = \frac{c}{6h^2}\,. \tag{2}$$

where A is the surface elevation above the undisturbed depth h. The KdV equation (1) is integrable (see, for instance, (1), (2), (4)) and many fascinating properties follow.

However, our main concern here is with the one-phase periodic travelling wave solutions of the KdV equation (1), the so-called cnoidal waves

$$A = a\{b(m) + \mathrm{cn}^2(\gamma\theta; m)\} + d\,, \tag{3}$$

$$\text{where} \quad \theta = k(x - Vt)\,, \tag{4}$$

$$b(m) = \frac{1-m}{m} - \frac{E(m)}{mK(m)}\,, \quad \mu a = 12m\delta\gamma^2 k^2\,, \tag{5}$$

$$\text{and} \quad V = c + \mu d + \frac{\mu a}{3}\left\{\frac{2-m}{m} - \frac{3E(m)}{mK(m)}\right\}. \tag{6}$$

Here $cn(x;m)$ is the Jacobian elliptic function of modulus $m, 0 < m < 1$ and $K(m), E(m)$ are the elliptic integrals of the first and second kind. The amplitude is a, the mean value of A over one period is d, while the spatial period is $2K(m)/\gamma k$. But since we can choose the phase θ so that A is 2π-periodic in θ we see that $\gamma = K(m)/\pi$. This periodic travelling wave (3) contains three free parameters; we take these to be the amplitude a, the mean level d and the modulus m, so that equations (5, 6) then determine k, V respectively.

As the modulus $m \to 1$, this becomes a solitary wave, since then $b \to 0$ and $\mathrm{cn}^2(x) \to \mathrm{sech}^2(x)$; in this limit $\gamma \to \infty, k \to 0$ with $\gamma k = K\Gamma$ held fixed. The outcome is the well-known solitary wave solution

$$A = a\,\mathrm{sech}^2(\Gamma(x - Vt))\,, \tag{7}$$

$$\text{where} \quad V = c + \frac{\mu a}{3} = 4\delta\Gamma^2\,. \tag{8}$$

On the other hand, as $m \to 0, \gamma \to 1/2$, and (3) reduces to sinusoidal waves of small amplitude $a \sim m$ and wavenumber k.

Many studies of weakly nonlinear long waves have used the KdV equation (1) or similar model equations with constant coefficents. However, particularly in the case of water waves in the coastal oceans, or for internal solitary waves propagating over the continental slope and shelf (see, for instance, (5)), (6), (16), (7)), there is a need to take account of the variation of the background medium in the wave propagation direction. In that case, the constant-coefficient equation (1) may be replaced by a variable-coefficient Korteweg-de Vries (vKdV) equation

$$A_t + cA_x + \frac{cQ_x}{2Q}A + \mu AA_x + \delta A_{xxx} = 0\,, \tag{9}$$

Here $A(x,t)$ is again the amplitude of the wave, but now the speed c and the coefficients μ, δ depend on x. The coefficient c is the relevant linear long wave speed, and Q is the linear magnification factor, defined so that QA^2 is the wave action flux. The vKdV equation was derived by Johnson (19) for water waves , where $Q = c$, and by Grimshaw (13) for internal waves (for recent reviews, see (5), (6)). The derivation assumes the usual KdV balance that the amplitude A has the same order as the dispersion, measured by $\partial^2/\partial x^2$, and in addition assumes that the waveguide properties (i.e. the coefficients c, Q, μ, λ) vary slowly so that Q_x/Q for instance is of the same order as the dispersion. In section 2 we shall give a brief description of the derivation of (9) for water waves.

As the first two terms in (9) are again the dominant terms, it is useful to make the transformation

$$\eta = \sqrt{Q}\,A\,, \quad \tau = \int^{x} \frac{dx}{c}\,, \quad \xi = \tau - t\,. \tag{10}$$

Substitution into (9) yields, to the same order of approximation as in the derivation of (9),

$$\eta_\tau + \alpha\eta\eta_\xi + \beta\eta_{\xi\xi\xi} = 0 \tag{11}$$

$$\alpha = \frac{\mu}{c\sqrt{Q}}\,, \quad \beta = \frac{\delta}{c^3}\,. \tag{12}$$

Here the coefficients α, β are functions of τ alone. Note that although τ is a variable along the spatial path of the wave, we shall subsequently refer to it as the "time". Similarly, although ξ is a temporal variable, in a reference frame moving with speed c, we shall subsequently refer to it as a "space" variable. The following sections are concerned with the derivation and with the solutions of equation (11)

2 Derivation of a variable coefficient Korteweg-de Vries equation

For simplicity, we shall describe the derivation of the variable coefficient KdV equation (9) for the main case of interest here, namely water waves propagating over variable topography. We consider a one-dimensional wave field so that the free-surface is represented by $z = \zeta(x,t)$ for an incompressible, inviscid fluid with constant density ρ, occupying the region $-h(x) < z < \zeta$ where the undisturbed depth $h(x)$ varies with x. The velocity field is $\mathbf{u} = (u, w)$ and can be assumed to be irrotational, so that $\mathbf{u} = \nabla\phi$, where $\phi(x,z,t)$ satisfies Laplace's equation

$$\phi_{xx} + \phi_{zz} = 0\,. \tag{13}$$

At the rigid bottom,

$$w + uh_x = 0 \quad \text{at} \quad z = -h. \tag{14}$$

At the free surface, the flow must satisfy two conditions

$$\zeta_t + u\zeta_x = w \quad \text{at} \quad z = \zeta\,, \tag{15}$$

$$\phi_t + \frac{|\mathbf{u}|^2}{2} + g\zeta = 0 \quad \text{at} \quad z = \zeta\,. \tag{16}$$

The first of these is the kinematic condition, and the second is the condition for constant pressure, where the Bernoulli relation has been used.

In order to obtain the vKdV equation (9) we shall use a multi-scale asymptotic expansion. This is a versatile approach and can be adapted to many other situations. Thus we introduce a small parameter, $\epsilon << 1$ measuring linear wave dispersion, and assume the usual KdV balance where the amplitude scales with ϵ^2. We then rescale the horizontal coordinate and the time, so that

$$X = \epsilon x\,, \quad T = \epsilon t\,, \tag{17}$$

and seek an asymptotic expansion of the form

$$\zeta = \epsilon^2 \zeta^{(1)}(X,T) + \epsilon^4 \zeta^{(2)}(X,T) + \cdots . \tag{18}$$

There is a similar expansion for the other fluid variables. At the same time, we assume that the depth varies slowly on a spatial scale of ϵ^{-3}, so that we may formally write $h = h(\chi)$ where $\chi = \epsilon^2 X$.

It is convenient to define the depth-averaged mean flow

$$U(X,T) = \frac{1}{h+\zeta} \int_{-h}^{\zeta} u(X,T,z)\, dz . \tag{19}$$

Then it is readily shown that conservation of mass implies that

$$\zeta_T + (U(h+\zeta))_X = 0 . \tag{20}$$

At the leading order we get

$$\zeta_T^{(1)} + hU_X^{(1)} = 0 , \tag{21}$$

$$U_T^{(1)} + g\zeta_X^{(1)} = 0 . \tag{22}$$

The general solution of this system is the sum of a wave propagating in the positive X-direction with speed c, and a wave propagating in the negative X-direction also with speed c, where we recall that $c = \sqrt{gh}$, which here depends on the slow variable χ, $c = c(\chi)$. We choose a wave propagating to the right, so that to leading order we get

$$\zeta^{(1)} = \frac{h}{c} U^{(1)} = A(\sigma, \theta) , \tag{23}$$

$$\text{where} \quad T^* = \int^X \frac{dX}{c} , \quad \theta = T^* - T , \quad \sigma = \epsilon^2 T^* . \tag{24}$$

Here we have anticipated that as the wave propagates to the right with speed c, it also evolves on the long spatial scale of ϵ^{-3}, and so we have introduced the slow variable σ. Note that the either of the slow variables σ or χ could be used here, but we have preferred the former as it has the dimensions of time.

At the next order we obtain the system of equations

$$\zeta_T^{(2)} + hU_X^{(2)} = F^{(2)} , \tag{25}$$

$$U_T^{(2)} + g\zeta_X^{(2)} = G^{(2)} , \tag{26}$$

$$\text{where} \quad F^{(2)} = -\zeta_\sigma^{(1)} - \frac{1}{c}(U^{(1)}\zeta^{(1)})_\theta - \frac{c_\sigma}{c}\zeta^{(1)} \tag{27}$$

$$G^{(2)} = -\frac{c}{h}\zeta_\sigma^{(1)} - \frac{1}{c}U^{(1)}U_\theta^{(1)} - \frac{h}{3c}\zeta_{\theta\theta\theta}^{(1)} . \tag{28}$$

Note that here, to leading order, $\zeta_X^{(1)} = \zeta_\theta^{(1)}/c$. From (23) the inhomogeneous terms are function of σ, θ, and so, to leading order this system of equations reduces to

$$-c\zeta_\theta^{(2)} + hU_\theta^{(2)} = F^{(2)} , \tag{29}$$

$$-cU_\theta^{(2)} + g\zeta_\theta^{(2)} = G^{(2)} . \tag{30}$$

The homogeneous version of the system (29, 30) has a non-trivial solution, namely the right-propagating wave $\zeta^{(1)}, U^{(1)}$ given by (23). Hence the inhomogeneous system

(29, 30) can have a solution only if the inhomogeneous terms on the left-hand side are orthogonal to the non-trivial solution of the homogeneous adjoint system. This is readily found to be (c, h) and so the required compatibility condition is

$$cF^{(2)} + hG^{(2)} = 0\,. \tag{31}$$

Next we substitute the expressions (23) into (31) and after some simplification get

$$A_\sigma + \frac{c_\sigma}{2c}A + \frac{3}{2h}AA_\theta + \frac{h^2}{6c^2}A_{\theta\theta\theta} = 0\,. \tag{32}$$

Using the transformations (17, 24) and replacing $\epsilon^2 A$ with A this becomes

$$A_t + cA_x + \frac{c_x}{2}A + \frac{3c}{2h}AA_x + \frac{hc^2}{6}A_{xxx} = 0\,, \tag{33}$$

This is just the vKdV equation (9) for the case of water waves. Finally, using the transformation (10) with $Q = c$ equation (33) becomes

$$\eta_\tau + \alpha^*\eta\eta_\xi + \beta^* A_{\xi\xi\xi} = 0 \tag{34}$$

$$\text{where here}\quad \eta = \sqrt{c}A \quad \alpha^* = \frac{3}{2h\sqrt{c}}\,, \quad \beta^* = \frac{h^2}{6c^2}\,. \tag{35}$$

The same type of multiscale asymptotic expansion can be used to derive a vKdV equation in many other physical systems. The key is the existence of a waveguide supporting a linear wave mode, whose dispersion relation for unidirectional sinusoidal waves, propagating along the waveguide (in the x-direction) with frequency ω and wavenumber k, has a long-wave expansion of the form.

$$\omega = ck - \delta k^3 + O(k^5)\,. \tag{36}$$

A typical fluid variable, say $u(x, t, z)$ can then be represented in the form

$$u = \epsilon^2 A(\sigma, \theta)\phi(z) + \epsilon u^{(2)} + \cdots\,. \tag{37}$$

Here the scaled variables σ, θ are again defined by (24), and $\phi(z, \sigma)$ is a known modal function in the z-direction, where z is a coordinate across the waveguide. For instance, for water waves and when u is the amplitude of the free surface elevation, $\phi(z) = z/h$ ($c = \sqrt{gh}$ in this case). But in most physical systems, the dependence on z is not so simple, and is determined by an associated eigenvalue problem, which also determines the linear long-wave speed c. For instance, this is the situation for internal waves (5). Note that in a slowly-varying inhomogeneous medium, the modal function also depends parametrically on σ (for water waves, through $h(\chi)$). It is immediately clear that for linearized waves in a homogeneous medium, the amplitude A will satisfy the linearization of the KdV equation (1). The next task is to find the magnification factor Q in (9) when the medium is inhomogeneous. This is most easily accomplished by finding the equation for conservation of wave action flux QA^2 in the linear long wave regime, Thus the main task of the multiscale aysmptotic expansion is the determination of the nonlinear coefficient μ. This is accomplished by constructing the equation for the second-order term in (37). This inevitably, as for water waves, takes the form of a linear inhomogeneous system, whose homogeneous part is just the defining equation for the linear long-wave mode being considered. Hence, the inhomogeneous system requires a compatibility condition, which yields the required KdV equation (9).

3 Slowly-varying waves

3.1 Cnoidal waves

Although our main concern will be with the behaviour of solitary waves, it is instructive to first review the asymptotic theory for slowly-varying periodic waves, namely here the cnoidal waves defined by (3). In this case the theory is analogous to the well-known WKB procedure for linear waves. One can either use a multi-scale asymptotic expansion, or make use of appropriate conservation laws (see (4) for instance). Here we shall use the former approach, and so we now suppose that the coefficients α, β in (11) are slowly-varying, and write

$$\alpha = \alpha(S)\,, \quad \beta = \beta(S)\,, \quad S = \nu\tau\,, \quad Y = \nu\xi\,, \quad \nu << 1\,. \tag{38}$$

Here the slow temporal variable S is introduced to provide an explicit description of the separation of scales between the variation of the coefficients and the more rapidly oscillating waves, while Y is an analogous slow spatial variable. Next we seek a multi-scale expansion for a modulated periodic wave, namely

$$\eta = \eta_0(\psi, S, Y) + \epsilon\eta_1(\psi, S, Y) + \cdots\,, \tag{39}$$

$$\text{where} \quad \psi = \frac{1}{\nu}\Psi(S, Y)\,. \tag{40}$$

$$\text{and} \quad k = \Psi_Y\,, \ kV = -\Psi_S\,. \tag{41}$$

It is assumed that η is periodic in ψ with a fixed period of 2π. Equation (41) defines the local wavenumber k, the local frequency kV, and the local phase speed V. Cross-differentiation yields the equation for conservation of waves

$$k_S + (kV)_Y = 0\,. \tag{42}$$

We should recall here, that although we have called k a wavenumber and kV a frequency, they are not the actual phsyical wavenumber and frequency, because of the transformation of variables defined in (10). Indeed the physical wavenumber is $k(1-W)/c$, the physical frequency is k, and the physical phase speed is $c/(1-W) \approx c(1+W)$, since $W << 1$ due to the scaling used to derive the KdV equation (11)

Substitution of (39) into (11) yields, at the leading orders

$$-V\eta_{0\psi} + \alpha\eta_0\eta_{0\psi} + \beta k^2\eta_{0\psi\psi\psi} = 0\,, \tag{43}$$

$$-V\eta_{1\psi} + \alpha(\eta_0\eta_1)_\psi + \beta k^2\eta_{1\psi\psi\psi} = \frac{1}{k}F_1\,, \tag{44}$$

$$\text{where} \quad F_1 = -\eta_{0S} - \alpha\eta_0\eta_{0Y} - 3\beta k^2\eta_{0\psi\psi Y} - 3\beta k k_Y\eta_{0\psi\psi}\,. \tag{45}$$

Each of these is essentially an ordinary differential equation with ψ as the independent variable, and with S, Y as parameters. The solution of (43) is the cnoidal wave (see (3))

$$\eta_0 = a\{b(m) + \mathrm{cn}^2(\gamma\psi; m)\} + d\,, \tag{46}$$

$$\text{where} \quad b(m) = \frac{1-m}{m} - \frac{E(m)}{mK(m)}\,, \quad \alpha a = 12m\beta\gamma^2k^2\,, \tag{47}$$

$$\text{and} \quad V = \alpha d + \frac{\alpha a}{3}\left\{\frac{2-m}{m} - \frac{3E(m)}{mK(m)}\right\}\,. \tag{48}$$

As before $\gamma = K(m)/\pi$, since η_0 is 2π-periodic in the phase ψ. In this solution the parameters a, m, γ, V, d each depend on the slow variables S, Y and now the task is to determine how they vary as functions of these slow variables. Note that there are three independent parameters, and hence three equations are needed. However, one of these is the conservation of waves equation (42). There are two main methods used to find the remaining two equations. One is the so-called Whitham averaging method, where one seeks two appropriate conservation laws for the vKdV equation (11), inserts the cnoidal wave (3) into these laws, and then averages over the phase ψ (see (3), (4)). Here we shall first describe the second approach which is to continue the asymptotic expansion to the next order, and then invoke the condition that A_1 is a periodic function of ψ. It is implicit in the Whitham averaging procedure that the higher-order terms in the expansion have this property. Although it can be shown that the presence of a suitable underlying Lagrangian usually ensures that this is so (see (4)), we shall nevertheless verify it directly here for the first-order term. This is given by (44) in which the right-hand side is now a known periodic function of ψ, given by (45). A necessary and sufficient condition for η_1 to be periodic in ψ is that the right-hand side of (44) should be orthogonal to the periodic solutions of the adjoint to the homogeneous operator on the left-hand side. This adjoint is

$$-V\eta_{1\theta} + \alpha\eta_0\eta_{1\psi} + \beta k^2\eta_{1\psi\psi\psi} = 0\,.. \tag{49}$$

It is readily seen that two solutions of (49) are $1, \eta_0$, both of which are periodic. A third solution can be found by the variation-of-parameters method, but it is not periodic. Hence there are two orthogonality conditons, given by

$$M_{0S} + (\alpha P_0)_Y = 0\,, \tag{50}$$
$$P_{0S} + (\alpha Q_0)_Y = 0\,, \tag{51}$$

$$M_0 = \frac{1}{2\pi}\int_0^{2\pi} \eta_0\, d\psi = d\,, \quad P_0 = \frac{1}{4\pi}\int_0^{2\pi} \eta_0^2\, d\psi\,, \tag{52}$$

$$Q_0 = \frac{1}{6\pi}\int_0^{2\pi} \eta_0^3\, d\psi - \frac{3\beta k^2}{4\alpha\pi}\int_0^{2\pi} \eta_{0\psi}^2\, d\psi\,. \tag{53}$$

As we discuss further below the first of these is the equation for conservation of "mass", and the second is an equation for conservation of "momentum". The final step is the substitution of (46) into (52, 53), which gives

$$P_0 = \frac{d^2}{2} + \frac{a^2}{2}\{C_4 - b^2\}\,, \tag{54}$$

$$Q_0 = \frac{d^3}{3} + da^2\{C_4 - b^2\} + \frac{a^3}{3}\{C_6 - 3bC_4 - 2b^3\} + \frac{3\beta k^2\gamma^2 a^2}{2\alpha}\{mC_6 - (2m-1)C_4 + (1-m)b\} \tag{55}$$

$$C_4 = \frac{1}{K}\int_0^K \mathrm{cn}^4(u;m)\, du\,, \tag{56}$$

$$C_6 = \frac{1}{K}\int_0^K \mathrm{cn}^6(u;m)\, du\,,\,. \tag{57}$$

Here we recall that $K(m)$ is the first elliptic integral, while $b(m)$ is defined in (47), and so $C_4(m), C_6(m)$ are functions of the modulus m alone, given by

$$3m^2 C_4(m) = (3m^2 - 5m + 2) + (4m - 2)\frac{E(m)}{K(m)}\,, \tag{58}$$

$$15m^3 C_6(m) = (15m^3 - 34m^2 + 27m - 8) + (23m^2 - 23m + 8)\frac{E(m)}{K(m)}\,. \tag{59}$$

The alternative approach to the derivation of (50, 51) is to make direct use of the conservations laws for "mass" and "momentum"

$$\eta_\tau + \{\frac{\alpha}{2}\eta^2 + \beta\eta_{\xi\xi}\}_\xi = 0\,, \tag{60}$$

$$(\frac{\eta^2}{2})_\tau + \{\frac{\alpha}{3}\eta^3 + \{\beta(\eta\eta_\xi)_\xi - \frac{3\beta}{2}\eta_\xi^2\}_\xi = 0\,. \tag{61}$$

Each of these is readily established from (11) and indeed (60) is just the vKdV equation (11) itself. Note that although we shall call these the laws for conservation of mass and momentum, the integrands do not necessarily correspond to the corresponding physical entities. Indeed, to leading order, (61) is usually the law for conservation of wave action flux. The conservation law (60) usually differs slightly from the actual law for conservation of mass . The issue has been explored by Miles (23) for water waves, where it can be shown that the difference is smaller than the error incurred in the derivation of the vKdV equation (33), and is due to reflected waves. The Whitham averaging procedure now consists of the substitution of (46) into (60, 61), and then averaging the results over the phase (see (3), (4)). The result is readily seen to be the derived equations (50, 51).

The equation set (42, 50, 51) are the three desired equations for the three chosen parameters, (k, d, m) say, and form a nonlinear hyperbolic system, provided that the underlying periodic wave is stable. They are quite complicated, and in general it is difficult to find explicit solutions. The issue is present even when the coefficients α, β are constant. But in that case, due to the integrability of the KdV equation, it can be shown that a subtle change of variables leads to a set of three nonlinear hyperbolic equations in Riemann form, see (3), (4) for instance. The resulting equations are also integrable, through a generalized hodograph tarnsformation. However, in the general case when the coefficients $\alpha = \alpha(S), \beta = \beta(S)$, no such reduction is available. Instead the system remains coupled, although it can be cast into a more transparent form using the same Riemann variables available in the integrable case, see (21).

However, one situation of interest can be solved explicitly. Let us suppose that the solution set for (k, d, m) depends only on S. Then equations (42, 50) readily show that $k, M_0 = d$ are respectively constants. The remaining variable m is then found from the remaining equation (51), whose solution is P_0 = constant. Then, using (54, 56) and the relation (47), we find that the amplitude a and m are related by the expression

$$\frac{a^2}{m^2}\{3m^2 - 5m + 2) + (4m - 2)\frac{E(m)}{K(m)} - 3m^2 b(m)^2\} = \text{constant}\,. \tag{62}$$

Finally, using the relation (47) we can determine the modulus m in terms of $\alpha(S), \beta(S)$,

$$F(m) = \text{constant}\,\frac{\alpha^2}{\beta^2}\,, \quad \text{where} \tag{63}$$

$$F(m) = K(m)^2\{(4 - 2m)E(m)K(m) - 3E(m)^2 - (1 - m)K(m)^2\}\,. \tag{64}$$

This expression for $F(m)$ was obtained by Ostrovsky and Pelinovsky (24), (25) and Miles (23) for the special case of water waves , where $\alpha/\beta = 9g^{3/4}/h^{9/4}$ (see (35)). $F(m)$ is plotted in Figure 1, which shows that it is a monotonically increasing function of m. It follows that as α/β increases so does m. Two limiting situations are of interest. First, if the nonlinear coefficient α decreases towards zero, then so does the modulus m where it can be shown that $F(m) \sim m^2$ as $m \to 0$; it follows that the modulus $m \sim \alpha$, but remarkably the amplitude a is finite in this limit. On the other hand, if the dispersive coefficient $\beta \to 0$, then $m \to 1$ and the waves become more like solitary waves . For water waves, we see that this situation arises as the waves propagate into shallow water, that is $h \to 0$. In this limit, the expressions (63, 64) show that $a \sim K(m) \sim h^{-3/4}$. Recalling the transformation (35), this leads to the well-known result that a solitary wave propagating in shallow water behaves as h^{-1}, see (11), (24), (20).

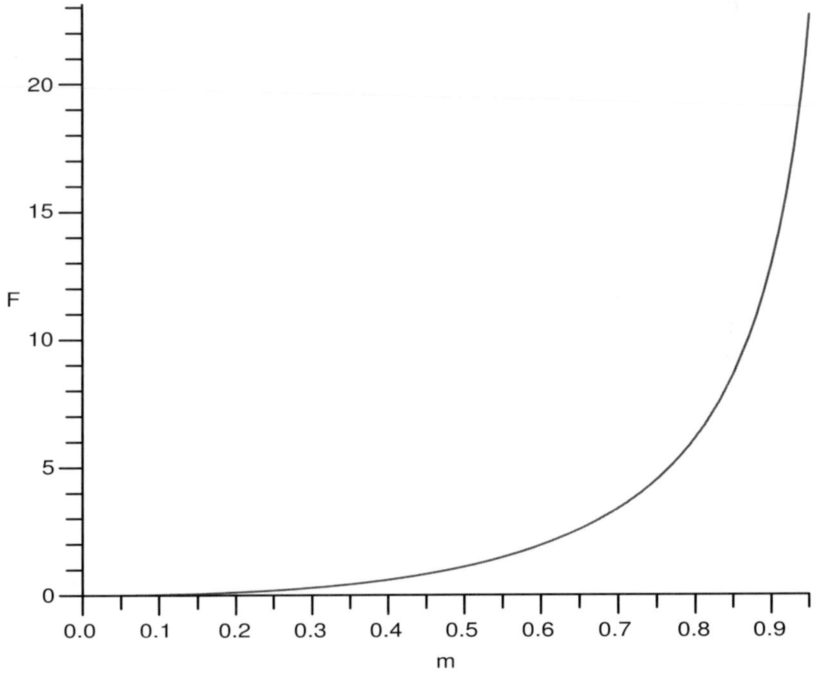

Fig. 1. A plot of $F(m)$ (64) versus m.

3.2 Solitary waves

The results obtained above for a slowly-varying periodic wave cannot immediately be extrapolated to a slowly-varying solitary wave , as the limits $m \to 1$ and $\nu \to 0$

do not commute. In physical terms, the basis for the validity of the slowly-varying periodic wave is that the local wavelength (i.e. $2\pi/k$) should be much less than the slowly-varying scale (i.e. $1/\nu$). The limit $m \to 1$ in (46, 47, 48) requires that $\gamma \to \infty, k \to 0$ with $\gamma k = \Gamma$ held constant, and so the wavelength technically becomes much larger than the slow scale. A new concept of slowly-varying is needed, which in physical terms is that the half-width (i.e. width of the solitary wave at the level of one half of the maximum amplitude) should be much less than $1/\nu$. We proceed as above and again invoke a multi-scale asymptotic expansion of the same form (38. 39) and we again obtain the equations (43, 44, 45) (see, for instance, (12) (14)). Note that k has a different meaning here, but that nevertheless equation (42) for conservation of waves still holds. but now we do not require that η is periodic in ψ with a fixed period. Instead we require that η is bounded in the limits $\psi \to \pm\infty$ (more strictly, we require that η grows at infinity as $\nu^{m+1}\psi^m$ at each order ν^{m+1} for $m = 0, 1, 2 \cdots$). We can suppose without loss of generality that $\delta > 0$, since the alternative case is recovered by replacing η, ξ with $-\eta, -\xi$ respectively. Then, small-amplitude waves will propagate in the negative ξ-direction, and we can suppose that $\eta \to 0$ as $\psi \to \infty$. However, it will transpire that we cannot impose this boundary condition as $\psi \to -\infty$. Note that it is possible to allow $\eta \to D^+(S, Y)$ as $\psi \to \infty$ where $D^+(S, Y)$ is known a priori, and satisfies the vKdV equation (11), that is

$$D_S^+ + \alpha D^+ D_Y^+ + \nu^2 \beta D_{YYY}^+ = 0 \,. \tag{65}$$

However, we shall not consider this extension here, although note that the basic procedure outlined below can be used also in this more general case.

Now the solution for η_0 is taken to be the solitary wave (see(7))

$$A = a\text{sech}^2(\Gamma\psi) \,, \tag{66}$$

$$\text{where} \quad V = \frac{\alpha a}{3} = 4\beta k^2 \Gamma^2 \,. \tag{67}$$

Here there is just a single free parameter, one of the set $(a(S, Y), V(S, Y), k\Gamma(S, Y))$, which depends on the slow variables. Its behaviour is determined by examining the next order equation (44), from which we seek a solution η_1 which is bounded as $\psi \to \pm\infty$: as above we can require that $\eta_1 \to 0$ as $\psi \to \infty$. As before, the adjoint equation to (44) is (49) for which two solutions are $1, \eta_0$; while both are bounded, only the second solution satisfies the condtion that $\eta_1 \to 0$ as $\psi \to \infty$. A third solution can be constructed using the variation-of-parameters method, but it is unbounded as $\psi \to \pm\infty$. Hence only one othogonality condition can now be imposed, namely that the right-hand side of (44) is orthogonal to η_0, which leads to (51) where now

$$P_0 = \frac{1}{2}\int_{-\infty}^{\infty} \eta_0^2 \, d\psi \,, \tag{68}$$

$$Q_0 = \frac{1}{3}\int_{-\infty}^{\infty} \eta_0^3 \, d\psi - \frac{3\beta k^2}{2\alpha}\int_{-\infty}^{\infty} \eta_{0\psi}^2 \, d\psi \,. \tag{69}$$

Substituting (66, 67) into (68, 69) yields

$$P_0 = \frac{2a^2}{3\Gamma} \,, \quad Q_0 = \frac{2a^3}{9\Gamma} \,. \tag{70}$$

Finally, substitution of these expressions into (51) yields

$$P_{0S} + (VP_0)_Y = o\,. \tag{71}$$

As the solitary wave (66) has just one free parameter (e.g. the amplitude a), this equation, together with (42) suffices to determine its variation.

We now recall that the vKdV equation (11) possesses two conservation laws, namely (60, 61) for mass and momentum respectively. But here we see that the equation (61), which reduces here to (71) for momentum is sufficient to determine the slowly-varying solitary wave. Hence we now see that the slowly-varying solitary wave cannot simultaneously conserve its mass. This is apparent when one examines the solution of (44) for η_1, from which it is readily shown that although $\eta_1 \to 0$ as $\psi \to \infty$, $\eta_1 \to D_1$ as $\psi \to -\infty$ where

$$-kVD_1 = M^s_{0S} + (\alpha P_0)_Y\,, \tag{72}$$

$$M^s_0 = \int_{-\infty}^{\infty} \eta_0 \, d\psi = \frac{2a}{\Gamma}\,. \tag{73}$$

Here M^s_0 can be interpreted as the mass of the solitary wave. It is readily verified that the right-hand side of (72) cannot vanish in general, and so the slowly-varying solitary wave generates a trailing shelf whose amplitude at the rear of the solitary wave is D_1. This non-uniformity has been recognized for some time, see, for instance, (14) and the references therein. The trailing shelf η_{sh} has small amplitude $O(\nu)$ but a long length-scale $O(1/\nu)$, and so carries $O(1)$ mass, but $O(\nu)$ momentum. It resides behind the solitary wave, and to leading order has a value independent of S, so that $\eta_{sh} = \nu A_1(Y)$ for $\Psi < 0$; it is determined by its value at the location $\Psi = 0$ of the solitary wave, namely $A_1(\Psi = 0) = D_1$ (72). It may readily be verified that the slowly-varying solitary wave and the trailing shelf together satisfy conservation of mass. The asymptotic expansion may be continued to second order (14). At higher orders in ν the shelf itself will evolve and may generate secondary solitary waves (10)

The two equations (42, 51) may be uncoupled by defining

$$\tilde{P}_0 = \frac{P_0}{k} = \frac{2a^3}{3\Gamma k} = \sqrt{\frac{\beta a^2}{27\alpha}}\,, \tag{74}$$

$$\text{so that} \quad \tilde{P}_{0S} + V\tilde{P}_{0Y} = 0\,. \tag{75}$$

Here we have used (67) to show that $\tilde{P}_0$, the momentum per unit distance, is a function of the amplitude a alone. Since $V = \alpha a/3$ (67) we see that equation (74) is a single equation for the amplitude. It show that the amplitude deforms to conserve $\tilde{P}_0$, which propagates with the solitary wave speed V.

An important special case arises when there is no Y-dependence. In that case (42) shows that k, $\tilde{P}_0$ are both constants, and so (75) reduces to

$$a^3 = \text{constant}\frac{\alpha}{\beta}\,, \quad \text{or} \quad (\Gamma k)^3 = \text{constant}\frac{\alpha^2}{\beta^2}\,. \tag{76}$$

In this same special case, (72) reduces to

$$Vk\Gamma D_1 = 2a_s\,. \tag{77}$$

The expression (76) shows that the amplitude increases (decreases) in absolute value as α/β increases (decreases). Note that the polarity of the wave is determined by the sign of α/β. Then, assuming without loss of generality that $\beta > 0$ so that $V > 0$, we see from (77) that a slowly-varying solitary wave of increasing (decreasing) amplitude, will generate a trailing shelf of the same (opposite) polarity. A particular case of interest is when the nonlinear coefficient α passes through zero, while β stays finite. Suppose this occurs at $S = 0$, where, without loss of generality, we may suppose that α passes from positive to negative values as S increases. Initially the solitary wave is located in $S < 0$ and has positive polarity. Then, near the transition point, the amplitude of the wave decreases to zero as $a \sim \alpha^{1/3}$, while $\Gamma \sim \alpha^{2/3}$; the momentum of the solitary wave is of course conserved (at least to leading order), the mass of the solitary wave increases as $1/\alpha^{1/3}$, its speed decreases as $\alpha^{4/3}$, and the amplitude D_1 of the trailing shelf just behind the solitary wave grows as $-1/\alpha^{8/3}$; the total mass of the trailing shelf grows as $-1/\alpha^{1/3}$, in balance with that of the solitary wave, while the total mass remains a positive constant. Thus the solitary wave itself is destroyed as the wave attempts to pass through the critical point $\alpha = 0$. The structure of the solution beyond this critical point has been examined numerically by Grimshaw et al (15), who showed that, in essence, the shelf passes through the critical point as a negative disturbance, which then being in an environment with $\alpha < 0$, can generate a train of solitary waves of negative polarity, riding on a positive pedestal. Of course, these conclusions may need to be modified when a cubic nonlinear term in is taken into account near the critical point (see Grimshaw et al (16)).

For the special case of water waves, $\alpha/\beta = 9g^{3/4}/h^{9/4}$ (see (35)) and so the amplitude behaves as $a \sim h^{-3/4}$, while the speed V behaves as h^{-2}, and the wavenumber Γ behaves as $h^{-3/2}$. Recalling the transformation (10 where $Q = c = (gh)^{1/2}$ for water waves, we recover the well-known result that the surface elevation amplitude of a surface solitary wave behaves as h^{-1} (see (11), (20), (24)). At the same time we see from (77) that the trailing shelf has negative polarity and behaves as $-h^2$ at the rear of the solitary wave.

4 Soliton fission

In the previous section we have considered the case when the variable coefficients in (11) vary slowly relative to a solitary wave . Here we consider the opposite case when the coefficients vary rapidly relative to a solitary wave. This scenario was considered by Johnson (19) for water waves and by Djordjevic and Redekopp (9) for internal waves . Let us suppose therefore that

$$\alpha = \alpha(Z)\,,\ \beta = \beta(Z)\,,\ \tau = \Delta Z\,,\ \Delta << 1\,, \tag{78}$$

$$\text{where}\quad \alpha(Z) \to \alpha_{a,b}\,,\ \beta(Z) \to \beta_{a,b}\,,\ \text{as } Z \to \pm\infty\,, \tag{79}$$

Thus, effectively the vKdV equation (11) has constant coefficients in $\tau < 0$ and in $\tau > 0$, while there is a small transition region in which τ is $O(\Delta)$ where the coefficients change their values from α_b, β_b in $\tau < 0$ to α_a, β_a in $\tau > 0$. In this transition zone, we may write $\eta = \eta(Z, \xi)$, and it is then readily shown that η_Z is $O(\Delta)$. Hence, to leading order, it follows that η is unchanged in the transition zone.

We now suppose that there is a single solitary wave in $\tau < 0$ (see (7, 8)),

$$A = a\,\mathrm{sech}^2(\Gamma(\xi - V\tau))\,, \tag{80}$$

$$\text{where}\quad V = \frac{\alpha_b a}{3} = 4\beta_b\Gamma^2\,. \tag{81}$$

This wave will pass through the transition zone unchanged, but, on arrival into the region $\tau > 0$ it is no longer a permissible solution of (11), which now has constant coefficients α_a, β_b. Instead, with $\tau = 0+$, the expression (80) forms an effective initial condition for the new constant-coefficient KdV equation. Using the inverse scattering transform, the solution in $\tau > 0$ can now be constructed; indeed in this case there is an explicit solution (e.g. (2)). The outcome is that the initial solitary wave *fissions* into N solitons , and some radiation. The number N of solitons produced is determined by the ratio of coefficients $R = \alpha_a\beta_b/\alpha_b\beta_a$. If $R > 0$ (i.e. there is no change in polarity for the solitary waves), then $N = 1+[(\sqrt{8R+1}-1)/2]$ ($[\cdots]$ denotes the integral part); as R increases from 0, a new soliton (initially of zero amplitude) is produced as R successively passes through the values $m(m+1)/2$ for $m = 1, 2, \cdots$. But if $R < 0$ (i.e. there is a change in polarity) no solitons are produced and the solitary wave decays into radiation.

For instance, for water waves $\alpha = 3/(2hc^{1/2}), \beta = h^2/(6c^2)$ (35) where h is the water depth. It can then be shown that a solitary water wave propagating from a depth h_b to a depth h_a will fission into N solitons where N is given as above with $R = (h_b/h_a)^{9/4}$. Here $R > 0$, and it follows that if $h_b > h_a$, so that the solitary wave moves into shallower water, $N \geq 2$ and at least one more soliton is produced; the initial soliton is said to have fissioned. But if $h_b < h_a$ so that the solitary wave moves into deeper water, then $N = 1$ and no further solitons are produced (19).

References

[1] Ablowitz, M. and H. Segur, H. (1981) Solitons and the Inverse Scattering Transform. SIAM Studies in Applied Mathematics **4**, Philadeplhia.

[2] Drazin, P.G. and Johnson, R.S. (1989). Solitons: an Introduction. Cambridge University Press, Cambridge.

[3] Kamchatnov, A.M. (2000) Nonlinear Periodic Waves and Their Modulations: An Introductory Course. World Scientific, Singapore.

[4] Whitham, G.B. (1974). Linear and Nonlinear Waves, Wiley, New York.

[5] Grimshaw, R. (2001) Internal solitary waves, in "Environmental Stratified Flows" . Kluwer, Boston, Chapter **1**, 1-28.

[6] P. Holloway, E. Pelinovsky and T. Talipova 2001. Internal tide transformation and oceanic internal solitary waves. In "Environmental Stratified Flows" (Kluwer, Boston, Chapter 2: 29-60, 2001).

[7] J. Rottman and R. Grimshaw. Atmospheric internal solitary waves. In "Environmental Stratified Flows" (Kluwer, Boston, Chapter 3: 61-88, 2001).

[8] Boussinesq. M.J. (1871) Theórie de l'intumescence liquide appellée onde solitaire ou de translation, se propageant dans un canal rectangulaire. *Comptes Rendus Acad. Sci (Paris)*, **72**, 755-759.

[9] Djordjevic, V. D. and Redekopp, L.G. (1978). The fission and disintegration of internal solitary waves moving over two-dimensional topography. *J. of Phys. Ocean.*, **8**, 1016-1024.

[10] El, G.A. and Grimshaw, R. (2002) Generation of undular bores in the shelves of slowly-varying solitary waves. *Chaos*, **12**, 1015-1026.

[11] Grimshaw, R. (1970). The solitary wave in water of variable depth. *J. Fluid Mech.*, **42**, 639-656.

[12] Grimshaw, R. (1979) Slowly varying solitary waves. I Korteweg-de Vries equation. *Proc. Roy. Soc.*, **368A**, 359-375.
[13] Grimshaw, R. (1981) Evolution equations for long nonlinear internal waves in stratified shear flows. *Stud. Appl. Math.*, **65**, 159-188.
[14] Grimshaw, R. and Mitsudera, H. (1993) Slowly-varying solitary wave solutions of the perturbed Korteweg-de Vries equation revisited. *Stud. Appl. Math.*, **90**, 75-86.
[15] Grimshaw, R., Pelinovsky, E. and Talipova, T. (1998) Solitary wave transformation due to a change in polarity. *Stud. Appl. Math.*, **101**, 357-388.
[16] Grimshaw, R., Pelinovsky, E. and Talipova, T. (1999). Solitary wave transformation in a medium with sign-variable quadratic nonlinearity and cubic nonlinearity. *Physica D*, **132**, 40-62.
[17] R. Grimshaw, E. Pelinovsky, T. Talipova, T. and A. Kurkin. Simulation of the transformation of internal solitary waves on oceanic shelves. *J. Phys. Ocean.*, **34**, 2774-2779 (2004).
[18] Gurevich, A. V. and Pitaevskii, L. P. (1974) Nonstationary structure of a collisionless shock wave. *Sov. Phys. JETP* **38**, 291–297.
[19] Johnson, R.S. (1973) On the development of a solitary wave moving over an uneven bottom. *Proc. Camb. Phil. Soc.*, **73**, 183-203.
[20] Johnson, R.S. (1973) On an asymptotic solution of the Korteweg - de Vries equation with slowly varying coefficients, *J. Fluid Mach.*, **60**, 813-824
[21] Kamchatnov, A.M. (2004). On Whitham theory for perturbed integrable equations, *Physica D*, **188**, 247–261 (2004).
[22] Korteweg, D.J. and de Vries, H. (1895) On the change of form of long waves advancing in a rectangular canal, and on a new type of long stationary waves. *Philosophical Magazine*, **39**, 422-443.
[23] Miles, J.W. (1979) On the Korteweg-de Vries equation for a gradually varying channel. *J. Fluid Mech.*, **91**, 181-190.
[24] Ostrovsky, L.A. and Pelinovsky, E.N. (1970) Wave transformation on the surface of a fluid of variable depth. *Akad. Nauk SSSR, Izv. Atmos. Ocean Phys.*, **6**, 552-555.
[25] Ostrovsky, L.A. and Pelinovsky, E.N. (1975). Refraction of nonlinear sea waves in a coastal zone. . *Akad. Nauk SSSR, Izv. Atmos. Ocean Phys.*, **11**, 37-41.
[26] Lord Rayleigh (1876) On waves. *Phil. Mag.* **1**, 257-279.
[27] Russell, J.S. (1844) Report on Waves. 14th meeting of the British Association for the Advancement of Science, 311-390.
[28] Zabusky, N.J. and Kruskal, M.D. (1965). Interactions of solitons in a collisionless plasma and the recurrence of initial states. *Physical Review Letters*, **15** 240-243.

Water waves generated by a moving bottom

Denys Dutykh, Frédéric Dias

Centre de Mathématiques et de Leurs Applications, Ecole Normale Supérieure de Cachan, 61 avenue du Président Wilson, 94235 Cachan cedex, France
dutykh@cmla.ens-cachan.fr, dias@cmla.ens-cachan.fr

1 Introduction

Waves at the surface of a liquid can be generated by various mechanisms: wind blowing on the free surface, wavemaker, moving disturbance on the bottom or the surface, or even inside the liquid, fall of an object into the liquid, liquid inside a moving container, etc. In this paper, we concentrate on the case where the waves are created by a given motion of the bottom. One example is the generation of tsunamis by a sudden seafloor deformation.

There are different natural phenomena that can lead to a tsunami. For example, one can mention submarine slumps, slides, volcanic explosions, etc. In this article we use a submarine faulting generation mechanism as tsunami source. The resulting waves have some well-known features. For example, characteristic wavelengths are large and wave amplitudes are small compared with water depth.

Two factors are usually necessary for an accurate modelling of tsunamis: information on the magnitude and distribution of the displacements caused by the earthquake, and a model of surface gravity waves generation resulting from this motion of the seafloor. Most studies of tsunami generation assume that the initial free-surface deformation is equal to the vertical displacement of the ocean bottom. The details of wave motion are neglected during the time that the source operates. While this is often justified because the earthquake rupture occurs very rapidly, there are some specific cases where the time scale of the bottom deformation may become an important factor. This was emphasized for example by Trifunac and Todorovska (1), who considered the generation of tsunamis by a slowly spreading uplift of the seafloor and were able to explain some observations. During the 26 December 2004 Sumatra-Andaman event, there was in the northern extent of the source a relatively slow faulting motion that led to significant vertical bottom motion but left little record in the seismic data. It is interesting to point out that it is the inversion of tide-gauge data from Paradip, the northernmost of the Indian east-coast stations, that led Neetu et al. (2) to conclude that the source length was greater by roughly 30% than the initial estimate of Lay et al. (3). Incidentally, the generation time is also longer for landslide tsunamis.

Our study is restricted to the water region where the incompressible Euler equations for potential flow can be linearized. The wave propagation away from the source can be investigated by shallow water models which may or may not take into account nonlinear effects and frequency dispersion. Such models include the Korteweg-de Vries equation (4) for unidirectional propagation, nonlinear shallow-water equations and Boussinesq-type models (5; 6; 7).

Several authors have modeled the incompressible fluid layer as a special case of an elastic medium (8; 9; 10; 11; 12). In our opinion it may be convenient to model the liquid by an elastic material from a mathematical point of view, but it is questionable from a physical point of view. The crust was modeled as an elastic isotropic half-space. This assumption will also be adopted in the present study.

The problem of tsunami generation has been considered by a number of authors: see for example (13; 14; 15). The models discussed in these papers lack flexibility in terms of modelling the source due to the earthquake. The present paper provides some extensions. A good review on the subject is (16).

Here we essentially follow the framework proposed by Hammack (17) and others (18). The tsunami generation problem is reduced to a Cauchy-Poisson boundary value problem in a region of constant depth. The main extensions given in the present paper consist in three-dimensional modelling and more realistic source models. This approach was followed recently in (1; 19), where the mathematical model was the same as in (17) but the source was different.

Most analytical studies of linearized wave motion use integral transform methods. The complexity of the integral solutions forced many authors (9; 20) to use asymptotic methods such as the method of stationary phase to estimate the far-field behaviour of the solutions. In the present study we have also obtained asymptotic formulas for integral solutions. They are useful from a qualitative point of view, but in practice it is better to use numerical integration formulas (21) that take into account the oscillatory nature of the integrals. All the numerical results presented in this paper were obtained in this manner.

One should use asymptotic solutions with caution since they approximate exact solutions of the linearized problem. The relative importance of linear and nonlinear effects can be measured by the Stokes (or Ursell) number (22):

$$U := \frac{a/h}{(kh)^2} = \frac{a}{k^2 h^3},$$

where k is a wave number, a a typical wave amplitude and h the water depth. For $U \gg 1$, the nonlinear effects control wave propagation and only nonlinear models are applicable. Ursell (22) proved that near the wave front U behaves like

$$U \sim t^{\frac{1}{3}}.$$

Hence, regardless of how small nonlinear effects are initially, they will become important.

Section 2 provides a description of the tsunami source when the source is an earthquake. In Section 3, we review the water-wave equations and provide the analytical solution to the linearized problem in the fluid domain. Section 4 is devoted to numerical results based on the analytical solution.

2 Source model

The inversion of seismic wave data allows the reconstruction of permanent deformations of the sea bottom following earthquakes. In spite of the complexity of the seismic source and of the internal structure of the earth, scientists have been relatively successful in using simple models for the source. One of these models is Okada's model (23). Its description follows.

The fracture zones, along which the foci of earthquakes are to be found, have been described in various papers. For example, it has been suggested that Volterra's theory of dislocations might be the proper tool for a quantitative description of these fracture zones (24). This suggestion was made for the following reason. If the mechanism involved in earthquakes and the fracture zones is indeed one of fracture, discontinuities in the displacement components across the fractured surface will exist. As dislocation theory may be described as that part of the theory of elasticity dealing with surfaces across which the displacement field is discontinuous, the suggestion makes sense.

As is often done in mathematical physics, it is necessary for simplicity's sake to make some assumptions. Here we neglect the curvature of the earth, its gravity, temperature, magnetism, non-homogeneity, and consider a semi-infinite medium, which is homogeneous and isotropic. We further assume that the laws of classical linear elasticity theory hold.

Several studies showed that the effect of earth curvature is negligible for shallow events at distances of less than 20° (25; 26; 27). The sensitivity to earth topography, homogeneity, isotropy and half-space assumptions was studied and discussed recently (28). A commercially available code, ABACUS, which is based on a finite element model (FEM), was used. Six FEMs were constructed to test the sensitivity of deformation predictions to each assumption. The author came to the conclusion that the vertical layering of lateral inhomogeneity can sometimes cause considerable effects on the deformation fields.

The usual boundary conditions for dealing with earth problems require that the surface of the elastic medium (the earth) shall be free from forces. The resulting mixed boundary-value problem was solved a century ago (29). Later, Steketee proposed an alternative method to solve this problem using Green's functions (24).

2.1 Volterra's theory of dislocations

In order to introduce the concept of dislocation and for simplicity's sake, this section is devoted to the case of an entire elastic space, as was done in the original paper by Volterra (29).

Let O be the origin of a Cartesian coordinate system in an infinite elastic medium, x_i the Cartesian coordinates ($i = 1, 2, 3$), and $\mathbf{e}_i$ a unit vector in the positive x_i−direction. A force $\mathbf{F} = F\mathbf{e}_k$ at O generates a displacement field $u_i^k(P, O)$ at point P, which is determined by the well-known Somigliana tensor

$$u_i^k(P,O) = \frac{F}{8\pi\mu}(\delta_{ik} r_{,\,nn} - \alpha r_{,\,ik}), \quad \text{with } \alpha = \frac{\lambda+\mu}{\lambda+2\mu}. \tag{1}$$

In this relation δ_{ik} is the Kronecker delta, λ and μ are Lamé's constants, and r is the distance from P to O. The coefficient α can be rewritten as $\alpha = 1/2(1-\nu)$, where ν is Poisson's ratio. Later we will also use Young's modulus E, which is defined as

$$E = \frac{\mu\,(3\lambda + 2\mu)}{\lambda + \mu}.$$

The notation $r,_i$ means $\partial r/\partial x_i$ and the summation convention applies.

The stresses due to the displacement field (1) are easily computed from Hooke's law:

$$\sigma_{ij} = \lambda\delta_{ij}u_{k,k} + \mu(u_{i,j} + u_{j,i}). \tag{2}$$

One finds

$$\sigma_{ij}^k(P,O) = -\frac{\alpha F}{4\pi}\left(\frac{3x_ix_jx_k}{r^5} + \frac{\mu}{\lambda+\mu}\frac{\delta_{ki}x_j + \delta_{kj}x_i - \delta_{ij}x_k}{r^3}\right).$$

The components of the force per unit area on a surface element are denoted as follows:

$$T_i^k = \sigma_{ij}^k\nu_j,$$

where the ν_j's are the components of the normal to the surface element. A Volterra dislocation is defined as a surface Σ in the elastic medium across which there is a discontinuity Δu_i in the displacement fields of the type

$$\Delta u_i = u_i^+ - u_i^- = U_i + \Omega_{ij}x_j, \tag{3}$$

$$\Omega_{ij} = -\Omega_{ji}. \tag{4}$$

Equation (3) in which U_i and Ω_{ij} are constants is the well-known Weingarten relation which states that the discontinuity Δu_i should be of the type of a rigid body displacement, thereby maintaining continuity of the components of stress and strain across Σ.

The displacement field in an infinite elastic medium due to the dislocation is then determined by Volterra's formula (29)

$$u_k(Q) = \frac{1}{F}\iint\limits_{\Sigma}\Delta u_i T_i^k\,dS. \tag{5}$$

Once the surface Σ is given, the dislocation is essentially determined by the six constants U_i and Ω_{ij}. Therefore we also write

$$u_k(Q) = \frac{U_i}{F}\iint\limits_{\Sigma}\sigma_{ij}^k(P,Q)\nu_j dS + \frac{\Omega_{ij}}{F}\iint\limits_{\Sigma}\{x_j\sigma_{il}^k(P,Q) - x_i\sigma_{jl}^k(P,Q)\}\nu_l dS, \tag{6}$$

where Ω_{ij} takes only the values Ω_{12}, Ω_{23}, Ω_{31}. Following Volterra (29) and Love (30) we call each of the six integrals in (6) an elementary dislocation.

It is clear from (5) and (6) that the computation of the displacement field $u_k(Q)$ is performed as follows. A force $F\mathbf{e}_k$ is applied at Q, and the stresses $\sigma_{ij}^k(P,Q)$ that this force generates are computed at the points $P(x_i)$ on Σ. In particular the components of the force on Σ are computed. After multiplication with prescribed weights of magnitude Δu_i these forces are integrated over Σ to give the displacement component in Q due to the dislocation on Σ.

2.2 Dislocations in elastic half-space

When the case of an elastic half-space is considered, equation (5) remains valid, but we have to replace σ_{ij}^k in T_i^k by another tensor ω_{ij}^k. This can be explained by the fact that the elementary solutions for a half-space are different from Somigliana solution (1).

The ω_{ij}^k can be obtained from the displacements corresponding to nuclei of strain in a half-space through relation (2). Steketee showed a method of obtaining the six ω_{ij}^k fields by using a Green's function and derived ω_{12}^k, which is relevant to a vertical strike-slip fault (see below). Maruyama derived the remaining five functions (31).

It is interesting to mention here that historically these solutions were first derived in a straightforward manner by Mindlin (32; 33), who gave explicit expressions of the displacement and stress fields for half-space nuclei of strain consisting of single forces with and without moment. It is only necessary to write the single force results since the other forms can be obtained by taking appropriate derivatives. The method consists in finding the displacement field in Westergaard's form of the Galerkin vector (34). This vector is then determined by taking a linear combination of some biharmonic elementary solutions. The coefficients are chosen to satisfy boundary and equilibrium conditions. These solutions were also derived by Press in a slightly different manner (35).

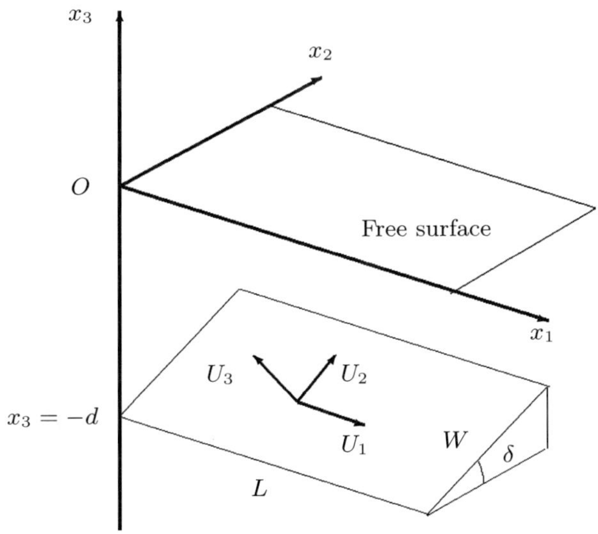

Fig. 1. Coordinate system adopted in this study and geometry of the source model

Here, we take the Cartesian coordinate system shown in Figure 1. The elastic medium occupies the region $x_3 \leq 0$ and the x_1-axis is taken to be parallel to the strike direction of the fault. In this coordinate system, $u_i^j(x_1, x_2, x_3; \xi_1, \xi_2, \xi_3)$ is the ith component of the displacement at (x_1, x_2, x_3) due to the jth direction point force of magnitude F at (ξ_1, ξ_2, ξ_3). It can be expressed as follows (23; 32; 35; 36):

$$u_i^j(x_1,x_2,x_3) = u_{iA}^j(x_1,x_2,-x_3) - u_{iA}^j(x_1,x_2,x_3) \\ +u_{iB}^j(x_1,x_2,x_3) + x_3 u_{iC}^j(x_1,x_2,x_3), \tag{7}$$

where

$$u_{iA}^j = \frac{F}{8\pi\mu}\left((2-\alpha)\frac{\delta_{ij}}{R} + \alpha\frac{R_iR_j}{R^3}\right),$$

$$u_{iB}^j = \frac{F}{4\pi\mu}\left(\frac{\delta_{ij}}{R} + \frac{R_iR_j}{R^3} + \frac{1-\alpha}{\alpha}\left[\frac{\delta_{ij}}{R+R_3} + \right.\right.$$

$$\left.\left. + \frac{R_i\delta_{j3} - R_j\delta_{i3}(1-\delta_{j3})}{R(R+R_3)} - \frac{R_iR_j}{R(R+R_3)^2}(1-\delta_{i3})(1-\delta_{j3})\right]\right),$$

$$u_{iC}^j = \frac{F}{4\pi\mu}(1-2\delta_{i3})\left((2-\alpha)\frac{R_i\delta_{j3} - R_j\delta_{i3}}{R^3} + \alpha\xi_3\left[\frac{\delta_{ij}}{R^3} - 3\frac{R_iR_j}{R^5}\right]\right).$$

In these expressions $R_1 = x_1 - \xi_1$, $R_2 = x_2 - \xi_2$, $R_3 = -x_3 - \xi_3$ and $R^2 = R_1^2 + R_2^2 + R_3^2$.

The first term in equation (7), $u_{iA}^j(x_1,x_2,-x_3)$, is the well-known Somigliana tensor, which represents the displacement field due to a single force placed at (ξ_1,ξ_2,ξ_3) in an infinite medium (30). The second term also looks like a Somigliana tensor. This term corresponds to a contribution from an image source of the given point force placed at $(\xi_1,\xi_2,-\xi_3)$ in the infinite medium. The third term, $u_{iB}^j(x_1,x_2,x_3)$, and $u_{iC}^j(x_1,x_2,x_3)$ in the fourth term are naturally depth dependent. When x_3 is set equal to zero in equation (7), the first and the second terms cancel each other, and the fourth term vanishes. The remaining term, $u_{iB}^j(x_1,x_2,0)$, reduces to the formula for the surface displacement field due to a point force in a half-space (23):

$$\begin{cases} u_1^1 = \frac{F}{4\pi\mu}\left(\frac{1}{R} + \frac{(x_1-\xi_1)^2}{R^3} + \frac{\mu}{\lambda+\mu}\left[\frac{1}{R-\xi_3} - \frac{(x_1-\xi_1)^2}{R(R-\xi_3)^2}\right]\right), \\ u_2^1 = \frac{F}{4\pi\mu}(x_1-\xi_1)(x_2-\xi_2)\left(\frac{1}{R^3} - \frac{\mu}{\lambda+\mu}\frac{1}{R(R-\xi_3)^2}\right), \\ u_3^1 = \frac{F}{4\pi\mu}(x_1-\xi_1)\left(-\frac{\xi_3}{R^3} - \frac{\mu}{\lambda+\mu}\frac{1}{R(R-\xi_3)}\right), \end{cases}$$

$$\begin{cases} u_1^2 = \frac{F}{4\pi\mu}(x_1-\xi_1)(x_2-\xi_2)\left(\frac{1}{R^3} - \frac{\mu}{\lambda+\mu}\frac{1}{R(R-\xi_3)^2}\right), \\ u_2^2 = \frac{F}{4\pi\mu}\left(\frac{1}{R} + \frac{(x_2-\xi_2)^2}{R^3} + \frac{\mu}{\lambda+\mu}\left[\frac{1}{R-\xi_3} - \frac{(x_2-\xi_2)^2}{R(R-\xi_3)^2}\right]\right), \\ u_3^2 = \frac{F}{4\pi\mu}(x_2-\xi_2)\left(-\frac{\xi_3}{R^3} - \frac{\mu}{\lambda+\mu}\frac{1}{R(R-\xi_3)}\right), \end{cases}$$

$$\begin{cases} u_1^3 = \frac{F}{4\pi\mu}(x_1-\xi_1)\left(-\frac{\xi_3}{R^3} + \frac{\mu}{\lambda+\mu}\frac{1}{R(R-\xi_3)}\right), \\ u_2^3 = \frac{F}{4\pi\mu}(x_2-\xi_2)\left(-\frac{\xi_3}{R^3} + \frac{\mu}{\lambda+\mu}\frac{1}{R(R-\xi_3)}\right), \\ u_3^3 = \frac{F}{4\pi\mu}\left(\frac{1}{R} + \frac{\xi_3^2}{R^3} + \frac{\mu}{\lambda+\mu}\frac{1}{R}\right). \end{cases}$$

In these formulas $R^2 = (x_1-\xi_1)^2 + (x_2-\xi_2)^2 + \xi_3^2$.

In order to obtain the displacements due to the dislocation we need to calculate the corresponding ξ_k-derivatives of the point force solution (7) and to insert them in Volterra's formula (5)

$$u_i = \frac{1}{F}\iint_\Sigma \Delta u_j\left[\lambda\delta_{jk}\frac{\partial u_i^n}{\partial\xi_n} + \mu\left(\frac{\partial u_i^j}{\partial\xi_k} + \frac{\partial u_i^k}{\partial\xi_j}\right)\right]\nu_k\, dS.$$

The ξ_k-derivatives are expressed as follows:

$$\frac{\partial u_i^j}{\partial \xi_k}(x_1,x_2,x_3) = \frac{\partial u_{iA}^j}{\partial \xi_k}(x_1,x_2,-x_3) - \frac{\partial u_{iA}^j}{\partial \xi_k}(x_1,x_2,x_3) + $$
$$+\frac{\partial u_{iB}^j}{\partial \xi_k}(x_1,x_2,x_3) + x_3\frac{\partial u_{iC}^j}{\partial \xi_k}(x_1,x_2,x_3),$$

with

$$\frac{\partial u_{iA}^j}{\partial \xi_k} = \frac{F}{8\pi\mu}\left((2-\alpha)\frac{R_k}{R^3}\delta_{ij} - \alpha\frac{R_i\delta_{jk}+R_j\delta_{ik}}{R^3} + 3\alpha\frac{R_iR_jR_k}{R^5}\right),$$

$$\frac{\partial u_{iB}^j}{\partial \xi_k} = \frac{F}{4\pi\mu}\left(-\frac{R_i\delta_{jk}+R_j\delta_{ik}-R_k\delta_{ij}}{R^3} + 3\frac{R_iR_jR_k}{R^5} + \right.$$
$$+\frac{1-\alpha}{\alpha}\left[\frac{\delta_{3k}R+R_k}{R(R+R_3)^2}\delta_{ij} - \frac{\delta_{ik}\delta_{j3}-\delta_{jk}\delta_{i3}(1-\delta_{j3})}{R(R+R_3)} + \right.$$
$$+\big(R_i\delta_{j3} - R_j\delta_{i3}(1-\delta_{j3})\big)\frac{\delta_{3k}R^2+R_k(2R+R_3)}{R^3(R+R_3)^2} +$$
$$\left.\left.+(1-\delta_{i3})(1-\delta_{j3})\Big(\frac{R_i\delta_{jk}+R_j\delta_{ik}}{R(R+R_3)^2} - R_iR_j\frac{2\delta_{3k}R^2+R_k(3R+R_3)}{R^3(R+R_3)^3}\Big)\right]\right),$$

$$\frac{\partial u_{iC}^j}{\partial \xi_k} = \frac{F}{4\pi\mu}(1-2\delta_{i3})\left((2-\alpha)\left[\frac{\delta_{jk}\delta_{i3}-\delta_{ik}\delta_{j3}}{R^3} + \frac{3R_k(R_i\delta_{j3}-R_j\delta_{i3})}{R^5}\right] + \right.$$
$$\left.+\alpha\delta_{3k}\left[\frac{\delta_{ij}}{R^3} - \frac{3R_iR_j}{R^5}\right] + 3\alpha\xi_3\left[\frac{R_i\delta_{jk}+R_j\delta_{ik}+R_k\delta_{ij}}{R^5} - \frac{5R_iR_jR_k}{R^7}\right]\right).$$

2.3 Finite rectangular source

Let us now consider a more practical problem. We define the elementary dislocations U_1, U_2 and U_3, corresponding to the strike-slip, dip-slip and tensile components of an arbitrary dislocation. In Figure 1 each vector represents the direction of the elementary faults. The vector $\mathbf{D}$ is the so-called Burger's vector, which shows how both sides of the fault are spread out: $\mathbf{D} = \mathbf{u}^+ - \mathbf{u}^-$.

A general dislocation can be determined by three angles: the dip angle δ of the fault ($0 \le \delta \le \pi$), the slip or rake angle θ ($0 \le \theta \le \pi$), and the angle ϕ between the fault plane and Burger's vector $\mathbf{D}$. When dealing with a geophysical application, an additional angle, the azimuth or strike, is introduced in order to provide an orientation of the fault. The general situation is schematically described in Figure 2.

For a finite rectangular fault with length L and width W occurring at depth d (Figure 2), the deformation field can be evaluated analytically by a change of variables and by integrating over the rectangle. This was done by several authors (23; 36; 37; 38; 39). Here we give the results of their computations. The final results are represented below in compact form, using Chinnery's notation $\|$ to represent the substitution

$$f(\xi,\eta)\| = f(x,p) - f(x,p-W) - f(x-L,p) + f(x-L,p-W),$$

where $p = y\cos\delta + d\sin\delta$. Next we introduce the notation

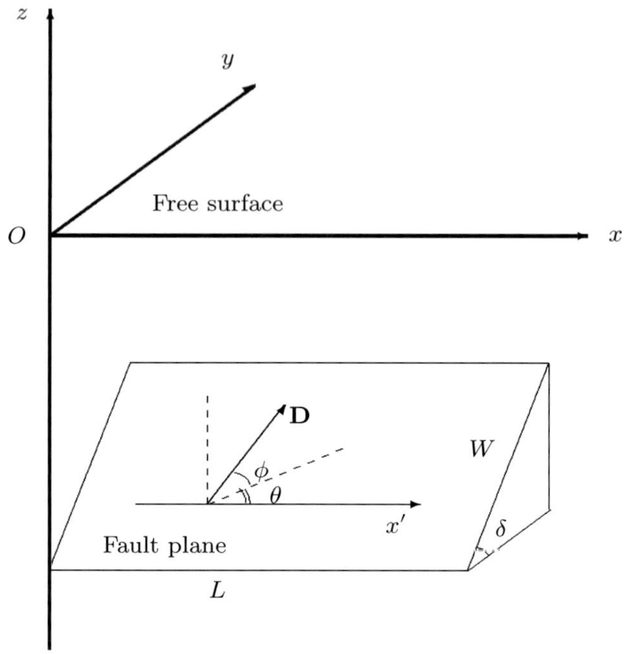

Fig. 2. Geometry of the source model and orientation of Burger's vector **D**

$$q = y \sin\delta - d\cos\delta, \quad \tilde{y} = \eta\cos\delta + q\sin\delta, \quad \tilde{d} = \eta\sin\delta - q\cos\delta$$

and

$$R^2 = \xi^2 + \eta^2 + q^2 = \xi^2 + \tilde{y}^2 + \tilde{d}^2, \quad X^2 = \xi^2 + q^2.$$

The quantities U_1, U_2 and U_3 are linked to Burger's vector through the identities

$$U_1 = |\mathbf{D}|\cos\phi\cos\theta, \quad U_2 = |\mathbf{D}|\cos\phi\sin\theta, \quad U_3 = |\mathbf{D}|\sin\phi.$$

For a strike-slip dislocation, one has

$$\begin{aligned}
u_1 &= -\frac{U_1}{2\pi}\left(\frac{\xi q}{R(R+\eta)} + \arctan\frac{\xi\eta}{qR} + I_1\sin\delta\right)\Bigg\|, \\
u_2 &= -\frac{U_1}{2\pi}\left(\frac{\tilde{y}q}{R(R+\eta)} + \frac{q\cos\delta}{R+\eta} + I_2\sin\delta\right)\Bigg\|, \\
u_3 &= -\frac{U_1}{2\pi}\left(\frac{\tilde{d}q}{R(R+\eta)} + \frac{q\sin\delta}{R+\eta} + I_4\sin\delta\right)\Bigg\|.
\end{aligned}$$

For a dip-slip dislocation, one has

$$u_1 = -\frac{U_2}{2\pi}\left(\frac{q}{R} - I_3 \sin\delta\cos\delta\right)\Bigg\|,$$
$$u_2 = -\frac{U_2}{2\pi}\left(\frac{\tilde{y}q}{R(R+\xi)} + \cos\delta \arctan\frac{\xi\eta}{qR} - I_1\sin\delta\cos\delta\right)\Bigg\|,$$
$$u_3 = -\frac{U_2}{2\pi}\left(\frac{\tilde{d}q}{R(R+\xi)} + \sin\delta \arctan\frac{\xi\eta}{qR} - I_5\sin\delta\cos\delta\right)\Bigg\|.$$

For a tensile fault dislocation, one has

$$u_1 = \frac{U_3}{2\pi}\left(\frac{q^2}{R(R+\eta)} - I_3\sin^2\delta\right)\Bigg\|,$$
$$u_2 = \frac{U_3}{2\pi}\left(\frac{-\tilde{d}q}{R(R+\xi)} - \sin\delta\left[\frac{\xi q}{R(R+\eta)} - \arctan\frac{\xi\eta}{qR}\right] - I_1\sin^2\delta\right)\Bigg\|,$$
$$u_3 = \frac{U_3}{2\pi}\left(\frac{\tilde{y}q}{R(R+\xi)} + \cos\delta\left[\frac{\xi q}{R(R+\eta)} - \arctan\frac{\xi\eta}{qR}\right] - I_5\sin^2\delta\right)\Bigg\|.$$

The terms $I_1, \ldots, I_5$ are given by

$$I_1 = -\frac{\mu}{\lambda+\mu}\frac{\xi}{(R+\tilde{d})\cos\delta} - \tan\delta I_5,$$
$$I_2 = -\frac{\mu}{\lambda+\mu}\log(R+\eta) - I_3,$$
$$I_3 = \frac{\mu}{\lambda+\mu}\left[\frac{1}{\cos\delta}\frac{\tilde{y}}{R+\tilde{d}} - \log(R+\eta)\right] + \tan\delta I_4,$$
$$I_4 = \frac{\mu}{\mu+\lambda}\frac{1}{\cos\delta}\left(\log(R+\tilde{d}) - \sin\delta\log(R+\eta)\right),$$
$$I_5 = \frac{\mu}{\lambda+\mu}\frac{2}{\cos\delta}\arctan\frac{\eta(X+q\cos\delta)+X(R+X)\sin\delta}{\xi(R+X)\cos\delta},$$

and if $\cos\delta = 0$,

$$I_1 = -\frac{\mu}{2(\lambda+\mu)}\frac{\xi q}{(R+\tilde{d})^2},$$
$$I_3 = \frac{\mu}{2(\lambda+\mu)}\left[\frac{\eta}{R+\tilde{d}} + \frac{\tilde{y}q}{(R+\tilde{d})^2} - \log(R+\eta)\right],$$
$$I_4 = -\frac{\mu}{\lambda+\mu}\frac{q}{R+\tilde{d}},$$
$$I_5 = -\frac{\mu}{\lambda+\mu}\frac{\xi\sin\delta}{R+\tilde{d}}.$$

Figures 3, 4, and 5 show the free-surface deformation due to the three elementary dislocations. The values of the parameters are given in Table 1.

2.4 Curvilinear fault

In the previous subsection analytical formulas for the free-surface deformation in the special case of a rectangular fault were given. In fact, Volterra's formula (5) allows to evaluate the displacement field that accompanies fault events with much

Table 1. Parameter set used in Figures 3, 4, and 5.

parameter	*value*
Dip angle δ	13°
Fault depth d, km	25
Fault length L, km	220
Fault width W, km	90
U_i, m	15
Young modulus E, GPa	9.5
Poisson's ratio ν	0.23

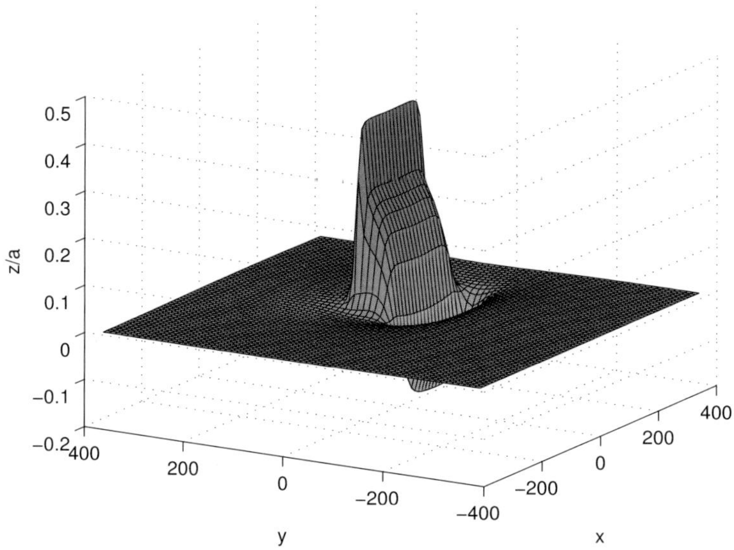

Fig. 3. Dimensionless free-surface deformation z/a due to dip-slip faulting: $\phi = 0$, $\theta = \pi/2$, $\mathbf{D} = (0, U_2, 0)$. Here a is $|\mathbf{D}|$ (15 m in the present application). The horizontal distances x and y are expressed in kilometers.

more general geometry. The shape of the fault and Burger's vector are suggested by seismologists and after numerical integration one can obtain the deformation of the seafloor for more general types of events as well.

Here we will consider the case of a fault whose geometry is described by an elliptical arc (see Figure 6). The parametric equations of this surface are given by

$$x(\xi,\eta) = \xi, \quad 0 \le \xi \le a, \qquad y(\xi,\eta) = \eta, \quad -\frac{c}{2} \le \eta \le \frac{c}{2},$$

$$z(\xi,\eta) = -(b+d) + \frac{b}{a}\sqrt{a^2 - \xi^2}.$$

Then the unit normal to this surface can be easily calculated:

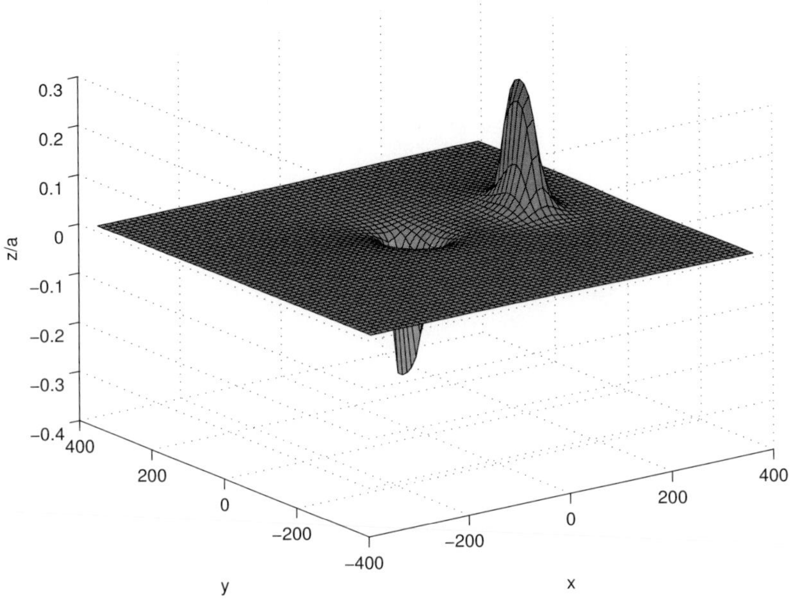

Fig. 4. Dimensionless free-surface deformation z/a due to strike-slip faulting: $\phi = 0$, $\theta = 0$, $\mathbf{D} = (U_1, 0, 0)$. Here a is $|\mathbf{D}|$ (15 m in the present application). The horizontal distances x and y are expressed in kilometers.

$$\mathbf{n} = \left(\frac{b\xi}{\sqrt{a^4 + (b^2 - a^2)\xi^2}}, 0, \frac{a\sqrt{a^2 - \xi^2}}{\sqrt{a^4 + (b^2 - a^2)\xi^2}} \right).$$

We also need to compute the coefficients of the first fundamental form in order to reduce the surface integral in (5) to a double Riemann integral. These coefficients are

$$E = \frac{a^4 + \xi^2(b^2 - a^2)}{a^2(a^2 - \xi^2)}, \quad F = 0, \quad G = 1$$

and the surface element dS is

$$dS = \sqrt{EG - F^2}\, d\xi d\eta = \frac{1}{a} \frac{\sqrt{a^4 + \xi^2(b^2 - a^2)}}{\sqrt{a^2 - \xi^2}}\, d\xi d\eta.$$

Since in the crust the hydrostatic pressure is very large, it is natural to impose the condition that $\mathbf{D} \cdot \mathbf{n} = 0$. The physical meaning of this condition is that both sides of the fault slide and do not detach. This condition is obviously satisfied if we take Burger's vector as

$$\mathbf{D} = D \left(\frac{a\sqrt{a^2 - \xi^2}}{\sqrt{a^4 + \xi^2(b^2 - a^2)}}, 0, -\frac{b\xi}{\sqrt{a^4 + \xi^2(b^2 - a^2)}} \right).$$

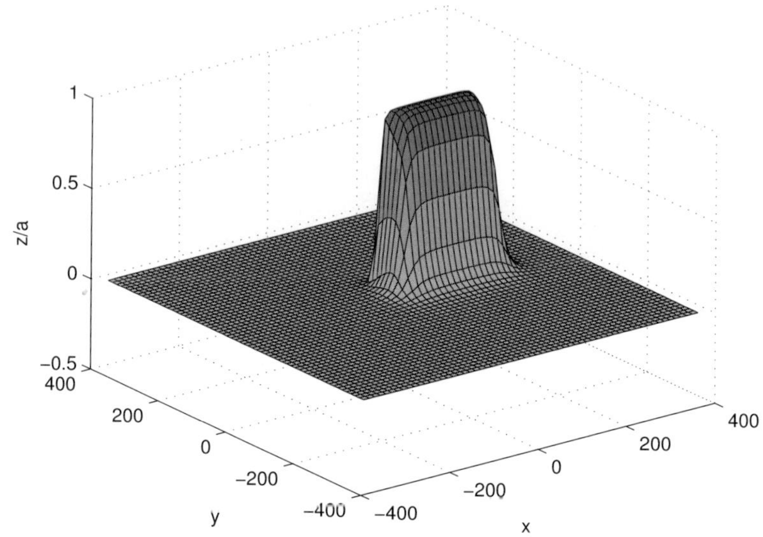

Fig. 5. Dimensionless free-surface deformation z/a due to tensile faulting: $\phi = \pi/2$, $\mathbf{D} = (0, 0, U_3)$. Here a is $|\mathbf{D}|$. The horizontal distances x and y are expressed in kilometers.

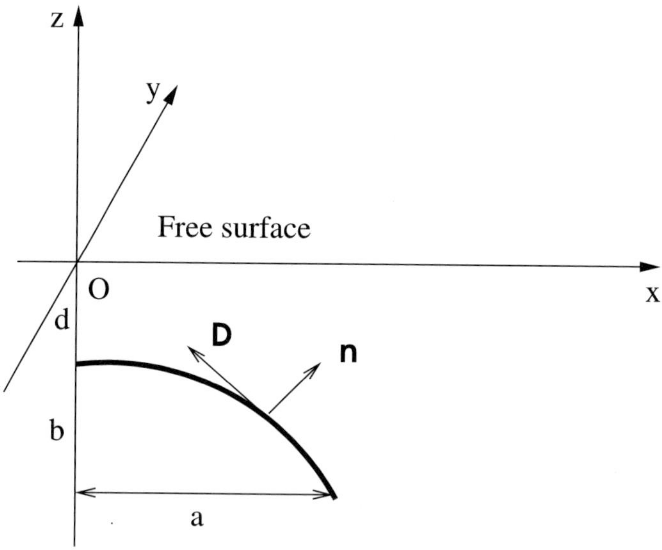

Fig. 6. Geometry of a fault with elliptical shape.

It is evident that $D = |\mathbf{D}|$.

The numerical integration was performed using a 9-point two-dimensional Gauss-type integration formula. The result is presented on Figure 7. The parameter values are given in Table 2.

Table 2. Parameter set used in Figure 7.

parameter	*value*
Depth event d, km	20
Ellipse semiminor axis a, km	17
Ellipse semimajor axis b, km	6
Fault width c, km	15
Young modulus E, GPa	9.5
Poisson's ratio ν	0.23

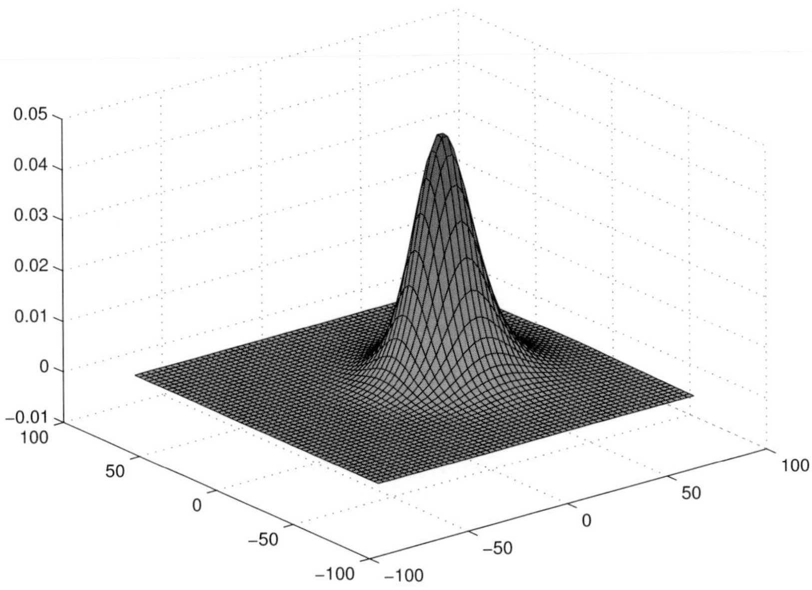

Fig. 7. Free-surface deformation due to curvilinear faulting. The horizontal distances x and y are expressed in kilometers.

The example considered in this subsection may not be physically relevant. However it shows how Okada's solution can be extended. For a more precise modeling of the faulting event we need to have more information about the earthquake source and its related parameters.

After having reviewed the description of the source, we now switch to the deformation of the ocean surface following a submarine earthquake. The traditional approach for hydrodynamic modelers is to use elastic models similar to the model we just described with the seismic parameters as input in order to evaluate the details of the seafloor deformation. Then this deformation is translated to the free surface of the ocean and serves as initial condition of the evolution problem described in the next section. Drawbacks of this traditional approach have recently been pointed out by Dutykh et al. (18).

3 Solution in fluid domain

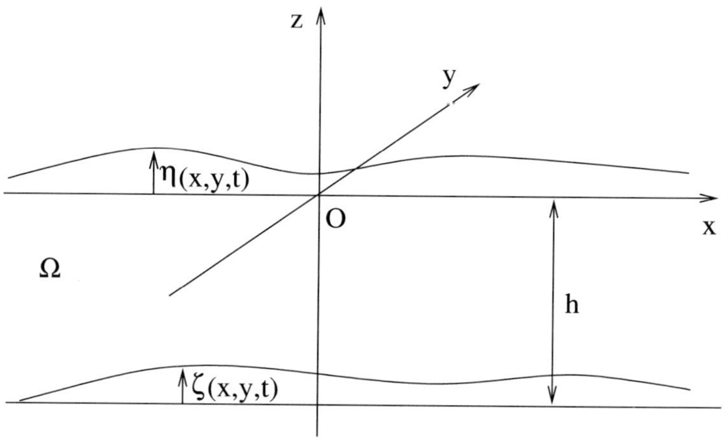

Fig. 8. Definition of the fluid domain and coordinate system

The fluid domain is supposed to represent the ocean above the fault area. Let us consider the fluid domain Ω shown in Figure 8. It is bounded above by the free surface of the ocean and below by the rigid ocean floor. The domain Ω is unbounded in the horizontal directions x and y, and can be written as

$$\Omega = \mathbb{R}^2 \times [-h + \zeta(x,y,t), \eta(x,y,t)]\,.$$

Initially the fluid is assumed to be at rest and the sea bottom to be horizontal. Thus, at time $t = 0$, the free surface and the sea bottom are defined by $z = 0$ and $z = -h$, respectively. For time $t > 0$ the bottom boundary moves in a prescribed manner which is given by

$$z = -h + \zeta(x,y,t).$$

The displacement of the sea bottom is assumed to have all the properties required to compute its Fourier transform in x, y and its Laplace transform in t. The resulting deformation of the free surface $z = \eta(x, y, t)$ must be found. It is also assumed that the fluid is incompressible and the flow is irrotational. The latter implies the existence of a velocity potential $\phi(x, y, z, t)$ which completely describes this flow. By definition of ϕ, the fluid velocity vector can be expressed as $\mathbf{q} = \nabla\phi$. Thus, the continuity equation becomes

$$\nabla \cdot \mathbf{q} = \Delta\phi = 0, \quad (x, y, z) \in \Omega. \tag{8}$$

The potential $\phi(x, y, z, t)$ must also satisfy the following kinematic boundary conditions on the free-surface and the solid boundary, respectively:

$$\frac{\partial\phi}{\partial z} = \frac{\partial\eta}{\partial t} + \frac{\partial\phi}{\partial x}\frac{\partial\eta}{\partial x} + \frac{\partial\phi}{\partial y}\frac{\partial\eta}{\partial y}, \qquad z = \eta(x, y, t), \tag{9}$$

$$\frac{\partial\phi}{\partial z} = \frac{\partial\zeta}{\partial t} + \frac{\partial\phi}{\partial x}\frac{\partial\zeta}{\partial x} + \frac{\partial\phi}{\partial y}\frac{\partial\zeta}{\partial y}, \qquad z = -h + \zeta(x, y, t). \tag{10}$$

Assuming that viscous effects as well as capillary effects can be neglected, the dynamic condition to be satisfied on the free surface reads

$$\frac{\partial\phi}{\partial t} + \frac{1}{2}|\nabla\phi|^2 + g\eta = 0, \qquad z = \eta(x, y, t). \tag{11}$$

As described above, the initial conditions are given by

$$\eta(x, y, 0) = 0 \quad \text{and} \quad \zeta(x, y, 0) = 0. \tag{12}$$

The significance of the various terms in the equations is more transparent when the equations are written in dimensionless variables. The new independent variables are

$$\widetilde{x} = \kappa x, \quad \widetilde{y} = \kappa y, \quad \widetilde{z} = \kappa z, \quad \widetilde{t} = \sigma t,$$

where κ is a wavenumber and σ is a typical frequency. Note that here the same unit length is used in the horizontal and vertical directions, as opposed to shallow-water theory.

The new dependent variables are

$$\widetilde{\eta} = \frac{\eta}{a}, \quad \widetilde{\zeta} = \frac{\zeta}{a}, \quad \widetilde{\phi} = \frac{\kappa}{a\sigma}\phi,$$

where a is a characteristic wave amplitude. A dimensionless water depth is also introduced:

$$\widetilde{h} = \kappa h.$$

In dimensionless form, and after dropping the tildes, equations (8)–(11) become

$$\Delta\phi = 0, \qquad (x, y, z) \in \Omega,$$

$$\frac{\partial\phi}{\partial z} = \frac{\partial\eta}{\partial t} + \kappa a\left(\frac{\partial\phi}{\partial x}\frac{\partial\eta}{\partial x} + \frac{\partial\phi}{\partial y}\frac{\partial\eta}{\partial y}\right), \qquad z = \kappa a\,\eta(x, y, t),$$

$$\frac{\partial\phi}{\partial z} = \frac{\partial\zeta}{\partial t} + \kappa a\left(\frac{\partial\phi}{\partial x}\frac{\partial\zeta}{\partial x} + \frac{\partial\phi}{\partial y}\frac{\partial\zeta}{\partial y}\right), \qquad z = -h + \kappa a\,\zeta(x, y, t),$$

$$\frac{\partial\phi}{\partial t} + \frac{1}{2}\kappa a|\nabla\phi|^2 + \frac{g\kappa}{\sigma^2}\eta = 0, \qquad z = \kappa a\,\eta(x, y, t).$$

Finding the solution to this problem is quite a difficult task due to the nonlinearities and the a priori unknown free surface. In this study we linearize the equations and the boundary conditions by taking the limit as $\kappa a \to 0$. In fact, the linearized problem can be found by expanding the unknown functions as power series of a small parameter $\varepsilon := \kappa a$. Collecting the lowest order terms in ε yields the linear approximation. For the sake of convenience, we now switch back to the physical variables. The linearized problem in dimensional variables reads

$$\Delta\phi = 0, \qquad (x, y, z) \in \mathbb{R}^2 \times [-h, 0], \tag{13}$$

$$\frac{\partial \phi}{\partial z} = \frac{\partial \eta}{\partial t}, \qquad z = 0, \tag{14}$$

$$\frac{\partial \phi}{\partial z} = \frac{\partial \zeta}{\partial t}, \qquad z = -h, \tag{15}$$

$$\frac{\partial \phi}{\partial t} + g\eta = 0, \qquad z = 0. \tag{16}$$

Combining equations (14) and (16) yields the single free-surface condition

$$\frac{\partial^2 \phi}{\partial t^2} + g\frac{\partial \phi}{\partial z} = 0, \qquad z = 0. \tag{17}$$

This problem will be solved by using the method of integral transforms. We apply the Fourier transform in (x, y):

$$\mathfrak{F}[f] = \widehat{f}(k, \ell) = \int\limits_{\mathbb{R}^2} f(x, y) e^{-i(kx+\ell y)}\, dxdy,$$

$$\mathfrak{F}^{-1}[\widehat{f}] = f(x, y) = \frac{1}{(2\pi)^2} \int\limits_{\mathbb{R}^2} \widehat{f}(k, \ell) e^{i(kx+\ell y)}\, dkd\ell,$$

and the Laplace transform in time t:

$$\mathfrak{L}[g] = \mathsf{g}(s) = \int\limits_0^{+\infty} g(t) e^{-st}\, dt.$$

For the combined Fourier and Laplace transforms, the following notation is introduced:

$$\mathfrak{FL}[F(x, y, t)] = \overline{F}(k, \ell, s) = \int\limits_{\mathbb{R}^2} e^{-i(kx+\ell y)}\, dxdy \int\limits_0^{+\infty} F(x, y, t) e^{-st}\, dt.$$

After applying the transforms, equations (13), (15) and (17) become

$$\frac{d^2\overline{\phi}}{dz^2} - (k^2 + \ell^2)\overline{\phi} = 0, \tag{18}$$

$$\frac{d\overline{\phi}}{dz}(k, \ell, -h, s) = s\overline{\zeta}(k, \ell, s), \tag{19}$$

$$s^2\overline{\phi}(k, \ell, 0, s) + g\frac{d\overline{\phi}}{dz}(k, \ell, 0, s) = 0. \tag{20}$$

The transformed free-surface elevation can be obtained from (16):

$$\overline{\eta}(k,\ell,s) = -\frac{s}{g}\overline{\phi}(k,\ell,0,s). \tag{21}$$

A general solution of equation (18) is given by

$$\overline{\phi}(k,\ell,z,s) = A(k,\ell,s)\cosh(mz) + B(k,\ell,s)\sinh(mz), \tag{22}$$

where $m = \sqrt{k^2+\ell^2}$. The functions $A(k,\ell,s)$ and $B(k,\ell,s)$ can be easily found from the boundary conditions (19) and (20):

$$A(k,\ell,s) = -\frac{gs\overline{\zeta}(k,\ell,s)}{\cosh(mh)[s^2+gm\tanh(mh)]},$$
$$B(k,\ell,s) = \frac{s^3\overline{\zeta}(k,\ell,s)}{m\cosh(mh)[s^2+gm\tanh(mh)]}.$$

From now on, the notation

$$\omega = \sqrt{gm\tanh(mh)} \tag{23}$$

will be used. The graphs of $\omega(m)$, $\omega'(m)$ and $\omega''(m)$ are shown in Figure 9.

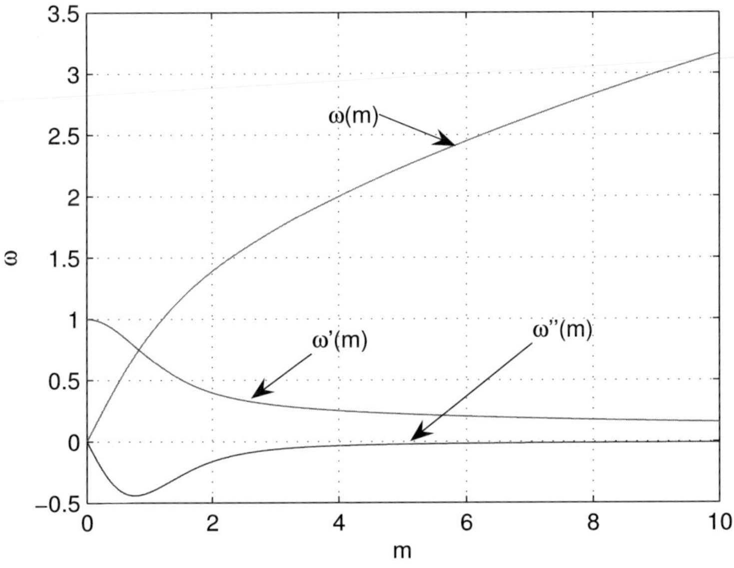

Fig. 9. Plot of the frequency $\omega(m) = \sqrt{gm\tanh(mh)}$ and its derivatives $d\omega/dm$, $d^2\omega/dm^2$. The acceleration due to gravity g and the water depth h have been set equal to 1.

Substituting the expressions for the functions A, B in (22) yields

$$\overline{\phi}(k,\ell,z,s) = -\frac{gs\overline{\zeta}(k,\ell,s)}{\cosh(mh)(s^2+\omega^2)}\left(\cosh(mz) - \frac{s^2}{gm}\sinh(mz)\right). \tag{24}$$

3.1 Free-surface elevation

From (21), the free-surface elevation becomes

$$\overline{\eta}(k,\ell,s) = \frac{s^2\overline{\zeta}(k,\ell,s)}{\cosh(mh)(s^2+\omega^2)}.$$

Inverting the Laplace and Fourier transforms provides the general integral solution

$$\eta(x,y,t) = \frac{1}{(2\pi)^2}\iint\limits_{\mathbb{R}^2} \frac{e^{i(kx+\ell y)}}{\cosh(mh)}\frac{1}{2\pi i}\int\limits_{\mu-i\infty}^{\mu+i\infty} \frac{s^2\overline{\zeta}(k,\ell,s)}{s^2+\omega^2}e^{st}\,ds\,dkd\ell. \qquad (25)$$

One can evaluate the Laplace integral in (25) using the convolution theorem:

$$\mathfrak{L}[f_1(t)*f_2(t)] = \mathsf{f}_1(s)\mathsf{f}_2(s).$$

It yields

$$\eta(x,y,t) = \frac{1}{(2\pi)^2}\iint\limits_{\mathbb{R}^2} \frac{e^{i(kx+\ell y)}}{\cosh(mh)}\int\limits_0^t (1-\omega\sin\omega\tau)\overline{\zeta}(k,\ell,t-\tau)d\tau\,dkd\ell.$$

This general solution contains as a special case the solution for an axisymmetric problem, which we now describe in detail. Assume that the initial solid boundary deformation is axisymmetric:

$$\zeta(x,y) = \zeta(r), \qquad r = \sqrt{x^2+y^2}.$$

The Fourier transform $\mathfrak{F}[\zeta(x,y)] = \widehat{\zeta}(k,\ell)$ of an axisymmetric function is also axisymmetric with respect to transformation parameters, i.e.

$$\widehat{\zeta}(k,\ell) = \widehat{\zeta}(m), \qquad m := \sqrt{k^2+\ell^2}.$$

In the following calculation, we use the notation $\psi = \arctan(\ell/k)$. One has

$$\widehat{\zeta}(k,\ell) = \iint\limits_{\mathbb{R}^2} \zeta(r)e^{-i(kx+\ell y)}\,dxdy = \int\limits_0^{2\pi} d\phi \int\limits_0^{\infty} \zeta(r)e^{-ir(k\cos\phi+\ell\sin\phi)}rdr =$$
$$= \int\limits_0^{2\pi} d\phi \int\limits_0^{\infty} r\zeta(r)e^{-irm\cos(\phi-\psi)}dr = \int\limits_0^{\infty} r\zeta(r)dr\int\limits_0^{\pi}(e^{-irm\cos\phi}+e^{irm\cos\phi})d\phi.$$

Using an integral representation of Bessel functions (40) finally yields

$$\widehat{\zeta}(k,\ell) = 2\pi\int\limits_0^{\infty} r\zeta(r)J_0(mr)dr \equiv \widehat{\zeta}(m).$$

It follows that

$$\eta(r,t) = \frac{1}{(2\pi)^2}\int_0^{2\pi} d\psi \int_0^{+\infty} \frac{me^{imr\cos(\phi-\psi)}}{\cosh(mh)} dm \int_0^t (1-\omega\sin\omega\tau)\overline{\zeta}(m,t-\tau)\,d\tau$$
$$= \frac{1}{2\pi}\int_0^{+\infty} m\frac{J_0(mr)}{\cosh(mh)} dm \int_0^t (1-\omega\sin\omega\tau)\overline{\zeta}(m,t-\tau)d\tau.$$

The last equation gives the general integral solution of the problem in the case of an axisymmetric seabed deformation. Below we no longer make this assumption since Okada's solution does not have this property.

In the present study we consider seabed deformations with the following structure:

$$\zeta(x,y,t) := \zeta(x,y)T(t). \tag{26}$$

Mathematically we separate the time dependence from the spatial coordinates. There are two main reasons for doing this. First of all we want to be able to invert analytically the Laplace transform. The second reason is more fundamental. In fact, dynamic source models are not easily available. Okada's solution, which was described in the previous section, provides the static sea-bed deformation $\zeta_0(x,y)$ and we will consider different time dependencies $T(t)$ to model the time evolution of the source. Four scenarios will be considered:

1. **Instantaneous**: $T_i(t) = H(t)$, where $H(t)$ denotes the Heaviside step function,
2. **Exponential**:
$$T_e(t) = \begin{cases} 0, & t<0, \\ 1-e^{-\alpha t}, & t \geq 0, \end{cases} \quad \text{with } \alpha > 0,$$
3. **Trigonometric**: $T_c(t) = H(t-t_0) + \frac{1}{2}[1-\cos(\pi t/t_0)]H(t_0-t)$,
4. **Linear**:
$$T_l(t) = \begin{cases} 0, & t<0, \\ t/t_0, & 0 \leq t \leq t_0, \\ 1, & t > t_0. \end{cases}$$

The typical graphs of $T_c(t)$ and $T_e(t)$ are shown in Figure 10. Inserting (26) into (25) yields

$$\eta(x,y,t) = \frac{1}{(2\pi)^2}\iint_{\mathbb{R}^2} \frac{\widehat{\zeta}(k,\ell)e^{i(kx+\ell y)}}{\cosh(mh)} \frac{1}{2\pi i}\int_{\mu-i\infty}^{\mu+i\infty} \frac{s^2\mathsf{T}(s)}{s^2+\omega^2}e^{st}ds\,dkd\ell. \tag{27}$$

Clearly, $\eta(x,y,t)$ depends continuously on the source $\zeta(x,y)$. Physically it means that small variations of ζ (in a reasonable space of functions such as L^2) yield small variations of η. Mathematically this problem is said to be well-posed, and this property is essential for modelling the physical processes, since it means that small modifications of the ground motion (for example, the error in measurements) do not induce huge modifications of the wave patterns.

Using the special representation (26) of seabed deformation and prescribed time-dependencies, one can compute analytically the Laplace integral in (27). To perform this integration, we first have to compute the Laplace transform of $T_{i,e,c,l}(t)$. The results are

$$\mathfrak{L}[T_i] = \frac{1}{s}, \qquad \mathfrak{L}[T_e] = \frac{\alpha}{s(\alpha+s)},$$

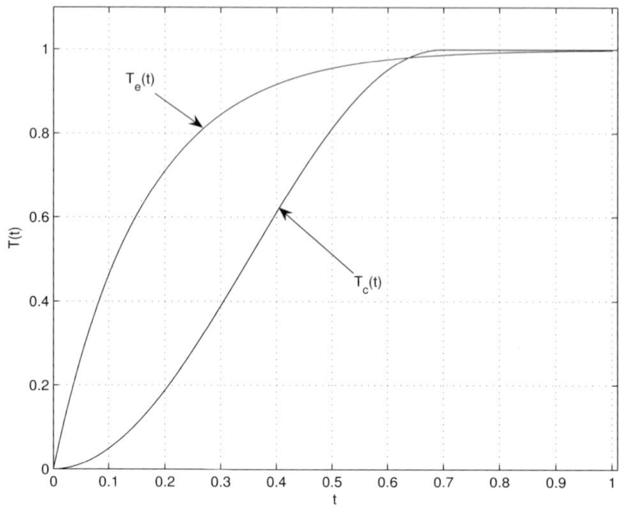

Fig. 10. Typical graphs of $T_e(t)$ and $T_c(t)$. Here we have set $\alpha = 6.2$, $t_0 = 0.7$.

$$\mathfrak{L}[T_c] = (1 + e^{-st_0})\frac{\gamma^2}{2s(s^2+\gamma^2)} \quad \text{with } \gamma = \frac{\pi}{t_0}, \qquad \mathfrak{L}[T_l] = \frac{1 - e^{-st_0}}{t_0 s^2}.$$

Inserting these formulas into the inverse Laplace integral yields

$$\frac{1}{2\pi i}\int\limits_{\mu-i\infty}^{\mu+i\infty} \frac{e^{st}s^2\mathsf{T}_{\mathsf{i}}(s)}{s^2+\omega^2}ds = \cos\omega t,$$

$$\frac{1}{2\pi i}\int\limits_{\mu-i\infty}^{\mu+i\infty} \frac{e^{st}s^2\mathsf{T}_{\mathsf{e}}(s)}{s^2+\omega^2}ds = -\frac{\alpha^2}{\alpha^2+\omega^2}\left(e^{-\alpha t} - \cos\omega t - \frac{\omega}{\alpha}\sin\omega t\right),$$

$$\frac{1}{2\pi i}\int\limits_{\mu-i\infty}^{\mu+i\infty} \frac{e^{st}s^2\mathsf{T}_{\mathsf{c}}(s)}{s^2+\omega^2}ds = \frac{\gamma^2}{2(\gamma^2-\omega^2)}$$

$$\left(\cos\omega t - \cos\gamma t + H(t-t_0)[\cos\omega(t-t_0) + \cos\gamma t]\right),$$

$$\frac{1}{2\pi i}\int\limits_{\mu-i\infty}^{\mu+i\infty} \frac{e^{st}s^2\mathsf{T}_{\mathsf{l}}(s)}{s^2+\omega^2}ds = \frac{\sin\omega t - H(t-t_0)\sin\omega(t-t_0)}{\omega t_0}.$$

The final integral formulas for the free-surface elevations with different time dependencies are as follows:

$$\eta_i(x,y,t) = \frac{1}{(2\pi)^2} \iint\limits_{\mathbb{R}^2} \frac{\widehat{\zeta}(k,\ell)e^{i(kx+\ell y)}}{\cosh(mh)} \cos\omega t \, dk d\ell,$$

$$\eta_e(x,y,t) = \frac{-\alpha^2}{(2\pi)^2} \iint\limits_{\mathbb{R}^2} \frac{\widehat{\zeta}(k,\ell)e^{i(kx+\ell y)}}{\cosh(mh)} \left(\frac{e^{-\alpha t} - \cos\omega t - \frac{\omega}{\alpha}\sin\omega t}{\alpha^2+\omega^2} \right) dk d\ell,$$

$$\eta_c(x,y,t) = \frac{\gamma^2}{(2\pi)^2} \iint\limits_{\mathbb{R}^2} \frac{\widehat{\zeta}(k,\ell)e^{i(kx+\ell y)}}{2(\gamma^2-\omega^2)\cosh(mh)}$$

$$(\cos\omega t - \cos\gamma t + H(t-t_0)[\cos\omega(t-t_0) + \cos\gamma t]) \; dk d\ell,$$

$$\eta_l(x,y,t) = \frac{1}{(2\pi)^2} \iint\limits_{\mathbb{R}^2} \frac{\widehat{\zeta}(k,\ell)e^{i(kx+\ell y)}}{\cosh(mh)} \left(\frac{\sin\omega t - H(t-t_0)\sin\omega(t-t_0)}{\omega t_0} \right) dk d\ell.$$

3.2 Velocity field

In some applications it is important to know not only the free-surface elevation but also the velocity field in the fluid domain. One of the goals of this work is to provide an initial condition for tsunami propagation codes. For the time being, tsunami modelers take initial seabed deformations and translate them directly to the free surface in order to obtain the initial condition $\eta(x,y,0)$. Since a priori there is no information on the flow velocities, they take a zero velocity field as initial condition for the velocity: $\nabla\phi(x,y,z,0) = 0$. The present computations show that it is indeed a very good approximation if the generation time is short.

In equation (24), we obtained the Fourier transform of the velocity potential $\phi(x,y,z,t)$:

$$\overline{\phi}(k,\ell,z,s) = -\frac{gs\widehat{\zeta}(k,\ell)\mathsf{T}(s)}{\cosh(mh)(s^2+\omega^2)} \left(\cosh(mz) - \frac{s^2}{gm}\sinh(mz) \right). \tag{28}$$

Let us evaluate the velocity field at an arbitrary level $z = \beta h$ with $-1 \le \beta \le 0$. In the linear approximation the value $\beta = 0$ corresponds to the free surface while $\beta = -1$ corresponds to the bottom. Next we introduce some notation. The horizontal velocities are denoted by $\mathbf{u}$. The horizontal gradient $(\partial/\partial x, \partial/\partial y)$ is denoted by ∇_h. The vertical velocity component is simply w. The Fourier transform parameters are denoted $\mathbf{k} = (k,\ell)$.

Taking the Fourier and Laplace transforms of

$$\mathbf{u}(x,y,t) = \nabla_h \phi(x,y,z,t)|_{z=\beta h}$$

yields

$$\begin{aligned} \overline{\mathbf{u}}(k,\ell,s) &= -i\overline{\phi}(k,\ell,\beta h,s)\mathbf{k} \\ &= i\frac{gs\widehat{\zeta}(k,\ell)\mathsf{T}(s)}{\cosh(mh)(s^2+\omega^2)} \left(\cosh(\beta mh) - \frac{s^2}{gm}\sinh(\beta mh) \right) \mathbf{k}. \end{aligned}$$

Inverting the Fourier and Laplace transforms gives the general formula for the horizontal velocities:

$$\mathbf{u}(x,y,t) = \frac{ig}{4\pi^2}\iint\limits_{\mathbb{R}^2} \frac{\mathbf{k}\widehat{\zeta}(k,\ell)\cosh(m\beta h)e^{i(kx+\ell y)}}{\cosh(mh)}\frac{1}{2\pi i}\int\limits_{\mu-i\infty}^{\mu+i\infty}\frac{s\mathsf{T}(s)e^{st}}{s^2+\omega^2}\,ds\,d\mathbf{k}$$
$$-\frac{i}{4\pi^2}\iint\limits_{\mathbb{R}^2} \frac{\mathbf{k}\widehat{\zeta}(k,\ell)\sinh(m\beta h)e^{i(kx+\ell y)}}{m\cosh(mh)}\frac{1}{2\pi i}\int\limits_{\mu-i\infty}^{\mu+i\infty}\frac{s^3\mathsf{T}(s)e^{st}}{s^2+\omega^2}\,ds\,d\mathbf{k}.$$

After a few computations, one finds the formulas for the time dependencies T_i, T_e and T_l. For simplicity we only give the velocities along the free surface ($\beta = 0$):

$$\mathbf{u}_i(x,y,t) = \frac{ig}{4\pi^2}\iint\limits_{\mathbb{R}^2}\frac{\mathbf{k}\widehat{\zeta}(k,\ell)e^{i(kx+\ell y)}}{\cosh(mh)}\frac{\sin\omega t}{\omega}\,d\mathbf{k},$$
$$\mathbf{u}_e(x,y,t) = \frac{ig\alpha}{4\pi^2}\iint\limits_{\mathbb{R}^2}\frac{\mathbf{k}\widehat{\zeta}(k,\ell)e^{i(kx+\ell y)}}{(\alpha^2+\omega^2)\cosh(mh)}\left(e^{-\alpha t} - \cos\omega t + \frac{\alpha}{\omega}\sin\omega t\right)d\mathbf{k},$$
$$\mathbf{u}_l(x,y,t) = \frac{ig}{4t_0\pi^2}\iint\limits_{\mathbb{R}^2}\frac{\mathbf{k}\widehat{\zeta}(k,\ell)e^{i(kx+\ell y)}}{\omega^2\cosh(mh)}$$
$$(1-\cos\omega t - H(t-t_0)[1-\cos\omega(t-t_0)])\,d\mathbf{k}.$$

Next we determine the vertical component of the velocity $w(x,y,z,t)$. It is easy to obtain the Fourier–Laplace transform $\overline{w}(k,\ell,z,s)$ by differentiating (28):

$$\overline{w}(k,\ell,z,s) = \frac{\partial\overline{\phi}}{\partial z} = \frac{sg\widehat{\zeta}(k,\ell)\mathsf{T}(s)}{\cosh(mh)(s^2+\omega^2)}\left(\frac{s^2}{g}\cosh(mz) - m\sinh(mz)\right).$$

Inverting this transform yields

$$w(x,y,z,t) = \frac{1}{4\pi^2}\iint\limits_{\mathbb{R}^2}\frac{\cosh(mz)\widehat{\zeta}(k,\ell)}{\cosh(mh)}e^{i(kx+\ell y)}\frac{1}{2\pi i}\int\limits_{\mu-i\infty}^{\mu+i\infty}\frac{s^3\mathsf{T}(s)e^{st}}{s^2+\omega^2}\,ds\,d\mathbf{k}$$
$$-\frac{g}{4\pi^2}\iint\limits_{\mathbb{R}^2}\frac{m\sinh(mz)\widehat{\zeta}(k,\ell)}{\cosh(mh)}e^{i(kx+\ell y)}\frac{1}{2\pi i}\int\limits_{\mu-i\infty}^{\mu+i\infty}\frac{s\mathsf{T}(s)e^{st}}{s^2+\omega^2}\,ds\,d\mathbf{k},$$

for $-h < z \leq 0$. One can easily obtain the expression of the vertical velocity at a given vertical level by substituting $z = \beta h$ in the expression for w.

The easiest way to compute the vertical velocity w along the free surface is to use the boundary condition (14). Indeed, the expression for w can be simply derived by differentiating the known formula for $\eta_{i,e,c,l}(x,y,t)$. Note that formally the derivative gives the distributions $\delta(t)$ and $\delta(t-t_0)$ under the integral sign. It is a consequence of the idealized time behaviour (such as the instantaneous scenario) and it is a disadvantage of the Laplace transform method. In order to avoid these distributions we can consider the solutions only for $t > 0$ and $t \neq t_0$. From a practical point of view there is no restriction since for any $\varepsilon > 0$ we can set $t = \varepsilon$ or $t = t_0 + \varepsilon$. For small values of ε this will give a very good approximation of the solution behaviour at these “critical” instants of time. Under this assumption we give the distribution-free expressions for the vertical velocity along the free surface:

$$w_i(x,y,t) = -\frac{1}{4\pi^2}\iint\limits_{\mathbb{R}^2} \frac{\widehat{\zeta}(k,\ell)e^{i(kx+\ell y)}}{\cosh(mh)}\omega\sin\omega t\, d\mathbf{k},$$

$$w_e(x,y,t) = \frac{\alpha^3}{4\pi^2}\iint\limits_{\mathbb{R}^2} \frac{\widehat{\zeta}(k,\ell)e^{i(kx+\ell y)}}{(\alpha^2+\omega^2)\cosh(mh)}\left(e^{-\alpha t}+\frac{\omega^2}{\alpha^2}\cos\omega t-\frac{\omega}{\alpha}\sin\omega t\right)\, d\mathbf{k},$$

$$w_c(x,y,t) = -\frac{\gamma^2}{4\pi^2}\iint\limits_{\mathbb{R}^2} \frac{\widehat{\zeta}(k,\ell)e^{i(kx+\ell y)}}{2(\gamma^2-\omega^2)\cosh(mh)}\big(\omega\sin\omega t-\gamma\sin\gamma t$$

$$+H(t-t_0)[\omega\sin\omega(t-t_0)+\gamma\sin\gamma t]\big)\, d\mathbf{k},$$

$$w_l(x,y,t) = \frac{1}{4t_0\pi^2}\iint\limits_{\mathbb{R}^2} \frac{\widehat{\zeta}(k,\ell)e^{i(kx+\ell y)}}{\cosh(mh)}[\cos\omega t-H(t-t_0)\cos\omega(t-t_0)]\, d\mathbf{k}.$$

3.3 Pressure on the bottom

Since tsunameters have one component that measures the pressure at the bottom (bottom pressure recorder or simply BPR (41)), it is interesting to provide as well the expression $p_b(x,y,t)$ for the pressure at the bottom. The pressure $p(x,y,z,t)$ can be obtained from Bernoulli's equation, which was written explicitly for the free surface in equation (11), but is valid everywhere in the fluid:

$$\frac{\partial\phi}{\partial t}+\frac{1}{2}|\nabla\phi|^2+gz+\frac{p}{\rho}=0. \tag{29}$$

After linearization, equation (29) becomes

$$\frac{\partial\phi}{\partial t}+gz+\frac{p}{\rho}=0. \tag{30}$$

Along the bottom, it reduces to

$$\frac{\partial\phi}{\partial t}+g(-h+\zeta)+\frac{p_b}{\rho}=0, \qquad z=-h. \tag{31}$$

The time-derivative of the velocity potential is readily available in Fourier space. Inverting the Fourier and Laplace transforms and evaluating the resulting expression at $z=-h$ gives for the four time scenarios, respectively,

$$\frac{\partial\phi_i}{\partial t} = -\frac{g}{(2\pi)^2}\iint\limits_{\mathbb{R}^2} \frac{\widehat{\zeta}(k,\ell)e^{i(kx+\ell y)}}{\cosh^2(mh)}\cos\omega t\, d\mathbf{k},$$

$$\frac{\partial\phi_e}{\partial t} = \frac{g\alpha^2}{(2\pi)^2}\iint\limits_{\mathbb{R}^2} \frac{\widehat{\zeta}(k,\ell)e^{i(kx+\ell y)}}{\alpha^2+\omega^2}\left(e^{-\alpha t}-\cos\omega t-\frac{\omega}{\alpha}\sin\omega t\right)\, d\mathbf{k}+\frac{\alpha^4}{(2\pi)^2}$$

$$\iint\limits_{\mathbb{R}^2} \frac{\widehat{\zeta}(k,\ell)\tanh(mh)e^{i(kx+\ell y)}}{m(\alpha^2+\omega^2)}\left(e^{-\alpha t}+\left(\frac{\omega}{\alpha}\right)^2\cos\omega t+\left(\frac{\omega}{\alpha}\right)^3\sin\omega t\right)\, d\mathbf{k},$$

$$\frac{\partial\phi_l}{\partial t} = -\frac{g}{t_0(2\pi)^2}\iint\limits_{\mathbb{R}^2} \frac{\widehat{\zeta}(k,\ell)e^{i(kx+\ell y)}}{\omega\cosh^2(mh)}[\sin\omega t-H(t-t_0)\sin\omega(t-t_0)]\, d\mathbf{k}.$$

The bottom pressure deviation from the hydrostatic pressure is then given by

$$p_b(x,y,t) = -\rho \frac{\partial \phi}{\partial t}\bigg|_{z=-h} - \rho g \zeta.$$

Plots of the bottom pressure will be given in Section 4.

3.4 Asymptotic analysis of integral solutions

In this subsection, we apply the method of stationary phase in order to estimate the far-field behaviour of the solutions. There is a lot of literature on this topic (see for example (42; 43; 44; 45; 46)). This method is a classical method in asymptotic analysis. To our knowledge, the stationary phase method was first used by Kelvin (47) in the context of linear water-wave theory.

The motivation to obtain asymptotic formulas for integral solutions was mainly due to numerical difficulties to calculate the solutions for large values of x and y. From equation (25), it is clear that the integrand is highly oscillatory. In order to be able to resolve these oscillations, several discretization points are needed per period. This becomes extremely expensive as $r = \sqrt{x^2+y^2} \to \infty$. The numerical method used in the present study is based on a Filon-type quadrature formula (21) and has been adapted to double integrals with $\exp[i(kx+\ell y)]$ oscillations. The idea of this method consists in interpolating only the amplitude of the integrand at discretization points by some kind of polynomial or spline and then performing exact integration for the oscillating part of the integrand. This method seems to be quite efficient.

Let us first obtain an asymptotic representation for integral solutions of the general form

$$\eta(x,y,t) = \frac{1}{4\pi^2} \iint_{\mathbb{R}^2} \frac{\widehat{\zeta}(k,\ell) e^{i(kx+\ell y)}}{\cosh(mh)} T(m,t)\, dk d\ell, \quad m = \sqrt{k^2+\ell^2}. \qquad (32)$$

Comparing with equation (27) shows that $T(m,t)$ is in fact

$$T(m,t) = \frac{1}{2\pi i} \int_{\mu-i\infty}^{\mu+i\infty} \frac{s^2 \mathsf{T}(s)}{s^2+\omega^2} e^{st}\, ds.$$

For example, we showed above that for an instantaneous seabed deformation $T(m,t) = \cos\omega t$, where $\omega^2 = gm\tanh mh$. For the time being, we do not specify the time behaviour $\mathsf{T}(s)$.

In equation (32), we switch to polar coordinates m and $\psi = \arctan(\ell/k)$:

$$\begin{aligned}\eta(x,y,t) &= \frac{1}{4\pi^2} \int_0^{\infty}\int_0^{2\pi} \frac{\widehat{\zeta}(m,\psi) e^{imr\cos(\varphi-\psi)}}{\cosh(mh)} T(m,t) m\, d\psi dm \\ &= \frac{1}{4\pi^2} \int_0^{\infty} \frac{mT(m,t)}{\cosh(mh)}\, dm \int_0^{2\pi} \widehat{\zeta}(m,\psi) e^{imr\cos(\varphi-\psi)}\, d\psi,\end{aligned}$$

where (r,φ) are the polar coordinates of (x,y). In the last expression, the phase function is $\Phi = mr\cos(\varphi-\psi)$. Stationary phase points satisfy the condition $\partial\Phi/\partial\psi =$

0, which yields two phases: $\psi_1 = \varphi$ and $\psi_2 = \varphi + \pi$. An approximation to equation (32) is then obtained by applying the method of stationary phase to the integral over ψ:

$$\eta(r,\phi,t) \simeq \frac{1}{\sqrt{8\pi^3 r}} \int_0^\infty \frac{\sqrt{m}T(m,t)}{\cosh(mh)} \left(\widehat{\zeta}(m,\varphi) e^{i(\frac{\pi}{4} - mr)} + \widehat{\zeta}(m,\varphi+\pi) e^{i(mr - \frac{\pi}{4})} \right) dm.$$

This expression cannot be simplified if we do not make any further hypotheses on the function $T(m,t)$.

Since we are looking for the far field solution behaviour, the details of wave formation are not important. Thus we will assume that the initial seabed deformation is instantaneous:

$$T(m,t) = \cos \omega t = \frac{e^{i\omega t} + e^{-i\omega t}}{2}.$$

Inserting this particular function $T(m,t)$ in equation (32) yields

$$\eta(r,\varphi,t) = \frac{1}{8\pi^2} (I_1 + I_2),$$

where

$$I_1 = \int_0^\infty \frac{m\widehat{\zeta}(m,\psi)}{\cosh(mh)} \int_0^{2\pi} e^{i(\omega t + mr\cos(\varphi - \psi))} \, d\psi dm,$$

$$I_2 = \int_0^\infty \frac{m\widehat{\zeta}(m,\psi)}{\cosh(mh)} \int_0^{2\pi} e^{i(-\omega t + mr\cos(\varphi - \psi))} \, d\psi dm.$$

The stationary phase function in these integrals is

$$\Phi(m,\psi) = mr\cos(\varphi - \psi) \pm \omega t, \qquad \omega^2(m) = gm \tanh mh.$$

The points of stationary phase are then obtained from the conditions

$$\frac{\partial \Phi}{\partial \psi} = 0, \quad \frac{\partial \Phi}{\partial m} = 0.$$

The first equation gives two points, $\psi_1 = \varphi$ and $\psi_2 = \varphi + \pi$, as before. The second condition yields

$$\frac{r}{t} \cos(\varphi - \psi_{1,2}) = \mp \frac{d\omega}{dm}. \tag{33}$$

Since $d\omega/dm$ decreases from $\sqrt{gh}$ to 0 as m goes from 0 to ∞ (see Figure 9), this equation has a unique solution for m if $|r/t| \leq \sqrt{gh}$. This unique solution will be denoted by m^*.

For $|r| > t\sqrt{gh}$, there is no stationary phase. It means physically that the wave has not yet reached this region. So we can approximately set $I_1 \approx 0$ and $I_2 \approx 0$. From the positivity of the function $d\omega/dm$ one can deduce that $\psi_1 = \varphi$ is a stationary phase point only for the integral I_2. Similarly, $\psi_2 = \varphi + \pi$ is a stationary point only for the integral I_1.

Let us obtain an asymptotic formula for the first integral:

$$
\begin{aligned}
I_1 &\approx \int\limits_0^\infty \frac{m}{\cosh(mh)} \left(\sqrt{\frac{2\pi}{mr}} \widehat{\zeta}(m, \varphi + \pi) e^{i(\omega t - mr)} e^{i\frac{\pi}{4}} \right) dm \\
&= \sqrt{\frac{2\pi}{r}} e^{i\frac{\pi}{4}} \int\limits_0^\infty \frac{\widehat{\zeta}(m, \varphi + \pi)}{\cosh(mh)} \sqrt{m} e^{i(\omega t - mr)} \, dm \\
&\approx \sqrt{\frac{2\pi}{r}} e^{i\frac{\pi}{4}} \left(\sqrt{\frac{2\pi m^*}{|\omega''(m^*)|\, t}} \frac{\widehat{\zeta}(m^*, \varphi + \pi)}{\cosh(m^* h)} e^{i(\omega(m^*)t - m^* r)} e^{-i\frac{\pi}{4}} \right) \\
&= \frac{2\pi}{t} \sqrt{\frac{m^*}{-\omega''\omega'}} \frac{\widehat{\zeta}(m^*, \varphi + \pi)}{\cosh(m^* h)} e^{i(\omega(m^*)t - m^* r)}.
\end{aligned}
$$

In this estimate we have used equation (33) evaluated at the stationary phase point (m^*, ψ_2):

$$
r = t \left. \frac{d\omega}{dm} \right|_{m=m^*}. \tag{34}
$$

Similarly one can obtain an estimate for the integral I_2:

$$
I_2 \approx \frac{2\pi}{t} \sqrt{\frac{m^*}{-\omega''\omega'}} \frac{\widehat{\zeta}(m^*, \varphi)}{\cosh(m^* h)} e^{-i(\omega(m^*)t - m^* r)}.
$$

Asymptotic values have been obtained for the integrals. As is easily observed from the expressions for I_1 and I_2, the wave train decays as $1/t$, or $1/r$, which is equivalent since r and t are connected by relation (34).

4 Numerical results

A lot of numerical computations based on the analytical formulas obtained in the previous sections have been performed. Because of the lack of information about the real dynamical characteristics of tsunami sources, we cannot really conclude which time dependence gives the best description of tsunami generation. At this stage it is still very difficult or even impossible.

Numerical experiments showed that the largest wave amplitudes with the time dependence $T_c(t)$ were obtained for relatively small values of the characteristic time . The exponential dependence has shown higher amplitudes for relatively longer characteristic times. The instantaneous scenario T_i gives at the free surface the initial seabed deformation with a slightly lower amplitude (the factor that we obtained was typically about $0.8 \sim 0.94$). The water has a high-pass filter effect on the initial solid boundary deformation (see Dutykh et al. (18) for more details). The linear time dependence $T_l(t)$ showed a linear growth of wave amplitude from 0 to also $\approx 0.9\zeta_0$, where $\zeta_0 = \max\limits_{(x,y)\in\mathbb{R}^2} |\zeta(x,y)|$.

In this section we provide several plots (Figure 11) of the free-surface deformation. For illustration purposes, we have chosen the instantaneous seabed deformation since it is the most widely used.

From Figure 12 it is clear that the velocity field is really negligible in the beginning of wave formation. Numerical computations showed that this situation does not change if one takes other time-dependences. The values of the parameters used in the

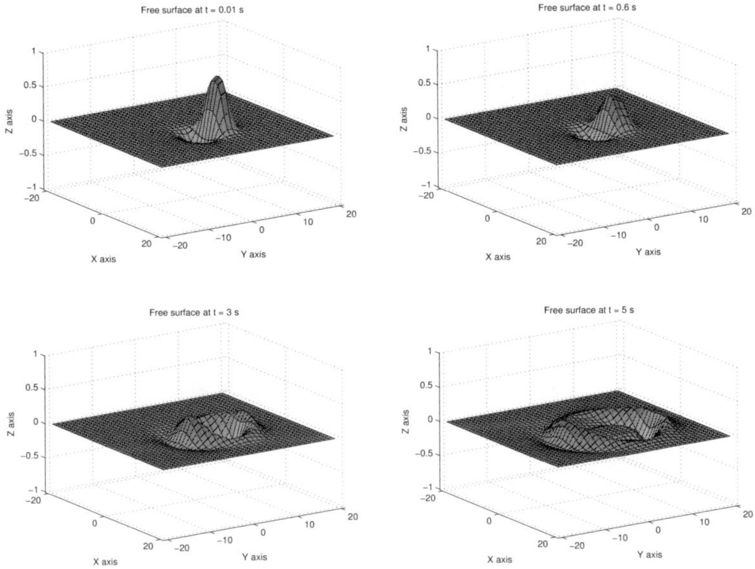

Fig. 11. Free-surface elevation at $t = 0.01, 0.6, 3, 5$ in dimensionless time. In physical time it corresponds to one second, one minute, five minutes and eight minutes and a half after the initial seabed deformation.

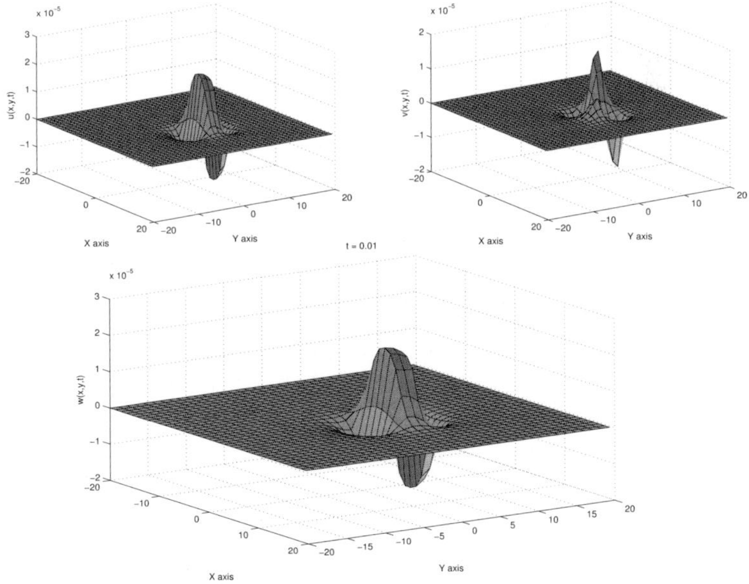

Fig. 12. Components u, v and w of the velocity field computed along the free surface at $t = 0.01$, that is one second after the initial seabed deformation.

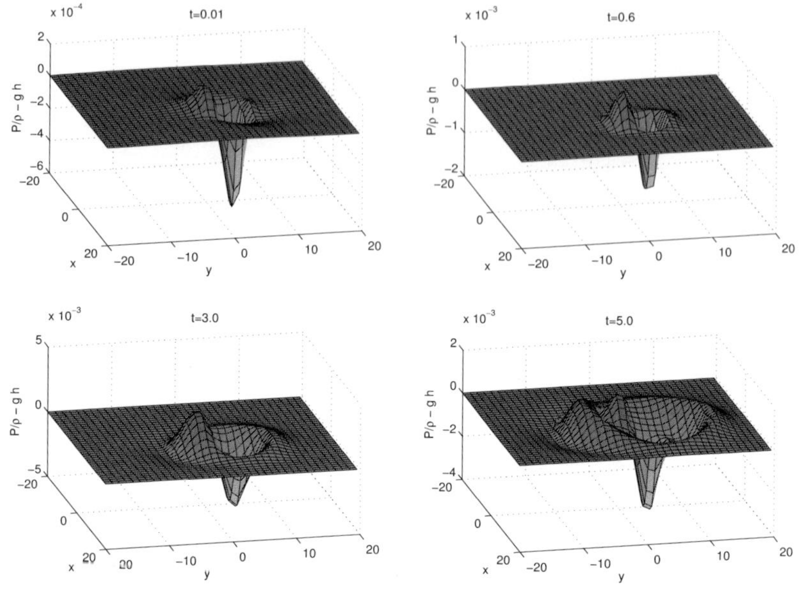

Fig. 13. Bottom pressure at $t = 0.01, 0.6, 3, 5$ in dimensionless time. In physical time it corresponds to one second, one minute, five minutes and eight minutes and a half after the initial seabed deformation.

computations are given in Table 3. We also give plots of the velocity components on the free surface a few seconds (physical) after the instantaneous deformation (Figure 12). Finally, plots of the bottom dynamic pressure are given in Figure 13.

The main focus of the present paper is the generation of waves by a moving bottom. The asymptotic behaviour of various sets of initial data propagating in a fluid of uniform depth has been studied in detail by Hammack and Segur (48; 49). In particular, they showed that the behaviours for an initial elevation wave and for an initial depression wave are different.

Table 3. Physical parameters used in the numerical computations

Parameter	**Value**
Young modulus, E, GPa	9.5
Poisson ratio, ν	0.27
Fault depth, d, km	20
Dip angle, δ, $^\circ$	13
Strike angle, θ, $^\circ$	90
Normal angle, ϕ, $^\circ$	0
Fault length, L, km	60
Fault width, W, km	40
Burger's vector length, $\lvert\mathbf{D}\rvert$, m	15
Water depth, h, km	4
Acceleration due to gravity, g, m/s^2	9.8
Wave number, k, $1/m$	10^{-4}
Angular frequency, ω, Hz	10^{-2}

References

[1] Todorovska MI, Trifunac MD (2001) Generation of tsunamis by a slowly spreading uplift of the sea-floor. Soil Dynamics and Earthquake Engineering 21:151–167

[2] Neetu S, Suresh I, Shankar R, Shankar D, Shenoi SSC, Shetye SR, Sundar D, Nagarajan B (2005) Comment on "The Great Sumatra-Andaman Earthquake of 26 December 2004". Science 310:1431a-1431b

[3] Lay T, Kanamori H, Ammon CJ, Nettles M, Ward SN, Aster RC, Beck SL, Bilek SL, Brudzinski MR, Butler R, DeShon HR, Ekstrom G, Satake K, Sipkin S (2005) The great Sumatra-Andaman earthquake of 26 December 2004. Science 308:1127–1133

[4] Korteweg DJ, de Vries G (1895) On the change of form of long waves advancing in a rectangular canal, and on a new type of long stationary waves. Phil. Mag. 39:422–443

[5] Boussinesq MJ (1871) Théorie de l'intumescence liquide appelée onde solitaire ou de translation se propageant dans un canal rectangulaire. C.R. Acad. Sci. Paris 72:755–759

[6] Peregrine DH (1966) Calculations of the development of an undual bore. J Fluid Mech 25:321–330

[7] Benjamin TB, Bona JL, Mahony JJ (1972) Model equations for long waves in nonlinear dispersive systems. Philos. Trans. Royal Soc. London Ser. A 272:47–78

[8] Podyapolsky GS (1968) The generation of linear gravitational waves in the ocean by seismic sources in the crust. Izvestiya, Earth Physics, Akademia Nauk SSSR 1:4–12, in Russian

[9] Kajiura K (1963) The leading wave of tsunami. Bull. Earthquake Res. Inst., Tokyo Univ. 41:535–571

[10] Gusyakov VK (1972) Generation of tsunami waves and ocean Rayleigh waves by submarine earthquakes. In: Mathematical problems of geophysics, vol 3, pages 250–272, Novosibirsk, VZ SO AN SSSR, in Russian

[11] Alekseev AS, Gusyakov VK (1973) Numerical modelling of tsunami and seismo-acoustic waves generation by submarine earthquakes. In: Theory of diffraction and wave propagation, vol 2, pages 194–197, Moscow-Erevan, in Russian

[12] Gusyakov VK (1976) Estimation of tsunami energy. In: Ill-posed problems of mathematical physics and problems of interpretation of geophysical observations, pages 46–64, Novosibirsk, VZ SO AN SSSR, in Russian

[13] Carrier GF (1971) The dynamics of tsunamis. In: Mathematical Problems in the Geophysical Sciences, Lectures in Applied Mathematics, vol 13, pages 157–187, American Mathematical Society, in Russian

[14] van den Driessche P, Braddock RD (1972) On the elliptic generating region of a tsunami. J. Mar. Res. 30:217–226

[15] Braddock RD, van den Driessche P, Peady GW (1973) Tsunami generation. J Fluid Mech 59:817–828

[16] Sabatier P (1986) Formation of waves by ground motion. In: Encyclopedia of Fluid Mechanics, pages 723–759, Gulf Publishing Company

[17] Hammack JL (1973) A note on tsunamis: their generation and propagation in an ocean of uniform depth. J Fluid Mech 60:769–799

[18] Dutykh D, Dias F, Kervella Y (2006) Linear theory of wave generation by a moving bottom. C. R. Acad. Sci. Paris, in press

[19] Todorovska MI, Hayir A, Trifunac MD (2002) A note on tsunami amplitudes above submarine slides and slumps. Soil Dynamics and Earthquake Engineering 22:129–141

[20] Keller JB (1961) Tsunamis: water waves produced by earthquakes. In: Proceedings of the Conference on Tsunami Hydrodynamics 24, pages 154–166, Institute of Geophysics, University of Hawaii

[21] Filon LNG (1928) On a quadrature formula for trigonometric integrals. Proc. Royal Soc. Edinburgh 49:38–47

[22] Ursell F (1953) The long-wave paradox in the theory of gravity waves. Proc. Camb. Phil. Soc. 49:685–694

[23] Okada Y (1985) Surface deformation due to shear and tensile faults in a half-space. Bull. Seism. Soc. Am. 75:1135–1154

[24] Steketee JA (1958) On Volterra's dislocation in a semi-infinite elastic medium. Can. J. Phys. 36:192–205

[25] Ben-Menahem A, Singh SJ, Solomon F (1969) Static deformation of a spherical earth model by internal dislocations. Bull. Seism. Soc. Am. 59:813–853

[26] Ben-Mehanem A, Singh SJ, Solomon F (1970) Deformation of an homogeneous earth model finite by dislocations. Rev. Geophys. Space Phys. 8:591–632

[27] Smylie DE, Mansinha L (1971) The elasticity theory of dislocations in real earth models and changes in the rotation of the earth. Geophys. J. Royal Astr. Soc. 23:329–354

[28] Masterlark, T (2003) Finite element model predictions of static deformation from dislocation sources in a subduction zone: Sensivities to homogeneous, isotropic, Poisson-solid, and half-space assumptions. J. Geophys. Res. 108(B11):2540

[29] Volterra V (1907) Sur l'équilibre des corps élastiques multiplement connexes. Annales Scientifiques de l'Ecole Normale Supérieure 24(3):401–517
[30] Love AEH (1944) A treatise on the mathematical theory of elasticity. Dover Publications, New York
[31] Maruyama T (1964) Static elastic dislocations in an infinite and semi-infinite medium. Bull. Earthquake Res. Inst., Tokyo Univ. 42:289–368
[32] Mindlin, RD (1936) Force at a point in the interior of a semi-infinite medium. Physics 7:195–202
[33] Mindlin RD, Cheng DH (1950) Nuclei of strain in the semi-infinite solid. J. Appl. Phys. 21:926–930
[34] Westergaard HM (1935) Bull. Amer. Math. Soc. 41:695
[35] Press F (1965) Displacements, strains and tilts at tele-seismic distances. J. Geophys. Res. 70:2395–2412
[36] Okada Y (1992) Internal deformation due to shear and tensile faults in a half-space. Bull. Seism. Soc. Am. 82:1018–1040
[37] Chinnery MA (1963) The stress changes that accompany strike-slip faulting. Bull. Seism. Soc. Am. 53:921–932
[38] Sato R, Matsu'ura M (1974) Strains and tilts on the surface of a semi-infinite medium. J. Phys. Earth 22:213–221
[39] Iwasaki T, Sato R (1979) Strain field in a semi-infinite medium due to an inclined rectangular fault. J. Phys. Earth 27:285–314
[40] Gradshteyn IS, Ryzhik M (2000) Tables of Integrals, Series, and Products, 6th edition, Academic Press, Orlando, Florida
[41] González FI, Bernard EN, Meinig C, Eble MC, Mofjeld HO, Stalin S (2005) The NTHMP tsunameter network. Natural Hazards 35:25–39
[42] Erdélyi A (1956) Asymptotic Expansions, Dover Publications
[43] Murray JD (1984) Asymptotic Analysis, Springer
[44] Petrashen GI, Latyshev KP (1971) Asymptotic Methods and Stochastic Models in Problems of Wave Propagation, American Mathematical Society
[45] Bleistein N, Handelsman RA (1986) Asymptotic Expansions of Integrals, Dover Publications
[46] Egorov YuV, Shubin MA (1994) Elements of the Modern Theory. Equations with Constant Coefficients. In: Partial Differential Equations, Encyclopedia of Mathematical Sciences, vol 2. Springer
[47] Kelvin Lord (W. Thomson) (1887) On the waves produced by a single impulse in water of any depth, or in a dispersive medium. Phil. Mag. 23(5):252–255
[48] Hammack JL, Segur H (1974) The Korteweg–de Vries equation and water waves. Part 2. Comparison with experiments. J Fluid Mech 65:289–314
[49] Hammack JL, Segur H (1978) The Korteweg–de Vries equation and water waves. Part 3. Oscillatory waves. J Fluid Mech 84:337–358

Tsunami surge in a river: a hydraulic jump in an inhomogeneous channel

Jean-Guy Caputo[1 2] and Y. A. Stepanyants[3]

[1] Laboratoire de Mathématiques, INSA de Rouen, B.P. 8, 76131 Mont-Saint-Aignan cedex, France.
caputo@insa-rouen.fr

[2] Laboratoire de Physique théorique et modelisation, Université de Cergy-Pontoise and C.N.R.S.

[3] Reactor Operations, ANSTO, PMB 1, Menai (Sydney), NSW, 2234, Australia.
Yury.Stepanyants@ansto.gov.au

1 Introduction

A tsunami or storm surge propagating in a river is a very energetic phenomenon and not many human constructions can resist it. However studying its propagation can help identify dangerous regions where hopefully human habitat will be avoided. As in a wave guide the tsunami or bore will be influenced by the variable section (depth or width) of the channel. We will not consider here the case of the ocean where the bathymetry plays a crucial role in fixing the wave speed but a channel or river where the depth and width vary slowly compared with the wave length.

For a typical estuary as one moves towards the mouth of the river, the depth increases and the width increases. When a tidal wave, tsunami or storm surge hits such an estuary, it can be seen as a hydraulic jump in the water height (and speed) that will propagate upstream. Far less dangerous but very similar is the well known bore (mascaret in french), a tidal wave that propagates in a river for considerable distances. To understand how the wave will evolve in the channel, it is important to understand what is the dominating parameter in the river profile, the depth or the width. These simple dependencies will help predict flood occurrence for a tsunami but also for a storm surge such that happened in New Orleans in 2005.

Consider two examples, the Seine river in France and the Hugli river near Calcutta in India. The Seine is a river in the northwest part of France (2) where a bore existed up to 1960 and disappeared when the river was dredged. Its tidal features are given in the table below.
This tide amplitude is one of the largest in the world and even though the bore has disappeared, this implies that boats should be released of their moorings when the tide reaches the seaport town of Rouen.

The Hugli river is a branch of the Ganges and flows through the city of Calcutta in the eastern part of India. Here a bore is present. From the data in (1) we can build the following table.

Table 1. Parameters for the Seine river in France

	Le Havre	Caudebec	Rouen
distance	$x = 0$	$x = 80km$	$x = 150km$
time	$t = 0$	$t = 2h$	$t = 4h$
Tide amplitude	8-10m		1-2m

Table 2. Parameters for the Hugli river in India

	Balari	Panchpara	Calcutta	Bansberia
distance	$x = 0$	$x = 30km$	$x = 50km$	$x = 90km$
Tide amplitude	4.5	4.7m	4.9m	4.4m
Bore height		1.m	1.2 m	0.4 m

Here the tide amplitude decreases very little as one moves upstream, a bore exists and can reach 1 m depending on the tide. As we see below the main reason for this is the shallowness of the river.

This article contains results from (4) and completes the picture by presenting it in a geophysics context. In particular in section 2, we justify our approach using an inhomogeneous Koreteweg de Vries equation by connecting it to classical shallow water models. The conservation laws, realistic types of dissipation and normalizations are detailed in section 3. Section 4 describes the breaking of Riemann waves and section 5 the disintegration of a bore into KdV solitons. Detailed numerical results are presented in section 6.

2 Shallow water models: long wave limit

The first steps in the derivation of shallow water equations are standard and can be found for example in (5). We reproduce these steps for the sake of completeness.

The equations describing surface (gravity) waves on a fluid are the conservation of mass and momentum

$$\nabla v = 0 \tag{1}$$

$$\rho v_t + v \nabla v = -\nabla p + \rho g \mathbf{z}, \tag{2}$$

where the subscripts are partial derivatives and v, p, ρ, g and $\mathbf{z}$ are respectively the 3D velocity field, pressure, density, gravity constant and unit vector along the fluid depth. Assuming curlv $= 0$ we introduce the velocity potential ϕ so that $v = \nabla\phi$. In the following we drop the 2nd horizontal direction for simplicity. Then the equations describing the evolution of the potential ϕ of the fluid in the region $-h(x) < z < \eta(x)$ where $z = \eta(x)$ is the free surface are (5)

$$\phi_{xx} + \phi_{zz} = 0 \quad -h(x) < z < \eta(x) \tag{3}$$
$$\phi_x h_x + \phi_z = 0 \quad z = -h(x) \tag{4}$$
$$\eta_t + \phi_x \eta_x - \phi_z = 0 \quad z = \eta(x) \tag{5}$$
$$\phi_t + \frac{1}{2}(\phi_x^2 + \phi_z^2) + g\eta = 0 \quad z = \eta(x). \tag{6}$$

The second and third equation state the continuity of the normal velocity field at the bottom and at the interface respectively. The last equation is Bernouilli's law obtained by integrating the second equation of (1) along z. At this level there are no approximations on the wave length.

Before going further, let us discuss briefly the influence of the bathymetry. When h depends on x, the conservation of mass takes the form

$$h'_t + \partial_x(hu) = 0, \tag{7}$$

where $h' = h(x) + \eta(x,t)$ is the absolute altitude of the water surface. Taking the x derivative of the fourth equation in the system (3) and dropping nonlinear terms we obtain

$$u_t + g\eta_x = 0. \tag{8}$$

Combining the two equations we get the evolution of the surface in the linear limit

$$\eta_{tt} - (gh\eta_x)_x = 0. \tag{9}$$

This inhomogeneous wave equation shows the dominant role of the bathymetry in the propagation of tsunamis which are very long ocean waves. In particular this equation allows propagation in both directions and therefore reflections of the wave. This is the main effect that will give the amplitude and direction of the waves as they hit the coastline (see the contributions of H. Segur and K. Fujima in this volume). Consider as an example the city of Pondicherry that was hit by the tsunami of December 2004. In the deep ocean $h \approx 3km$ so the waves travel at around 600 km/hr. When they reach the coastline h goes from that depth to about 100 m in a $60km$ interval because of the presence of underwater canyons. This causes a considerable slowing down and "pile-up" of the waves which will then increase in amplitude. A systematic study of how the east coast of India was affected would provide information as to where to expect coastal damage in future events, tsunamis or storm surges (see the contribution of K. S. R. Murthy in this volume). In Fig. 1 we present a picture of how the wave piles up as it reaches the shallow regions. The plot shows the solution of (9) as a function of x for three successive times $t = 0, 7$ and $t = 10$ respectively in continuous, continuous and dashed lines for a right going $\tanh(x - vt + 12)$ profile with the adapted velocity $v = 3.24$. The bathymetry profile $h(x) = 0.5h_0(\tanh(x/w_x) - 1 - \epsilon) \;\; h_0 = 10, \epsilon = 0.1 w_x = 2$ is shown in the bottom with a reduced scale. Notice how the wave slows down and its amplitude doubles as its front reaches the cliff. This plot illustrates how the continental shelf affects a long wave tsunami impinging on it. The calculations were done using the Femlab (6) finite element software with homogeneous Neuman boundary conditions on both ends of the domain.

In the rest of the article we will assume that the depth h varies very slowly with respect to the wave length so that we are in a coastal region or an estuary ie on the right hand side of Fig. 1. The dispersion relation can be obtained by linearizing the above system of equations and assuming traveling wave solutions in x of the form

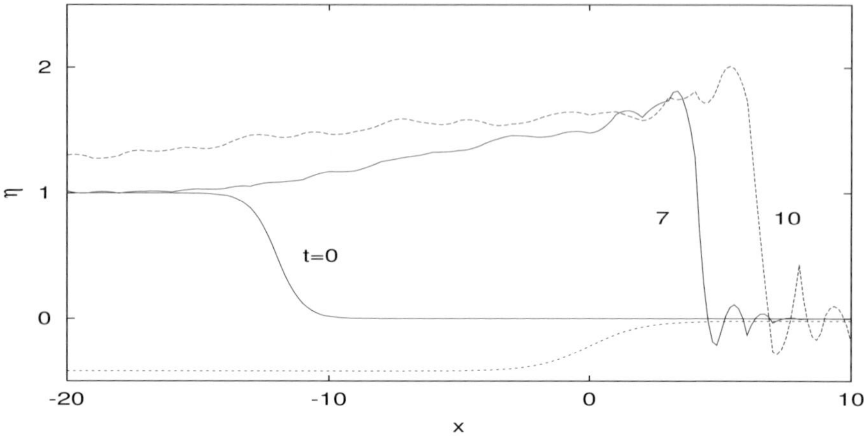

Fig. 1. Evolution of a right going wave disturbance as it hits an underwater cliff as a function of x for times $t = 0, 7$ and 10. The cliff profile is shown in the bottom of the graph. The initial condition is a rightgoing tanh profile with the adapted velocity.

$\eta = Ae^{i(kx-\omega t)}$, $\phi = B(z)e^{i(kx-\omega t)}$ where $B(z)$ should be determined. Doing this one finds

$$\omega^2 = gk\tanh(kh) \approx c_0^2k^2 - \frac{1}{3}c_0^2h^2k^4, \tag{10}$$

for small kh and where $c_0 = \sqrt{gh}$ and $B(z) = \cosh[k(z+h)]$.

Using the linear theory one can obtain the long wave Boussinesq model and then assuming unidirectional propagation, the Korteweg de Vries equation. Following Whitham (5), we introduce the normalized variables and fields

$$x' = x/l,\ Z' = (h+z)/h,\ t' = c_0t/l,\ \eta' = \eta/a,\ \phi' = \frac{c_0}{gla}\phi, \tag{11}$$

and the important nonlinear and dispersive parameters

$$\alpha = \frac{a}{h} \quad, \beta = \frac{h^2}{l^2}, \tag{12}$$

which are assumed small. We assume for simplicity the depth to be constant. Omitting the primes, the system of equations (3) becomes

$$\beta\phi_{xx} + \phi_{ZZ} = 0, \qquad 0 < Z < 1 + \alpha\eta \tag{13}$$

$$\phi_Z = 0 \quad Z = 0 \tag{14}$$

$$\eta_t + \alpha\phi_x\eta_x - \frac{\phi_Z}{\beta} = 0 \quad Z = 1 + \alpha\eta \tag{15}$$

$$\phi_t + \frac{1}{2}\alpha(\phi_x^2 + \frac{\phi_Z^2}{\beta}) + \eta = 0 \quad Z = 1 + \alpha\eta. \tag{16}$$

The key point is to solve the Laplace equation using the expansion in β

$$\phi(x, Z, t) = f(x, t) - \frac{Z^2}{2}\beta f_{xx} + \frac{Z^4}{4!}\beta f_{4x} \ldots, \tag{17}$$

which takes care of the $Z = 0$ boundary condition. Plugging (17) into the system (13) and keeping only terms $O(\alpha, \beta)$ we get a variant of the system proposed by Boussinesq a century ago

$$\eta_t + [(1 + \alpha\eta)u]_x - \frac{1}{6}\beta u_{3x} = 0 \tag{18}$$

$$u_t + \alpha u u_x - \frac{1}{2}\beta u_{xxt} = 0, \tag{19}$$

where $u \equiv f_x$ is the fluid horizontal velocity at the bottom $Z = 0$. Variants of these equations are obtained by evaluating the fluid velocity at different points in Z between 0 and η. Taking $\beta = 0$ in (18) yields the well known nonlinear (non dispersive) shallow water equations. At this point, waves can still propagate in both directions, we only assumed that they are long.

The Korteweg de Vries equation describes unidirectional waves of the system (18). Following Whitham (5) we assume right going waves such that

$$u = \eta \quad, \eta_t + \eta_x = 0,$$

and look for a solution

$$u = \eta + \alpha A + \beta B + O(\alpha^2 + \beta^2).$$

We find then the classical Korteweg de Vries equation (KdV)

$$\eta_t + \eta_x + \frac{3}{2}\alpha\eta\eta_x + \frac{\beta}{6}\eta_{xxx} = 0, \tag{20}$$

where the first two terms come from the wave equation (9) and describe right going waves. The last two terms are nonlinear and dispersive corrections because α and β were assumed small. The typical values of these parameters in the deep ocean are $\alpha = \frac{a}{h} = \frac{1m}{3000m} \approx 3 \times 10^{-4}$ and $\beta = (\frac{h}{l})^2 = (\frac{3000m}{100000m})^2 = 9 \times 10^{-4}$ so these corrections are very small. The nonlinearity α becomes significant near the coastline where h decreases and the wave amplitude a increases. The dispersion term is also important there because the wavelength is reduced due to the pile-up effect discussed above and illustrated in Fig. 2. For a tsunami reaching the shore typical values are $\alpha = \frac{10m}{30m} \approx 0.3$ and $\beta = (\frac{30m}{100m})^2 = 0.09$ so that these correction terms are important for the dynamics.

Another way of obtaining (20) is to introduce new coordinates $\xi = k(x - at)$, $\tau = k^3 t$ inspired by the dispersion relation for $k << 1$ and expand u, η in powers of k. Going back to the dimensional space and time coordinates and dimensional field variables from (11) the KdV equation becomes

$$\eta_t + c_0(1 + \frac{3}{2}\frac{\eta}{h})\eta_x + \frac{c_0 h^2}{6}\eta_{3x} = 0. \tag{21}$$

Despite the long wave and unidirectional approximations, the KdV equation describes quantitatively long waves propagating in narrow shallow channels. This was first established by John Scott Russel (7). A systematic comparison with experiments for different initial conditions was done by Joe Hammack and Harvey Segur in the series of articles (8; 9; 10). In all cases, the data showed remarkable agreement with the solutions of (21).

3 The model: inhomogeneous Korteweg-De Vries equation

For hydrological purposes, ie to determine the amount of water present at a given position for all times one needs to solve a boundary value problem. This is difficult for the KdV equation (13; 14). However going back to the derivation of the equation, one sees that at zero order $\eta_t + c_0\eta_x = 0$ so at that order $\rightarrow$ $\partial x = -\frac{1}{c_0}\partial t$. We can then transform the nonlinear and dispersive corrections.

This approach preserves the accuracy of the model and also allows to use the entire powerful kit of exact methods available for the KdV equation, including inverse scattering transform, Hirota method, Darboux-Matveev and Bäcklund transformations. The equation in signalling coordinates is (15)

$$\frac{\partial \eta}{\partial x} + \frac{1}{c_0}\frac{\partial \eta}{\partial t} - \frac{a_1}{c_0^2}\eta\frac{\partial \eta}{\partial t} - \frac{a_2}{c_0^4}\frac{\partial^3 \eta}{\partial t^3} = 0. \tag{22}$$

This equation can describe channel inhomogeneities that are long compared to the typical wave length. If the channel is curved then x is a curvilinear ray coordinate directed along the channel axis (see Fig. 2) and orthogonal at each point to the other axes. Such an approach was used in (11) to describe sea waves in a coastal zone. Another important effect is the change of the depth $h(x)$ or width $l(x)$ of the channel which we will assume to of rectangular cross-section for simplicity throughout the paper. This leads not only to an x-dependence of the coefficients of the Eq. (22) but also to the appearance of an additional term. To derive this term, we note first that the constructing generalized KdV equation contains additively the terms describing different effects such as nonlinearity, dispersion, dissipation, inhomogeneity, etc, because all these effects are assumed small and *the same order of smallness.* This assumption allows us to derive the corresponding terms in the equation separately from the others. To derive the inhomogeneous term we use the energy flux conservation law in the channel cross-section. Time averaged density of wave energy integrated over the channel depth can be readily obtained in the linear approximation $\mathcal{E} = \rho g \eta^2/2$ (16; 17), where ρ is fluid density. Hence, the total energy flux across the channel cross-section is

$$Q = c_0(x)\mathcal{E}l(x) = \frac{1}{2}\rho g c_0(x) l(x) \eta^2(x) = \text{const}. \tag{23}$$

Differentiating this expression with respect to x we get

$$\frac{dQ}{dx} = 0 \quad \rightarrow \quad \eta_x + \frac{\eta}{2c_0 l}(c_0 l)_x = 0. \tag{24}$$

In addition to inhomogeneities of the channel, an incoming wave will be damped in a real river, we will discuss this in the next section. Then using relation (24), we obtain the final inhomogeneous Korteweg de Vries equation

$$\eta_x + \frac{1}{c_0(x)}\left[1 - \alpha(x)\eta\right]\eta_t - \beta(x)\eta_{3t} + \frac{1}{2}\frac{\Delta_x}{\Delta}\eta = -F(\eta), \tag{25}$$

where $F(\eta)$ is the friction term and

$$\Delta \equiv c_0 l \ , \ \alpha = \frac{3}{2h} \ , \ \beta = \frac{h^2}{6c_0^3}. \tag{26}$$

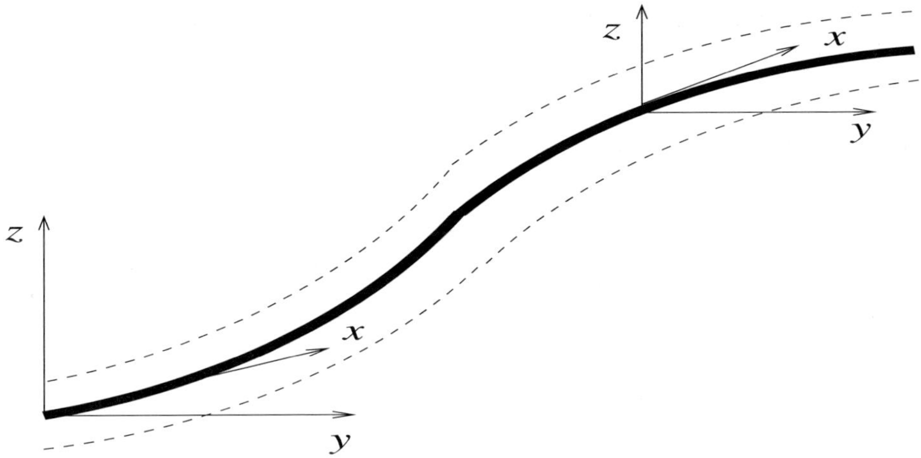

Fig. 2. Sketch of a curved channel and its associated curvilinear coordinate system.

The nonlinear and dispersive coefficients α and β are the dimensional versions of the coefficients discussed in the previous sections. We use the same names to indicate this and will use this form throughout the article. The channel will be considered of rectangular cross-section, depth $h(x)$ and width $l(x)$ and we will assume these parameters to be slowly varying functions of the ray coordinate x. Specifically we will assume

$$h = h_0(1 - \kappa_h x), \quad h_0 e^{-\kappa_h x} \quad \text{or} \quad l = l_0(1 - \kappa_l x), \quad l_0 e^{-\kappa_l x}, \tag{27}$$

where h_0, l_0 are the depth and width at $x = 0$. Equations (25-27) will be the main model that we will study in the article. We will consider the evolution of initial ($x = 0$) perturbations as they propagate along the channel. The analysis is impossible in the general case so we will consider the most typical and important wave processes separately.

3.1 Conservation laws in the absence of dissipation $F \equiv 0$

In the absence of dissipation and for a localized solution, (25) has two conserved quantities, the momentum

$$\Delta^{1/2}(x) \int \eta \, dt = \text{const}, \tag{28}$$

and the energy

$$\Delta \int \eta^2 \, dt = \text{const}. \tag{29}$$

For a linear wave, assuming a perturbation in the form

$$\eta(x,t) = Af\left[\frac{1}{T}\left(t - \int_0^x \frac{dx'}{c_0(x')}\right)\right], \tag{30}$$

where f is a dimensionless function describing wave profile and plugging it into (28) or (29), one gets the general linear Green's law

$$A(x) \sim \Delta^{-1/2}(x) \sim h^{-1/4}(x) l^{-1/2}(x). \tag{31}$$

For a (nonlinear) KdV soliton of the form

$$\eta(x,t) = A \operatorname{sech}^2 \left[\frac{1}{T} \left(t - \int_0^x \frac{dx'}{V(x')} \right) \right] \tag{32}$$

we should use the energy relation (29) and not the momentum because the perturbation generates a shelf in the wave and this should not be taken into account. One obtains then the nonlinear Greens law instead of (31)

$$l(x) h^{3/2}(x) A^{3/2}(x) = \text{const} \quad \text{or} \quad A(x) \sim l^{-2/3}(x) h^{-1}(x). \tag{33}$$

3.2 Different types of dissipations $F(\eta)$

Geophysical flows are different from laboratory flows in particular for the friction mechanisms involved. For example the boundary layer which is the main component of friction in laboratory experiments plays little role in a tsunami or turbulent bore. For these the dissipation terms are not well known so authors use very general models. The main dissipation terms used in geophysical flows are Rayleigh, Chezy and Reynolds damping. Rayleigh damping is $F(\eta) = \gamma_0(x)\eta$ where $\gamma_0(x) = \frac{3\nu}{4c_0(x)h^2(x)}$ and $0.2m^2 s < \nu < 30m^2 s$ is the turbulent viscosity(11).
Chezy damping is $F(\eta) = \gamma_c(x)|\eta|\eta$ where $\gamma_c(x) = \frac{k}{h^2(x)}$ and the the coefficient $k \approx 1.510^{-3}$ is related to friction on the bottom.
Finally Reynolds damping is $F(\eta) = -\gamma_2(x)\frac{\partial^2 \eta}{\partial t^2}$, where $\gamma_2(x) = \frac{\nu}{c_0^3}$

4 Normalizations of the model

The first step is to eliminate the independent term by introducing $u = s(x)\eta$; $s(x) = \sqrt[2]{\frac{\Delta(x)}{\Delta(0)}}$ Equation (25) becomes

$$\frac{\partial u}{\partial x} + \left[\frac{1}{c_0(x)} - p(x)u \right] \frac{\partial u}{\partial t} - \beta(x)\frac{\partial^3 u}{\partial t^3} = -s(x) F\left[\frac{u}{s(x)} \right], \tag{34}$$

where $p(x) = \frac{\alpha(x)}{c_0(x)s(x)}$ The boundary condition is $u(x=0,t) = U\Phi(t/T)$
To eliminate the galilean term and reduce the nonlinearity we introduce the new fields and coordinates

$$v = \frac{u}{U} \quad ; \quad \zeta = \frac{U}{T}\int_0^x p(\xi)d\xi \quad ; \quad \theta = \frac{1}{T}(t - \int_0^x \frac{d\xi}{c_0(\xi)}) \tag{35}$$

, and obtain the final simplified KdV equation

$$\frac{\partial v}{\partial \zeta} - v\frac{\partial v}{\partial \theta} - \frac{1}{\sigma^2(\zeta)}\frac{\partial^3 v}{\partial \theta^3} = -\frac{1}{R_d}\mathcal{F}(v). \tag{36}$$

Equation (36) features the Ursell parameter $\sigma^2(\zeta) = p(\zeta)UT^2/\beta(\zeta)$ which gives the importance of the dispersion vs nonlinearity. The dissipation term is $R_d = p(\zeta)U/T$ and

$$\mathcal{F}(v) = \left\{ \frac{3\nu}{4c_0(\zeta)h^2(\zeta)}v \quad ; \quad \frac{kU}{s(\zeta)h^2(\zeta)}|v|v \quad ; \quad \frac{\nu}{T^2c_0^3(\zeta)}\frac{\partial^2 v}{\partial \theta^2} \right\}. \tag{37}$$

This normalization is very useful to compare the relative strength of the different dissipations. First consider the case of variable depth $h = h_0(1-\kappa x)$ $x \ll \frac{1}{\kappa}$. Then the normalized coefficients of the Rayleigh, Chezy and Reynolds damping terms are respectively $\tilde{g}_0 = 1 + W\zeta$, $\tilde{g}_c = (1+W\zeta)^{2/3}$, $\tilde{g}_2 = (1+W\zeta)^{-1/3}$, where $W = \kappa T\sqrt{gh_0^3}/(2U)$ and $\tilde{\sigma}^2 = (1+W\zeta)^3$.
For the variable width case $l = l_0(1-\kappa x)$ $x \ll 1/\kappa$, we get similarly $\tilde{g}_0 = 1 - \frac{2}{3}W\zeta$; $\tilde{g}_c = 1$; $\tilde{g}_2 = 1 - \frac{2}{3}W\zeta$; $\tilde{\sigma}^2 = (1-\frac{2}{3}W\zeta)^{-1}$. The dissipative coefficients are plotted as a function of the reduced space variable $W\zeta$ in the left panel of Fig. 3 for variable depth and on the right panel of Fig. 3 for variable width. Notice how, for linearly decreasing depth, the Reynolds dissipation coefficient decreases as one moves upstream. On the contrary the Chezy and Rayleigh coefficients increase. When the width of the channel is decreased, the Chezy coefficient is constant while the other two types of damping decrease.

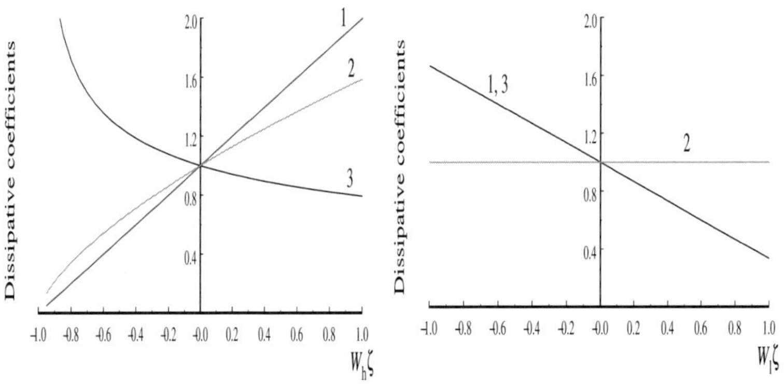

Fig. 3. Evolution of the Rayleigh, Chezy and Reynolds normalized dissipation coefficients g_0, g_c and g_2 as a function of $W\zeta$ for linearly decreasing depth (left panel) and linearly decreasing width (right panel).

5 Breaking of Riemann waves:

When the dispersion and dissipation terms are small compared to the nonlinearity ie for very large amplitude waves like a tsunami, they can be neglected and the problem reduced to the quasilinear wave equation

$$\frac{\partial u}{\partial x} + \frac{1}{V(u,x)}\frac{\partial u}{\partial t} = 0, \tag{38}$$

with $V(u,x) = \frac{c_0(x)}{1-\alpha(x)u/s(x)}$. Then the solution is given in implicit form

$$u(x,t) = \Phi\left(t - \int\limits_0^x \frac{dx'}{V[u,x']}\right). \tag{39}$$

It is well known that the wave breaks at the distance $x = X_b$ such that(5) $\frac{\partial}{\partial \tau}\left(t - \tau - \int\limits_0^{X_b} \frac{dx'}{V[\Phi(\tau),x']}\right) = 0.$

When the channel depth or width decreases, the breaking distance X_b is changed from $X_b^0 = \frac{2h_0 c_0}{3\max(\Phi')}$ its unperturbed value and the change can be estimated as shown in the table below for the typical numbers. $h_0 = 10m,\ g = 9.8ms^{-2} \ \rightarrow \ \alpha = 0.15, \ \ \beta = 0.017\ \Phi(\tau) = A\sin(\omega\tau),\ A = 1\ ,\ \omega = 0.5s^{-1}$
$X_b^0 = 131m$

Table 3. Breaking distance for inhomogeneous channels

Variable	depth $h = h_0(1-\kappa x)$	width $l = l_0(1-\kappa x)$
X_b	$\approx X_b^0(1-\frac{7}{6}\kappa X_b^0)$	$= X_b^0(1-\frac{1}{4}\kappa X_b^0)$
$\kappa = 5\ 10^{-3}$	31 m	109 m

As one can see breaking occurs for a much shorter distance upstream when the depth is decreased. Decreasing the width has a much smaller effect. Dissipation of Rayleigh and Chezy types will increase X_b (4). On the contrary Reynolds dissipation leads to Burgers equation where no breaking occurs. This is a turbulent bore where energy is dissipated at a constant rate(3). One can think that vegetation at the water front (mangroves..) causes this type of damping and thus limits the destruction caused by the large release of kinetic energy due to breaking.

6 Bore disintegration into KdV solitons:

In general, dispersion is present so that any initial condition will decompose into KdV solitons if the inhomogeneous and damping terms are small enough. In the homogeneous and inviscid case $s \equiv 1,\ F \equiv 0$ (25) becomes

$$\frac{\partial \eta}{\partial x} + \frac{1}{c_0}(1-\alpha_1\eta)\frac{\partial \eta}{\partial t} - \beta\frac{\partial^3 \eta}{\partial t^3} = 0, \tag{40}$$

with the initial (boundary) condition $\eta(x=0,t) = UH(t)$. Introducing the change of variables $v = \frac{\eta}{U}$; $\xi = x\sqrt{\frac{p_0^3 U^3}{\beta}}$; $\tau = \sqrt{\frac{p_0 U}{\beta}}\left(t - \frac{x}{c_0}\right)$, we get the dimensionless form

$$\frac{\partial v}{\partial \xi} - v\frac{\partial v}{\partial \tau} - \frac{\partial^3 v}{\partial \tau^3} = 0. \tag{41}$$

From this one of the authors estimated the distance of first soliton emergence from a step profile as $\xi_s \approx 45$ (12). In the real units the distance of soliton emerging is $X_s = \xi_s \sqrt{\frac{p_0^3 U^3}{\beta}} = 10h^{5/2}U^{-3/2}$ As example we take $h = 10$ and obtain $X_s = 3160m$ for $U = 1$, $X_s = 1120m$ for $U = 2$ and $X_s = 560m$ for $h = 5m$, $U = 1m$.

In the inhomogeneous case, this analysis is not valid, however the scaling is instructive to understand the role of depth vs width of the channel. The reduced field and variables are $v = \frac{u}{U}$; $\xi = U^{3/2}\sqrt{\frac{p(0)}{\beta(0)}} \int\limits_0^x p(x')\,dx'$; $\tau = \sqrt{\frac{Up(0)}{\beta(0)}}\left(t - \int\limits_0^x \frac{dx'}{c_0(x')}\right)$, and the inhomogeneous KdV equation is

$$\frac{\partial v}{\partial \xi} - v\frac{\partial v}{\partial \tau} - B(\xi)\frac{\partial^3 v}{\partial \tau^3} = 0, \tag{42}$$

where the generalized Ursell parameter $B(\xi) = \frac{\beta(\xi)}{\beta(0)}\frac{p(0)}{p(\xi)}$ depends on the position ξ along the channel. The table below gives the behavior of $B(\xi)$ as a function of the reduced coordinate along the channel $\xi/\bar{\xi}$ with $\bar{\xi} = \frac{9}{2}\frac{U^{3/2}h_0^{-5/2}}{\kappa}$. It shows that decreasing the channel depth gives a smaller dispersion than decreasing the channel width so that the distance of soliton emerging will be smaller for variable depth.

Table 4. Generalized Ursell parameter for different channel profiles

Variable	depth	width
Linear	$B_{dl}(\xi) = \left(1 + \frac{3}{4}\frac{\xi}{\bar{\xi}}\right)^{-3}$	$B_{wl}(\xi) = 1 - \frac{1}{2}\frac{\xi}{\bar{\xi}}$
Exponential	$B_{de}(\xi) = \left(1 + \frac{7}{4}\frac{\xi}{\bar{\xi}}\right)^{-9/7}$	$B_{we}(\xi) = \left(1 + \frac{1}{2}\frac{\xi}{\bar{\xi}}\right)^{-1}$

6.1 Numerical results: variable depth or width, no damping

We have solved the inhomogeneous KdV equation (36)

$$v_\xi - \alpha_0(\xi) v v_\tau + \beta_0(\xi) v_{3\tau} = 0$$

using the following finite difference scheme in space and time $O(dt^2, dx^2)$

$$v_\xi = \frac{v_j^{n+1} - v_j^{n-1}}{2dx} + O(dx^2) \quad , \quad v_\tau = \frac{v_{j+1}^n - v_{j-1}^n}{2dt} + O(dt^2)$$

$v_{3\tau} = \frac{v_{j+2}^n - 2v_{j+1}^n + 2v_{j-1}^n - v_{j-2}^n}{2dt^3}$. The scheme is stable if $dx \sim dt^3$. In the homogeneous case it conserves the mass and energy $C_1 = \int v d\tau$, $C_2 = \int v^2 d\tau$. On a practical side, since it is a 3 step iteration, it has to be initialized with a simple 2 level scheme $O(dt, dx^2)$. The Reynolds dissipation term is treated with the DuFort Frankel discretisation,

$$v_{\tau\tau} \approx \frac{v_{j+1}^n - v_j^{n+1} - v_j^{n-1} + v_{j-1}^n}{dt^2},$$

which us absolutely stable when $\alpha_0 = \beta_0 = 0$.

6.2 Numerical results: geometry

We first consider the role of geometry in the absence of dissipation. The initial condition is a tanh profile

$$\eta(0,t) = \frac{U_0}{2}\left(1 + \tanh\frac{t - t_0}{T_0}\right), \tag{43}$$

where we took $U_0 = 1m$, $t_0 = 940s$, $T_0 = 20s$. We assumed a linearly decreasing depth such that $\kappa = 10^{-3}$. Fig. 4 shows the decomposition of the bore (43) into solitons in the reduced frame (left panel) and the laboratory frame (right panel). Notice the main features of the transformations (35), the amplification of the wave and the time shift due to the Galilean transformation. In Fig. 5 we compare the solutions

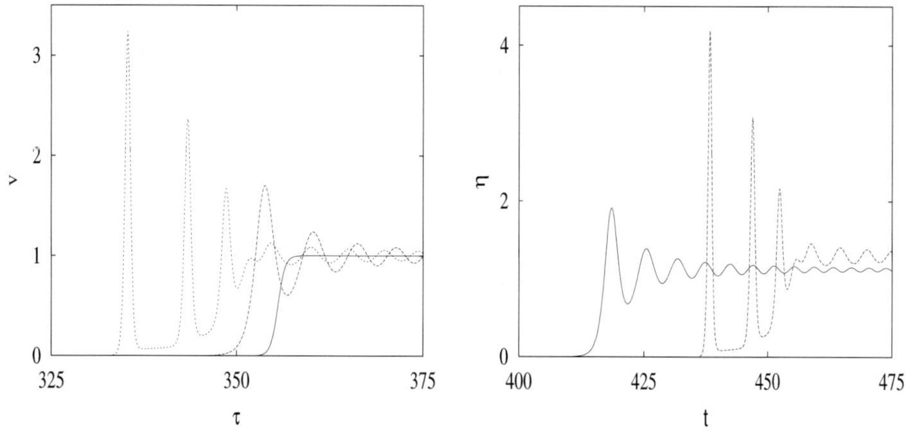

Fig. 4. Bore decomposition into solitons for a channel of linearly decreasing depth. The left panel shows the solution in the reduced coordinates $v(\xi, \tau)$ for $\xi = 0$ (initial condition) and $\xi = 18, 30$. The right panel presents the solution $\eta(x,t)$ in the laboratory frame for $\xi = 18, 30$ corresponding respectively to $x = 532m$ and $x = 644m$.

obtained for a linearly decreasing depth (continuous line) and width (dashed line) at the same location $x = 1000m$. Solitons have appeared for the former while they are just emerging from the latter as expected from the reduced Ursell parameter $B(\xi)$ (Table 6). Once solitons are created,they evolve practically independently of each other and their amplitude follows the nonlinear Greens law (33) as seen in Fig. 6.

6.3 Numerical results: effect of damping

To illustrate how different dissipations act we present in Fig. 7 $\eta(x = 2710m, t)$ for the three types of damping and a linear decreasing depth $\kappa = 10^{-4}$ for the bore initial condition (43). One sees that the Chezy dissipation reduces most the wave. As expected from the results in Fig. 4, the Reynolds dissipation affects it very little and the wave gets amplified by the inhomogeneity.

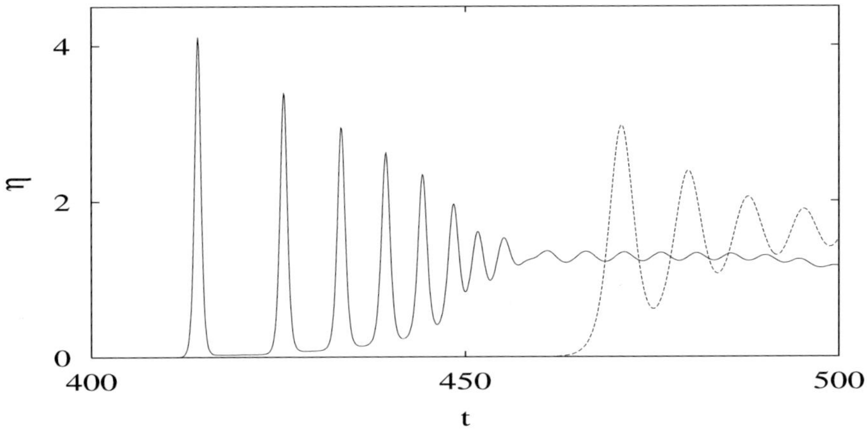

Fig. 5. Bore evolution $\eta(x,t)$ for $x = 1000m$ for a channel of linearly decreasing depth (continuous line) and width (dashed line).

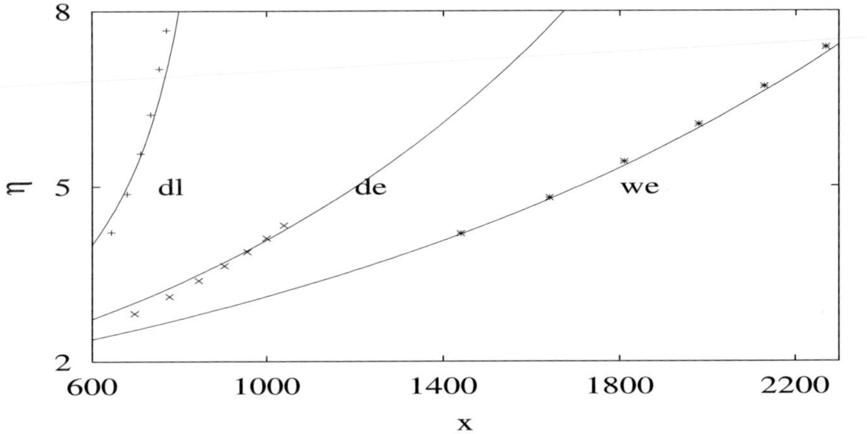

Fig. 6. Evolution of the leading pulse of Fig. 5 as a function of x computed numerically (symbols) for a channel of linearly decreasing depth (dl), exponentially decreasing depth (de) and exponentially decreasing width (we). The corresponding Green's laws (33) are given for each case in solid lines.

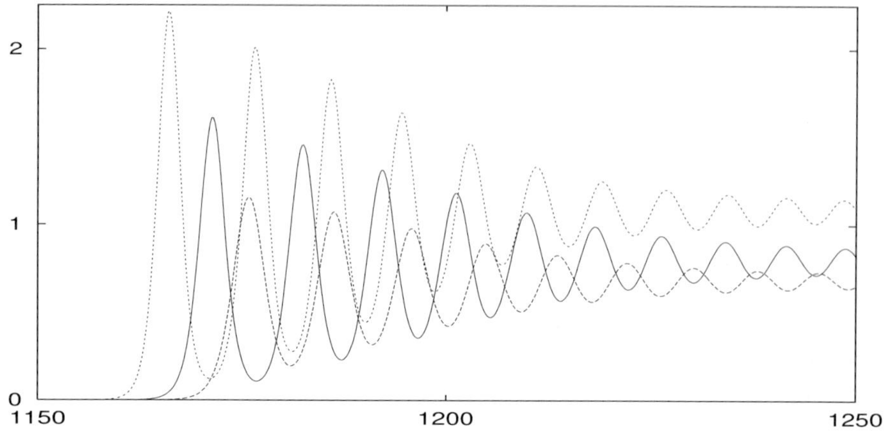

Fig. 7. Solution $\eta(x = 2710m, t)$ for a channel of linearly decreasing depth and Rayleigh dissipation (continuous line), Chezy dissipation (long dash) and Reynolds dissipation (short dash).

6.4 Adiabatic decay of KdV solitons

Solitons once formed, evolve practically independently so it is useful to understand how dissipation and inhomogeneities affect them. This is true for small perturbations for which the parameters of the soliton can be assumed to be slowly varying. In that case and for the Rayleigh damping, the energy balance equation applied to (25) yields

$$\frac{1}{2}\frac{d}{dx} < \eta^2 > + \frac{1}{2}\frac{< \eta^2 >}{\Delta}\frac{d\Delta}{dx} = -\gamma_0(x) < \eta^2 > \tag{44}$$

where $< \eta^2 > \equiv \int_{-\infty}^{\infty} \eta^2(x,t)dt$. The solution can be obtained as

$$< \eta^2 > \Delta(x) = < \eta_0^2 > \Delta_0 \exp\left(-2\int\limits_{x_0}^{x} \gamma_0(x')\, dx'\right),$$

where $\eta_0 = \eta(x = 0)$, $\Delta_0 = \Delta(x = 0)$. Assuming a soliton ansatz yields a generalization of Green's law in the presence of dissipation

$$A(x) = A_0 \frac{h_0}{h(x)}\left(\frac{l_0}{l(x)}\right)^{2/3} \exp\left(-\int\limits_{x_0}^{x} \frac{\nu\, dx'}{c_0(x')h^2(x')}\right) \tag{45}$$

For example in the case of a channel of linearly decreasing depth, the amplitude is unchanged if the depth $h(x)$ is

$$h(x) = h_0 \left[1 - \frac{5}{2}\frac{\nu}{\sqrt{gh_0^5}}(x - x_0)\right]^{2/5} . \tag{46}$$

These considerations can be for the other types of damping as well but may have to be done numerically.

7 Conclusion

We examined the propagation of a tsunami from the deep sea where nonlinearity and dispersion are small to the coastline where these parameters become significant. In particular we showed on the inhomogeneous linear wave equation the importance of bathymetry in the wave amplification.

We introduced a very simple model which allows in depth analysis of a tsunami or storm surge propagating in an inhomogeneous channel or river of slowly varying depth or width. This allows to separate the effects of the different parameters and showed the dominant role of depth variation as opposed to width variation in the breaking of Riemann waves and the bore decomposition into solitons.

Since a reliable model for wave dissipation is not known, we considered the main dissipations used for geophysical flows, the Rayleigh, Chezy and Reynolds dampings and compared their effects as one moves along the channel. These damping terms have a different dependence on the channel inhomogeneity and it would be useful to compare these predictions with the results of a large scale experiment to see which one dominates. One could then solve the inverse problem and estimate the parameters.

Acknowledgements

JGC thanks Koji Fujima, K S. R. Murthy and Harvey Segur for useful discussions.

References

[1] Mazumder N. C. and Bose S. (1995). J. of waterway, port, coastal and ocean engineering, 167-175
[2] Le Hir P. et Silva Jacinto R. (2001). Programme Scientifique Seine-Aval Vol. 2, Editions Ifremer, (2001).
[3] Tsuji Y., Yanuma T. , Murata I. and Fujiwara C. (1991) Natural Hazards 4:257–266
[4] Caputo J. G. and Stepanyants Y. (2003) Nonlinear processes in geophysics, 10:1–18
[5] Whitham G. B. (1974) Linear and Nonlinear Waves. J. Wiley & Sons
[6] http://www.comsol.com
[7] Russel J. S., Report on waves, 14th meeting of the British association for the advancement of science (BAAS), 1844.
[8] J. Hammack and H. Segur, *J. Fluid Mech.*, 1974, v. 65, 239.
[9] J. Hammack and H. Segur, *J. Fluid Mech.*, 1978, v. 84, 337.
[10] J. Hammack and H. Segur, *J. Fluid Mech.*, 1978, v. 84, 359.
[11] Pelinovsky E. N., Stepanyants Yu. A., Talipova T. G. Nonlinear dispersion model of sea waves in the coastal zone. – *J. Korean Soc. Coastal Ocean Eng.*, 1993, v. 5, n. 4, 307–317.
[12] Pelinovsky E. N., Stepanyants Yu. A. On the estimation of the time of solitons creations from the initial perturbations in the framework of Korteweg–de Vries equation. – *Izv. VUZov, Radiofizika*, 1981, v. 24, n. 7, 908–911 (in Russian).

[13] Marchant T. R., Smyth N. F. Initial–boundary problems for the Korteweg–de Vries equations. – *IMA J. Appl. Math.*, 1991, v. 47, 247–264.
[14] Marchant T. R., Smyth N. F. The initial boundary problem for the Korteweg–de Vries equation on negative quarter-plane. – *Proc. R. Soc. Lond. A*, 2002, v. 458, 857–871.
[15] Osborne A. R. The inverse scattering transform: tools for the nonlinear Fourier analysis and filtering of ocean surface waves. – *Chaos, solitons and fractals*, 1995, v. 5, 2623–2637.
[16] Mei C. C. *The Applied Dynamics of Ocean Surface Waves.* – John Wiley & Sons, New York, 1983.
[17] Stepanyants Yu. A. Relationship between kinetic and potential energies in internal waves in the presence of shear currents. – *Sov. Phys. Izvestia. Atmospheric and Oceanic Physics*, 1985, v. 21, n. 6, 518–520 (English translation of the Russian journal).

On the modelling of huge water waves called rogue waves

Christian Kharif

Institut de Recherche sur les phénomènes Hors Equilibre, Marseille, France
kharif@irphe.univ-mrs.fr

Summary. The chapter focuses on the physics and modelling of the extreme water wave events called rogue waves. A particular attention is paid to their formation in presence of strong wind. Two mechanisms producing the giant waves are considered: The dispersive spatio-temporal focusing and the modulational instability. In both cases an amplification of the height and duration of the rogue wave event is observed under wind action.

1 Introduction

Extreme wave events such as rogue waves correspond to large-amplitude waves occurring suddenly on the sea surface. As it has been emphasized by (13) these huge water waves have been part of marine folklore for centuries. Since the seventies of the last century, oceanographers have started to believe them. In situ observations provided by oil and shipping industries and capsizing of giant vessels confirm the existence of such events. Up to now there is no definitive consensus about a unique definition of rogue wave event. The definition based on height criterion is often used. When the height of the wave exceeds twice the significant height it is considered as a rogue wave. Owing to the non-Gaussian and non-stationary character of the water wave fields on sea surface, it is a very tricky task to compute the probability density function of rogue waves. So, the approaches presented herein are rather deterministic than statistical. Recently (7) and (11) provided reviews on the physics of these events when the direct effect of the wind is not considered. Rogue waves can occur far away from storm areas where wave fields are generated. In that case huge waves are possible on quasi still water. Hence, our approach to the problem is aimed at describing the deterministic mechanisms responsible for the occurrence of these huge waves in presence of strong wind, that is in storm areas.

There are a number of physical mechanisms producing the occurrence of rogue waves. Extreme wave events can be due to refraction (presence of variable currents or bottom topography), dispersion (frequency modulation), wave instability (Benjamin-Feir instability), soliton interactions, etc. that may focus the wave energy into a small area. All these different mechanisms can work without direct effect of wind on waves. For more details see the paper (11).

The most popular example of rogue waves is that corresponding to abnormal waves appearing suddenly off the south-east coast of South Africa when the dominant wind-generated waves meet a counter-current (Agulhas Current). Theoretical and numerical studies have been developed to understand this wave-current interaction ((23), (12), (26)). Recently, (29) reported experimental results of limiting rogue waves on currents.

Refraction of surface waves can be due to underwater topography as well. The result is spatial variations of the kinematic (frequency and wavenumber) and dynamic (amplitude or energy) properties of the wave packets. The geometrical focusing of wave energy can generate huge waves. Note that rogue waves can arise from wave-current interactions in water of varying depth. It means that refraction effects due to sea bottom and current both are working.

Rogue wave events due to spatio-temporal focusing phenomenon can be described as follows. If initially short wave packets are located in front of longer wave packets having larger group velocities, then during the stage of evolution, longer waves will overtake short waves. A large-amplitude wave can occur at some fixed time because of the superposition of all the waves merging at the same location (the focus point). Afterward, the longer waves will be in front of the short waves, and the amplitude of the wave train will decrease. This focusing-defocusing cycle was described by (18) within the framework of the shallow water theory and later by (22) using the Davey-Stewartson system for three-dimensional water waves propagating in finite depth. More recently, this technique was also used in the experiments on rogue waves conducted by (8) and (25).

Another mechanism generating extreme wave events is the modulational instability or the Benjamin-Feir instability. Due to this instability uniform wave trains suffer modulation-demodulation cycles (the Fermi-Pasta-Ulam recurrence). At the maximum of modulation a frequency downshifting is observed and very steep waves occur.

Soliton interaction as a possible model for extreme waves in shallow water has been suggested by (19) and (20). They considered the interaction of two long-crested shallow water waves within the framework of the two-soliton solution of the Kadomstev-Petviashvili equation. It was found that extreme surface elevation exceeds several times the amplitude of the incoming waves over a small area. In deep water (3) and (4) showed that strong interactions between envelope-solitons may produce rogue wave event. They performed long time simulations based on fully nonlinear equations.

The present study focuses on the two main mechanisms producing rogue waves : The spatio-temporal focusing mechanism and the modulational instability mechanism. In section 2 is defined the criterion which characterizes a rogue wave event. Section 3 presents the equations of water waves and the sheltering theory. The experimental study of extreme wave event is presented in section 4 while section 5 is devoted to the numerical modelling and simulations.

2 Rogue wave definition

In the first approximation, the sea elevation is considered as a summation of sinusoidal waves of different frequencies with random phases and amplitudes. The random wave field is considered as a stationary random Gaussian process with the following probability density distribution

$$f(\eta) = \frac{1}{\sqrt{2\pi}\sigma} \exp(-\frac{\eta^2}{2\sigma^2}), \tag{1}$$

where η is the sea surface elevation with zero mean level, $<\eta>= 0$, and σ^2 is the variance computed from the frequency spectrum, $S(\omega)$

$$\sigma^2 =< \eta^2 >= \int_0^\infty S(\omega) d\omega. \tag{2}$$

The wind wave spectrum is assumed to be narrow, thus the cumulative probability function of the wave heights will be given by the Rayleigh distribution

$$P(H) = \exp(-\frac{H^2}{8\sigma^2}). \tag{3}$$

The probability that the wave heights will exceed a threshold value, H, is given by (3).
One specific wave height frequently used in oceanography and ocean engineering is the significant wave height, H_s. This concept was introduced by (24) who defined H_s as the average of the highest one-third of wave heights. This wave height is close to the mean wave height estimated by human eye. Using the Rayleigh distribution (15) showed that H_s is given by

$$H_s = (3\sqrt{2\pi}\text{erfc}(\sqrt{\ln 3} + 2\sqrt{2\ln 3})\sigma \approx 4\sigma, \tag{4}$$

where erfc(.) is the complementary error function. So H_s is four times the standard deviation σ. Equation (3) can be rewritten as follows

$$P(H) = \exp(-\frac{2H^2}{{H_s}^2}). \tag{5}$$

Mathematically, a wave is considered to be a rogue wave if its height, H_f, satisfies the condition

$$H_f > 2H_s. \tag{6}$$

3 Mathematical formulation

3.1 Water wave equations

The fluid is assumed to be inviscid and the motion irrotational, such that the velocity $\mathbf{u}$ may be expressed as the gradient of a potential $\phi(x, y, z, t)$: $\mathbf{u} = \nabla\phi$. If the fluid is assumed to be incompressible, such that $\nabla.\mathbf{u} = \mathbf{0}$, the equation that holds throughout the fluid is the Laplace's equation

$$\nabla^2\phi = 0 \quad \text{for} \quad -h < z < \eta(x, y, t). \tag{7}$$

where $\eta(x, y, t)$ is the surface elevation.

The x and y coordinates are taken to be horizontal plane, the z axis vertically upwards. The bottom is located at $z = -h(x, y)$. The bottom condition is

$$\frac{\partial\phi}{\partial x}\frac{\partial h}{\partial x} + \frac{\partial\phi}{\partial y}\frac{\partial h}{\partial y} + \frac{\partial\phi}{\partial z} = 0 \quad \text{on} \quad z = -h(x, y). \tag{8}$$

The kinematic requirement that a particle on the sea surface, $z = \eta(x, y, t)$, remains on it is expressed by

$$\frac{\partial\eta}{\partial t} + \frac{\partial\phi}{\partial x}\frac{\partial\eta}{\partial x} + \frac{\partial\phi}{\partial y}\frac{\partial\eta}{\partial y} - \frac{\partial\phi}{\partial z} = 0 \quad \text{on} \quad z = \eta(x, y, t). \tag{9}$$

Since surface tension effects are ignored, the dynamic boundary condition which corresponds to pressure continuity trough the interface, can be written

$$\frac{\partial\phi}{\partial t} + \frac{1}{2}(\nabla\phi)^2 + g\eta + \frac{p_a}{\rho_w} = 0 \quad \text{on} \quad z = \eta(x, y, t). \tag{10}$$

where g is the gravitational acceleration, p_a the pressure at the sea surface and ρ_w the density of the water. The atmospheric pressure at the sea surface can vary in space and time. In water of infinite depth, the kinematic boundary equation (8) is replaced by $\nabla\phi \to 0$ as $z \to -\infty$.

The mathematical formulation of the water wave problem has been presented for general 3D flows. The present study is confined to 2D flows and the corresponding equations can be derived from the previous system by using $\partial/\partial y = 0$.

3.2 Wind modelling: The Jeffreys' sheltering theory

Previous works on rogue wave have not considered the direct effect of wind on their dynamics. It was assumed that they occur independently of wind action, that is far away from storm areas where wind wave fields are formed. Herein the Jeffreys' theory (see (9)) is invoked for the modelling of the pressure, p_a. Jeffreys proposed a plausible mechanism to explain the phase shift of the atmospheric pressure, p_a, needed for an energy transfer from wind to the water waves. He suggested that the energy transfer was due to the form drag associated with the flow separation occurring on the leeward side of the crests. The air flow separation would cause a pressure asymmetry with respect to the wave crest resulting in a wave growth. This mechanism can be invoked only if the waves are sufficiently steep to produce air flow separation. (1) have shown that separation occurs over near breaking waves. For weak or moderate steepness of the waves this phenomenon cannot apply and the Jeffreys' sheltering mechanism becomes irrelevant.
Following (9), the pressure at the interface $z = \eta(x, t)$ is related to the local wave slope according to the following expression

$$p_a = \rho_a s(U - c)^2 \frac{\partial\eta}{\partial x}, \tag{11}$$

where the constant, s is termed the sheltering coefficient, U is the wind speed, c is the wave phase velocity and ρ_a is atmospheric density. The sheltering coefficient, $s = 0.5$, has been calculated from experimental data. In order to apply the relation (11) for only very steep waves we introduce a threshold value for the slope $(\partial\eta/\partial x)_c$. When the local slope of the waves becomes larger than this critical value, the pressure is given by equation (11) otherwise the pressure at the interface is taken equal to a constant which is chosen equal to zero without loss of generality. This means that wind forcing is applied locally in time and space. Figure 1 [1] shows the pressure distribution at the interface in the vicinity of the crest, given by equation (11), for a threshold value close to the slope corresponding to the Stokes' corner.

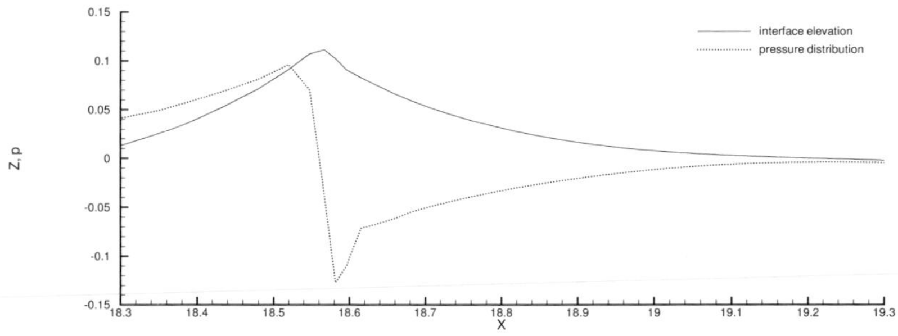

Fig. 1. Pressure at the interface given in $10^{-1}HPa$ (dotted line) and surface elevation given in m (solid line) as a function of x .

4 Rogue wave observation in presence of wind

Rogue waves have been generated in the large wind-wave tank of IRPHE which is $40m$ long, $1m$ deep and $2.6m$ wide. The wind tunnel above the water surface is $40m$ long, $3.2m$ wide and $1.6m$ high. Figure 2 gives a schematic view of the facility. The blower can produce a wind velocity up to $14m/s$ and a computer-controlled submerged wave maker located under the upstream beach can generate regular or random waves in a frequency range from $0.5Hz$ to $2Hz$. Particular care has been taken to obtain a pure logarithmic mean wind velocity profile with a constant shear layer over the water surface. Several wave gauges are installed on a trolley to measure the water surface elevation at different fetches (distance measured from the upstream beach). For additional details on experiments see the paper by (25)

Rogue waves are generated using the spatio-temporal focusing mechanism based on the dispersive nature of water waves. Within the framework of a linear approach the sea surface can be considered as the superposition of linear waves of frequency

[1] Figures 1, 4, 5 and 6 have been reprinted from (25).

$\omega(x,t)$. According to (28), the spatio-temporal evolution of the frequency of these components is governed by the following hyperbolic equation

$$\frac{\partial \omega}{\partial t} + c_g(\omega)\frac{\partial \omega}{\partial x} = 0, \tag{12}$$

where c_g is the group velocity. This equation can be solved by using the method of characteristics. The solution is given by

$$\omega(x,t) = \omega_0(\tau), \qquad v_g(\tau) = c_g(\omega_0(\tau)) \quad \text{on} \quad t = \tau + x/v_g(\tau), \tag{13}$$

where ω_0 corresponds to the temporal frequency distribution of the wave train at $x = 0$. The temporal partial derivative of the frequency is

$$\frac{\partial \omega}{\partial t} = \frac{\frac{d\omega_0}{d\tau}}{1 - \frac{x}{v_g^2}\frac{dv_g}{d\tau}}. \tag{14}$$

One can notice that the case $dv_g/d\tau > 0$, which corresponds to short waves emitted before longer waves, leads to a singularity. This singularity corresponds to the focusing of several waves at $t = T_{f_{th}}$ and $x = X_{f_{th}}$. For infinite depth, the frequency to impose to the wave maker located at $x = 0$ is given by

$$\omega(0,t) = \frac{g}{2}\frac{T_{f_{th}} - t}{X_{f_{th}}}, \tag{15}$$

where $X_{f_{th}}$ and $T_{f_{th}}$ are the coordinates of the point of focus in the $(x - t)$ plane. Using $\omega = 2\pi f$ the coordinates of the focus point reads

$$T_{f_{th}} = \Delta T \frac{f_{max}}{f_{max} - f_{min}}$$

$$X_{f_{th}} = \frac{g\Delta T}{4\pi}\frac{1}{f_{max} - f_{min}}$$

where f_{max} and f_{min} are the maximal and minimal values of the frequency imposed to the wave maker during a period of time equal to ΔT and g is the acceleration due to gravity.

The wave amplitude, a, satisfies the following equation

$$\frac{\partial a^2}{\partial t} + \frac{\partial}{\partial x}(c_g a^2) = 0. \tag{16}$$

This equation corresponds to the conservation of wave energy, and its solution is found explicitly

$$a(x,t) = \frac{a_0(\tau)}{\sqrt{1 - \frac{x}{v_g^2}\frac{dv_g}{d\tau}}}, \tag{17}$$

where $a_0(\tau)$ is the temporal distribution of the wave amplitude at $x = 0$. Within the framework of the linear theory focus points are singular points where the amplitude becomes infinite and behaves like $(X_{f_{th}} - x)^{-1/2}$.

The wave train generated at the wave maker is uniform in amplitude and frequency modulated. The experimental data are: $f_{max} = 1.3Hz$, $f_{min} = 0.8Hz$ and $\Delta T = 10s$. Experiments are performed with and without wind. Figure 3 shows the surface

elevation at a distance of $1m$ from the upstream beach for two values of the wind velocity. It can be seen that the initial group of waves is almost uniform in amplitude and unaffected by wind. From the data we find that $T_{\mathrm{f_{th}}} = 26s$ and $X_{\mathrm{f_{th}}} = 17m$ while the experimental values are $T_{\mathrm{f_{exp}}} = 26s$ and $X_{\mathrm{f_{exp}}} = 20m$ (see Figure 4 corresponding to $U = 0m/s$). Experimental data are in close agreement with the linear theory. The difference observed between the theoretical and experimental values of the focus point is mainly due to the nonlinearity of the experimental wave train. Note that the wave train generated at the wave maker is uniform in amplitude, hence the short waves are steeper than the longer waves, and the result is a downstream shift of the focusing location. From Figure 4 it can be seen that dispersion leads short waves to propagate slower than long waves, and as a result, the waves focus at a given position in the wave tank leading to the occurrence of a large amplitude wave. Downstream the point of focus, the amplitude of the group decreases rapidly (defocusing).
The same initial wave train is generated and propagated under the action of wind for several values of the wind velocity ranging from $U = 4m/s$ to $10m/s$ respectively. In presence of wind, the focusing wave train is generated once wind waves have developed. For each value of the mean wind velocity U the water surface elevation is measured at $1m$ fetch and at different fetches between $3m$ and $35m$. The fetch is measured from the entrance of the wave tank where the air flow meets the water surface i.e. at the end of the upstream beach. The wave maker is totally submerged, to avoid any perturbation of the air flow which could be induced by its displacement. Figure 5 shows the same time series of $\eta(x,t)$, at several values of the fetch x, and for a wind speed $U = 6m/s$. The wave groups mechanically generated by the wave maker are identical to those used in the experiments without wind (see Figure 3). Nevertheless, short wind waves can be observed. In any case, some differences appear in the time-space evolution of the focusing wave train. One can observe that the group of the rogue wave event is sustained longer.

The amplification factor $A(x,U)$ of the group between fetches x and $1m$ is defined as follows

$$A(x,U) = \frac{H_{\max}(x,U)}{H_{\mathrm{ref}}}, \tag{18}$$

where U is the mean wind velocity and $H_{\max}(x,U)$ is the maximal height between two consecutive crest and through in the transient group. The height, H_{ref}, of the quasi uniform wave train generated at the entrance of the tank is measured at $1m$. The mean height crest to through is $H_{\mathrm{ref}} = 6.13cm$. Figure 6 shows how evolves in space the amplification factor for three values of the wind velocity. For $U = 0m/s$ (without wind) as expected the amplification is maximal at the focus point and the curve exhibits a symmetry with respect to the straight line of equation $x = X_{\mathrm{f_{exp}}}$. Focusing and defocusing stages present a symmetrical behaviour. For $U = 4m/s$ and $U = 6m/s$ this symmetry is broken and the focus point is shifted downstream. One can notice that the amplification factor increases as the wind velocity increases. Another important feature is observed: The rogue wave criterion, $A > 2$, is satisfied on a longer distance (or period of time) while the wind velocity increases.
The experiments suggest that the physical mechanism which could be responsible of the persistence of rogue events is the occurrence of air flow separation over steep waves.

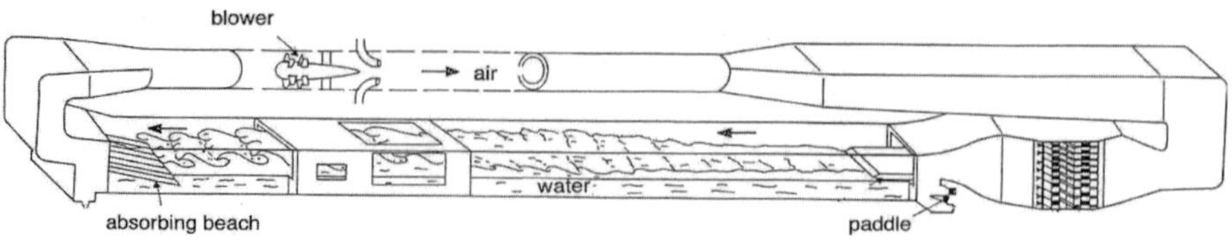

Fig. 2. Schematic view of the wind-wave facility

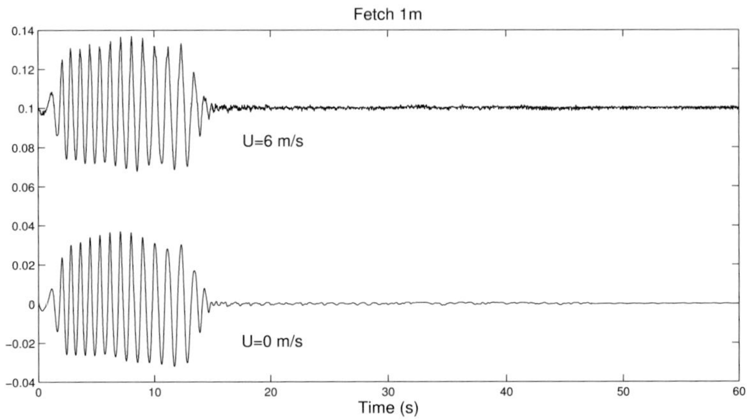

Fig. 3. Surface elevation (in cm) at fetch $1m$ for wind velocities $U = 0m/s$ and $U = 6m/s$.

5 Numerical modelling

5.1 Focusing due to linear dispersion

Herein we considered a numerical wave tank simulating the experimental water wave tank briefly described in the previous section. The gravity wave train is generated by a piston-type wave maker. An absorbing beach located at the end of the wave tank dissipates the incident wave energy.

The Laplace equation (7) is solved within a domain bounded by the water surface and solid boundaries of the numerical wave tank. The condition on the solid boundary writes

$$\nabla\phi \cdot \mathbf{n} = \mathbf{v} \cdot \mathbf{n}, \quad \text{on} \quad \partial\boldsymbol{\Omega}_{\mathrm{S}}, \tag{19}$$

where $\partial\Omega_{\mathrm{S}}$ corresponds to solid boundaries, $\mathbf{v}$ is the velocity of the solid boundaries, set equal to zero on the horizontal bottom and downstream wall of the wave tank and equal to the velocity of the piston at any point of the wave maker, and $\mathbf{n}$ is the unit normal vector to the boundaries.

A Lagrangian description of the water surface is used

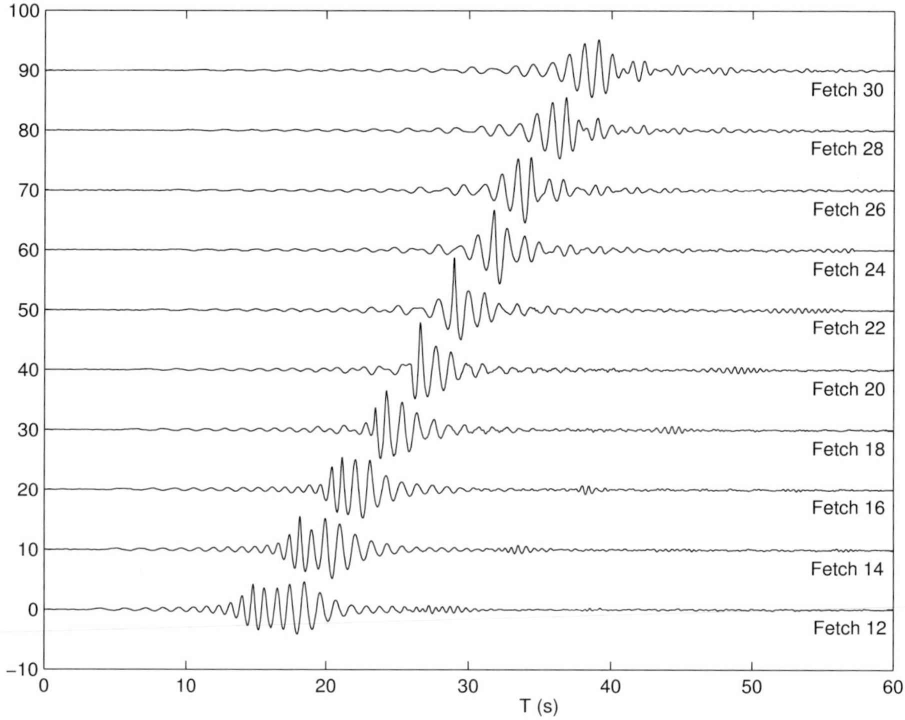

Fig. 4. Surface elevation (in cm) at several fetches (in m) for wind velocity $U = 0m/s$ as a function of time

$$\frac{D\eta}{Dt} = \frac{\partial\phi}{\partial z}, \tag{20}$$

$$\frac{Dx}{Dt} = \frac{\partial\phi}{\partial x}, \tag{21}$$

where x is the abscissa of the water surface and $D/Dt = \partial/\partial t + \nabla\phi \cdot \nabla$.

Equation (20) is an alternative form of equation (9). The kinematic boundary condition writes as well

$$\frac{DS}{Dt} = 0, \tag{22}$$

where $S(x, z, t) = \eta(x, t) - z = 0$ is the water surface equation.

The dynamic boundary condition (10) is rewritten as follows

$$\frac{D\phi}{Dt} = \frac{1}{2}(\nabla\phi)^2 - g\eta - \frac{p_a}{\rho_w}, \tag{23}$$

where the pressure, $p_a(x, t)$, at the water surface is given by equation (11), i.e. the Jeffreys' theory presented in subsection 3.2 is used for modelling wind effect on the extreme waves. Over waves presenting slopes less than a threshold value, the

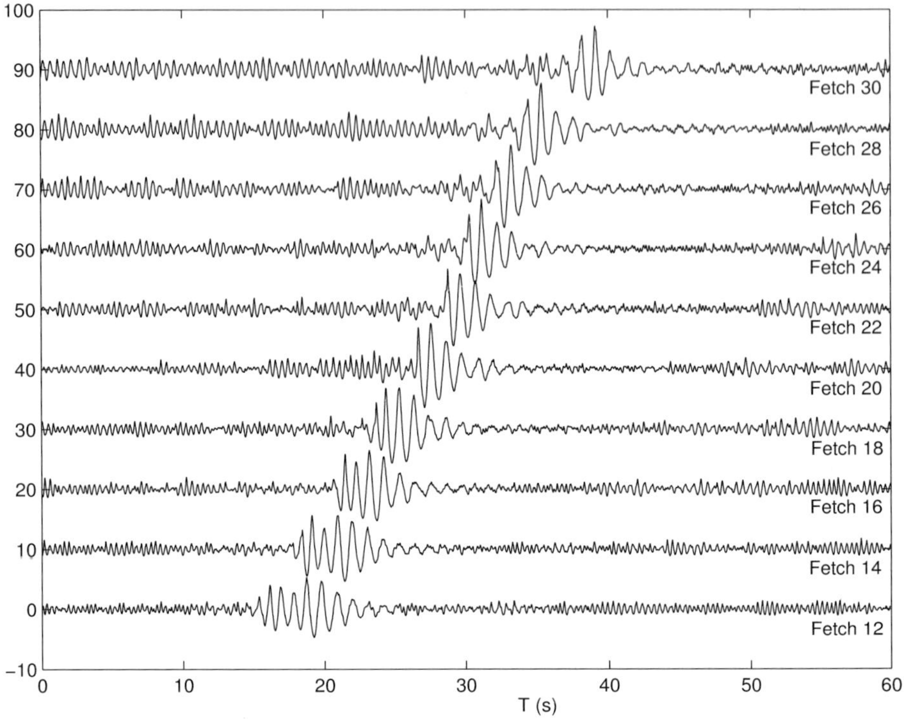

Fig. 5. Surface elevation (in cm) at several fetches (in m) for wind velocity $U = 6m/s$ as a function of time

atmospheric pressure is uniform, set equal to zero without loss of generality.

The system of equations to solve is (7), (19), (20), (21), and (23). The method to integrate numerically this system is a boundary integral equation method (BIEM) with a mixed Euler-Lagrange (MEL) time marching scheme. The numerical method is based on the Green's second identity. For more details see the paper (25).

A focusing wave train is generated by the piston wave maker, leading during the focusing stage to the generation of a extreme wave followed by a defocusing stage. The water surface and the solid boundaries (downstream wall, bottom and wave maker) are discretised by 2000 and 1000 meshes respectively, uniformly distributed. The time integration is performed using a RK4 scheme, with a constant time step of 0.01 s. To avoid numerical instability the grid spacing Δx and time increment Δt have been chosen to the satisfy the following Courant criterion derived from the linearized surface conditions

$$(\Delta t)^2 \leq \frac{8\Delta x}{\pi g}. \tag{24}$$

The focusing mechanism is investigated with and without wind as well.

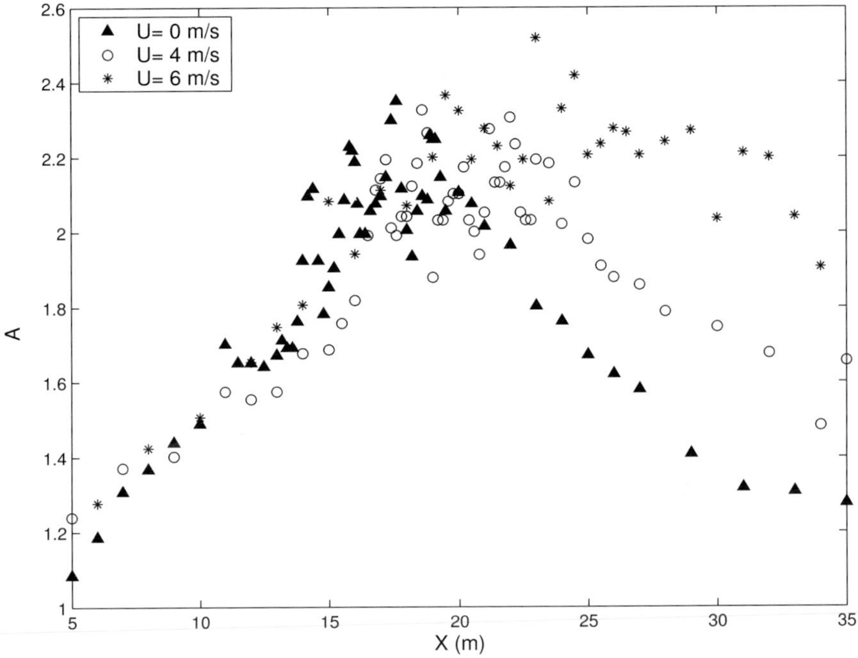

Fig. 6. Experimental amplification factor $A(x, U)$ as a function of the distance (in m) for several values of the wind velocity

A series of numerical simulations have been run for three values of the wind velocity: $U = 0\ m/s$, $4\ m/s$, $6\ m/s$. Using equation (18), Figure 7 describes the spatial evolution of the amplification factor computed numerically. It can be observed that the numerical curves behave similarly to those plotted in Figure 6 and thus emphasize the asymmetry found in the experiments. This asymmetry results in an increase of the life time of the rogue wave event which increases with the wind velocity. The threshold value of the slope beyond which the wind forcing is applied is $(\partial\eta/\partial x)_c = 0.5$. This value corresponds to a wave close to the limiting form for which the Jeffreys' theory applies. Hence the duration of the wind effect is relatively short to increase the amplification of the rogue wave event significantly. However a very weak increase of the amplification factor is observed in presence of wind which is significantly weaker than in the experiments. The main effect of Jeffreys' sheltering mechanism is to sustain the coherence of the short group involving the rogue wave event. Inspection of Figures 6 and 7 shows that the numerical maxima of the amplification factor are larger than those obtained experimentally. This can be due to spilling breaking events which where observed in the experiments, resulting in energy dissipation and in saturation in the growth of amplitude. Notice that the present model which is based on the assumption of inviscid fluid cannot describe energy dissipation. In our model, the transfer of energy from the wind to the water waves depends on the wind velocity and threshold wave slope value. If the latter value is low the energy transferred becomes high and breaking occurs.

To summarize, within the framework of the spatio-temporal focusing both experimental and numerical results are in qualitative good agreement even if some quantitative difference have been observed for the height of the rogue wave.

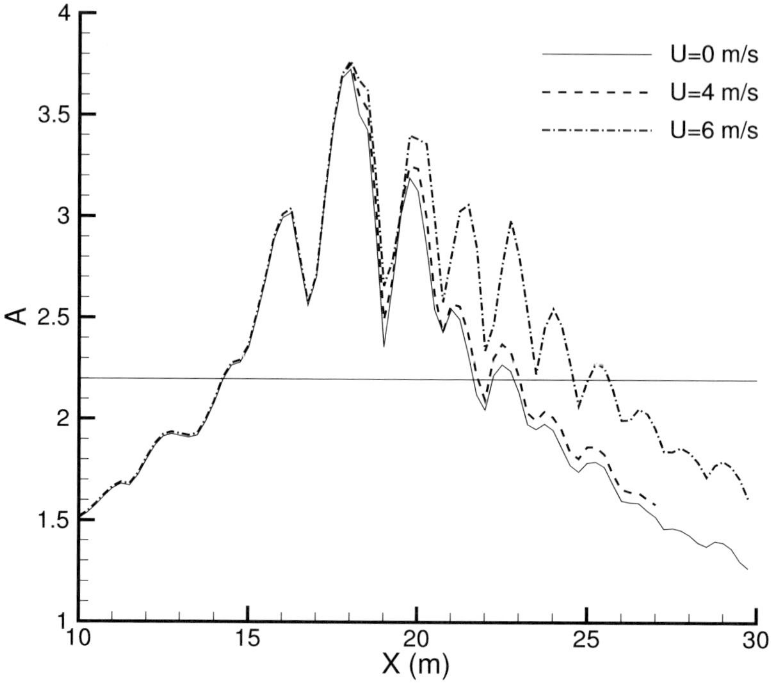

Fig. 7. Numerical amplification factor $A(x, U)$ as a function of the distance (in m) for several values of the wind velocity

5.2 Focusing due to modulational instability

Beside the focusing due to dispersion of a chirped wave group, another mechanism, the modulational instability or Benjamin-Feir instability (see the paper (2)) of uniform wave trains, can generate extreme wave events. This periodic phenomenon is investigated numerically using a high-order spectral method (HOSM) without experimental counterpart. The question is to know how evolve extreme wave events due to modulational instability under strong wind action. How is modified their amplification and time duration under wind effect? Are these effects similar or different from those observed in the case of extreme wave due to dispersive focusing?

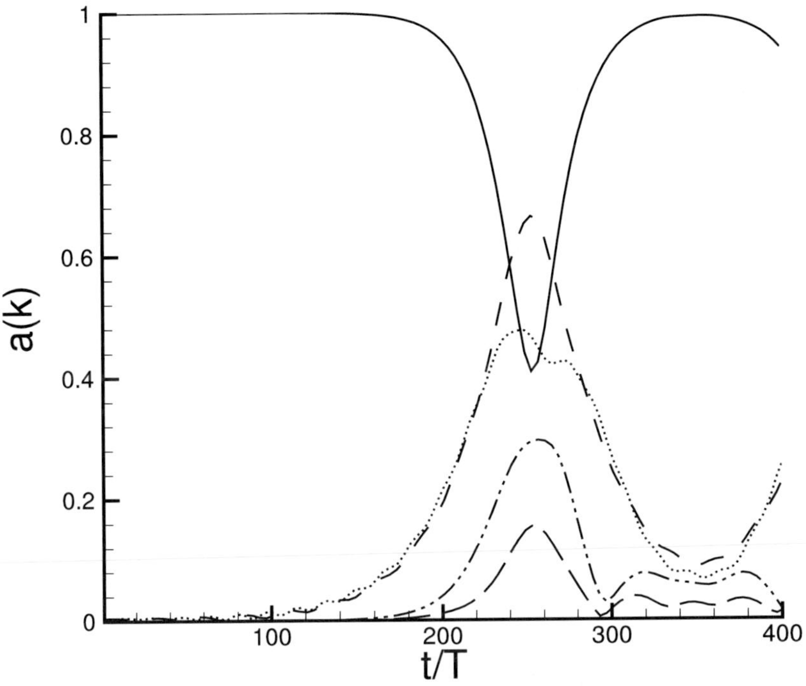

Fig. 8. Time histories of the amplitude of the fundamental, $k_0 = 5$ (solid line), subharmonic, $k_1 = 4$ (dashed line), and superharmonic, $k_2 = 6$ (dotted line), modes for an evolving perturbed Stokes wave of initial wave steepness $\epsilon = 0.11$ and fundamental wave period T, without wind action. The two lowest curves (dashed-dotted lines) correspond to the modes $k_3 = 3$ and $k_4 = 7$.

Introducing the potential velocity at the free surface $\phi^s(x,t) = \phi(x, \eta(x,t), t)$, equations (9) and (10) write

$$\frac{\partial \phi^s}{\partial t} = -\eta - \frac{1}{2}\nabla\phi^s \cdot \nabla\phi^s + \frac{1}{2}W^2[1 + (\nabla\eta)^2] - p_a, \tag{25}$$

$$\frac{\partial \eta}{\partial t} = -\nabla\phi^s \cdot \nabla\eta + W[1 + (\nabla\eta)^2], \tag{26}$$

where

$$W = \frac{\partial \phi}{\partial z}(x, y, \eta(x,y,t), t). \tag{27}$$

Equations (25) and (26) are given in dimensionless form. Reference length, reference velocity and reference pressure are, $1/k_0$, $\sqrt{g/k_0}$ and $\rho_w g/k_0$ respectively.

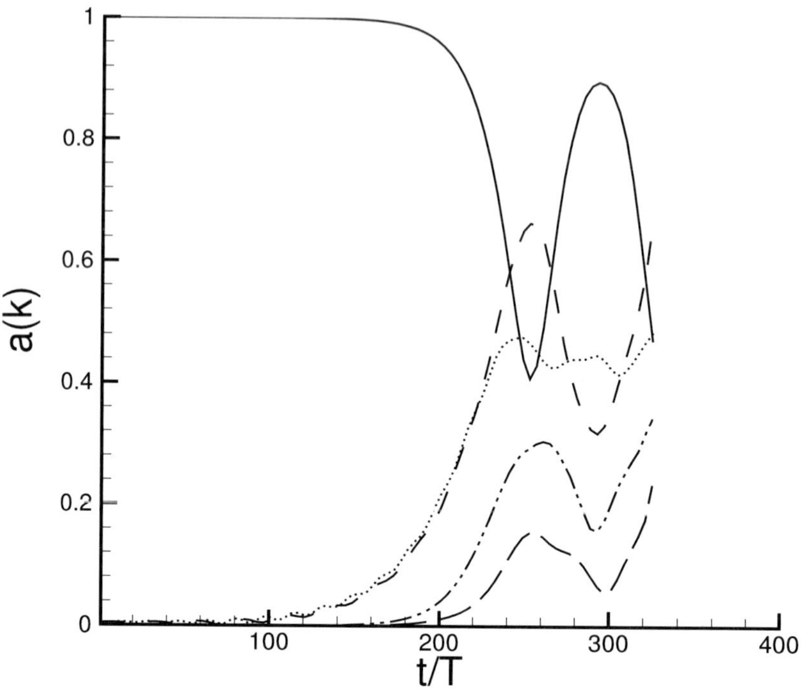

Fig. 9. Time histories of the amplitude of the fundamental, $k_0 = 5$ (solid line), subharmonic, $k_1 = 4$ (dashed line), and superharmonic, $k_2 = 6$ (dotted line), modes for an evolving perturbed Stokes wave of initial wave steepness $\epsilon = 0.11$ and fundamental wave period T, with wind action ($U = 1.75c$). The two lowest curves (dashed-dotted lines) correspond to the modes $k_3 = 3$ and $k_4 = 7$.

The numerical method used to solve the evolution equations is based on a pseudo-spectral treatment with a fourth-order Runge-Kutta integrator with constant time step, similar to the method developed by (6). For more details see the paper (21).

It is well known that uniformly-traveling wave train of Stokes' waves are unstable to the Benjamin-Feir instability (or modulational instability) which results from a quartet resonance, that is, a resonance interaction between four components of the wave field. This instability corresponds to a quartet interaction between the fundamental component (the carrier) $\mathbf{k_0} = \mathbf{k_0(1, 0)}$ counted twice and two satellites $\mathbf{k_1} = \mathbf{k_0(1 + p, q)}$ and $\mathbf{k_2} = \mathbf{k_0(1 - p, -q)}$ where p and q are the longitudinal wavenumber and transversal wavenumber respectively of the modulation. Instability occurs when the following resonance conditions are fulfiled.

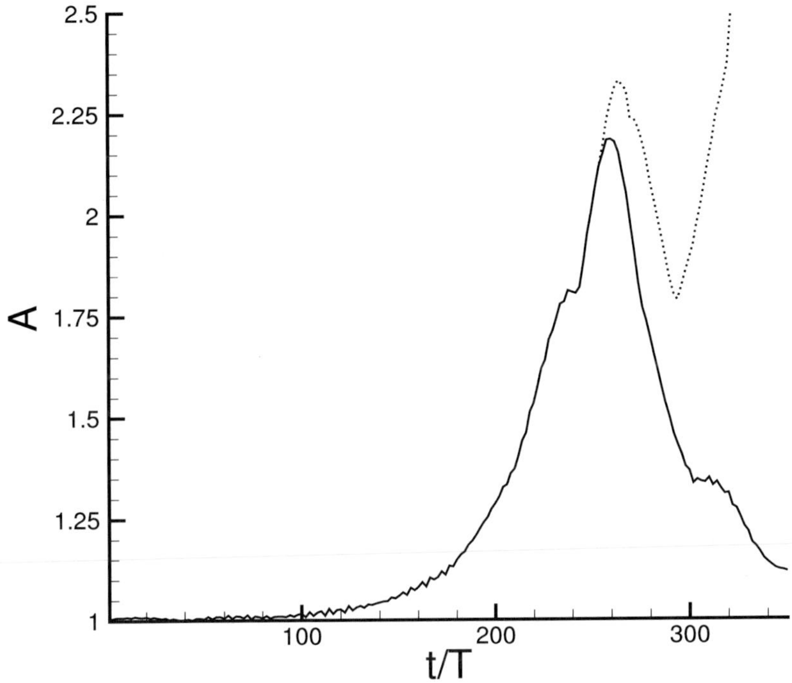

Fig. 10. Numerical amplification factor as a function of time without wind (solid line) and with wind (dotted line) for $U = 1.75c$

$$\mathbf{k_1} + \mathbf{k_2} = \mathbf{2k_0}. \tag{28}$$

$$\omega_1 + \omega_2 = 2\omega_0. \tag{29}$$

where ω_i with $i = 0, 1, 2$ are frequencies of the carrier and satellites.
A presentation of the different classes of instability of Stokes waves is given in the review paper (5).

The procedure used to calculate the linear stability of Stokes waves is similar to the method described by (10). Let $\eta = \bar{\eta} + \eta'$ and $\phi = \bar{\phi} + \phi'$ be the perturbed elevation and perturbed velocity potential where $(\bar{\eta}, \bar{\phi})$ and (η', ϕ') correspond respectively to the unperturbed Stokes wave and infinitesimal perturbative motion ($\eta' \ll \bar{\eta}, \phi' \ll \bar{\phi}$). Following (14), the Stokes wave of amplitude a_0 and wavenumber k_0 is computed iteratively. Substitute these decomposition in the boundary conditions linearized about the unperturbed motion and using the following forms for a two-dimensional flow:

$$\eta' = \exp(\lambda t + ipx) \sum_{-\infty}^{\infty} a_j \exp(ijx), \tag{30}$$

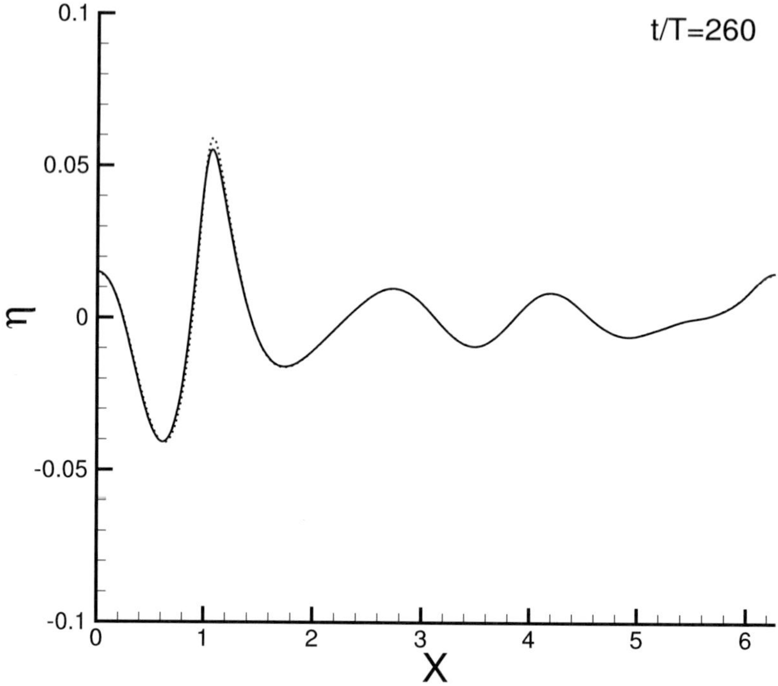

Fig. 11. Surface wave profile at $t = 260T$: without wind (solid line) and with wind (dotted line).

$$\phi' = \exp(\lambda t + ipx) \sum_{-\infty}^{\infty} b_j \exp(ijx + \gamma_j z)), \tag{31}$$

where λ, a_j and b_j are complex numbers and $\gamma_j = \mid p + j \mid$.

Equations (30) and (31) correspond to an eigenvalue problem for λ with eigenvector $\mathbf{u} = (\mathbf{a_j}, \mathbf{b_j})^{\mathbf{t}}$:

$$(\mathbf{A} - \lambda \mathbf{B})\mathbf{u} = \mathbf{0}, \tag{32}$$

where $\mathbf{A}$ and $\mathbf{B}$ are complex matrices depending on the unperturbed wave steepness of the basic wave, $\epsilon = a_0 k_0$, and the arbitrary real number p. The eigenvalue, λ, satisfies

$$\det(\lambda \mathbf{B} - \mathbf{A}) = \mathbf{0}. \tag{33}$$

The physical disturbances are obtained from the real part of the complex expressions η' and ϕ' at $t = 0$.

(17) and (16) showed that the dominant instability of a uniformly-traveling train of Stokes' waves in deep water is the two-dimensional modulational instability (class I) provided its steepness is less than $\epsilon = 0.30$. For higher values of the wave steepness

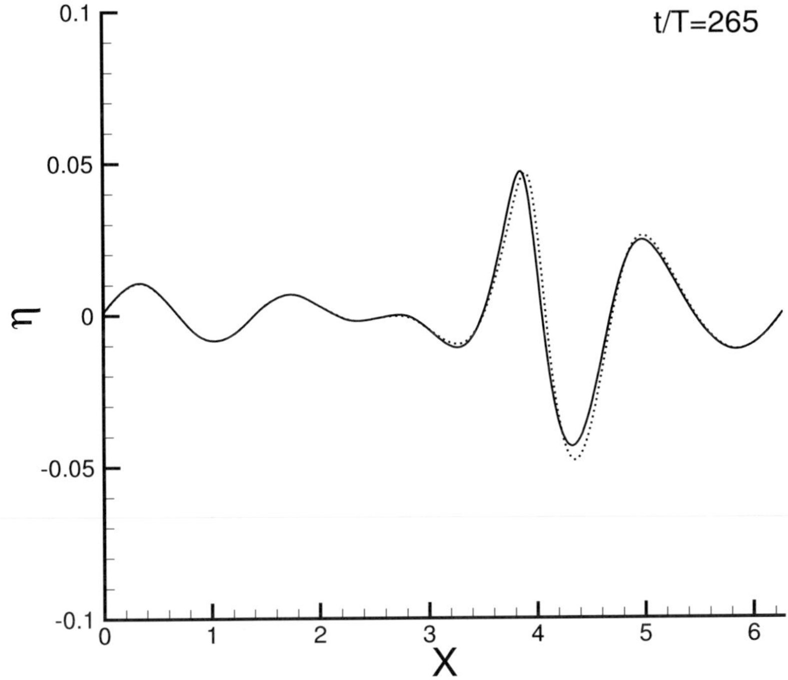

Fig. 12. Surface wave profile at $t = 265T$: without wind (solid line) and with wind (dotted line).

three-dimensional instabilities (class II) become dominant, phase locked to the unperturbed wave. Herein we shall focus on the two-dimensional nonlinear evolution of a Stokes' wave train suffering modulational instability with and without wind action.

Numerical simulation without wind action:

The initial condition is a Stokes wave of steepness $\epsilon = 0.11$, disturbed by its most unstable perturbation which corresponds to $p \approx 0.20 = 1/5$. The fundamental wavenumber of the Stokes wave is chosen so that integral numbers of the sidebands perturbation (satellites) can be fitted into the computational domain. For $p = 1/5$ the fundamental wave harmonic of the Stokes wave is $k_0 = 5$ and the dominant sidebands are $k_1 = 4$ and $k_2 = 6$ for subharmonic and superharmonic part of the perturbation respectively. The wave parameters have been re-scaled in order to have the wavelength of the perturbation equal to 2π. There exists higher harmonics present in the interactions which are not presented here. The normalized amplitude of the perturbation relative to Stokes wave amplitude is initially taken

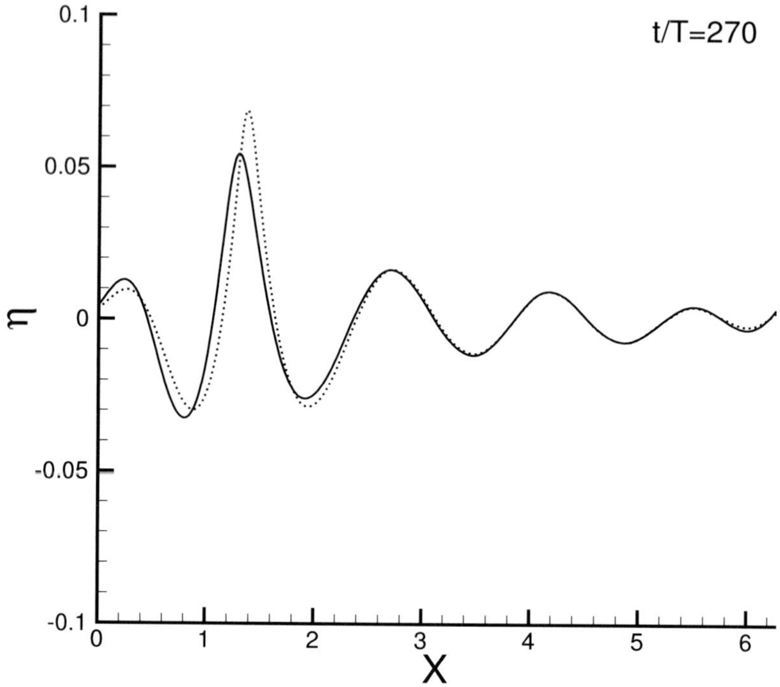

Fig. 13. Surface wave profile at $t = 270T$: without wind (solid line) and with wind (dotted line).

equal to 10^{-3}. The order of nonlinearity is $M = 6$, the number of mesh points is $N > (M + 1)k_{\max}$ where $k_{\max}$ is the highest harmonic taken into account in the simulation. The latter criterion concerning N is introduced to avoid aliasing errors. To compute the long time evolution of the wave packet the time step Δt is chosen equal to $T/100$ where T is the fundamental period of the basic wave. This temporal discretisation satisfies the CFL condition.

For the case without wind, the time histories of the normalized amplitude of the carrier, lower sideband and upper sideband of the most unstable perturbation are plotted in Figure 8. Another perturbation which was initially linearly stable becomes unstable in the vicinity of maximum of modulation resulting in the growth of the sidebands $k_3 = 3$ and $k_4 = 7$. The nonlinear evolution of the two-dimensional wave train exhibits the Fermi-Pasta-Ulam recurrence phenomenon. This phenomenon is characterized by a series of modulation-demodulation cycles in which initially uniform wave trains become modulated and then demodulated until they are again uniform. Herein one cycle is reported. At $t \approx 360T$ the initial condition is more or less recovered. At the maximum of modulation $t = 260T$, one can observe a temporary frequency (and wavenumber) downshifting since the subharmonic mode $k_1 = 4$

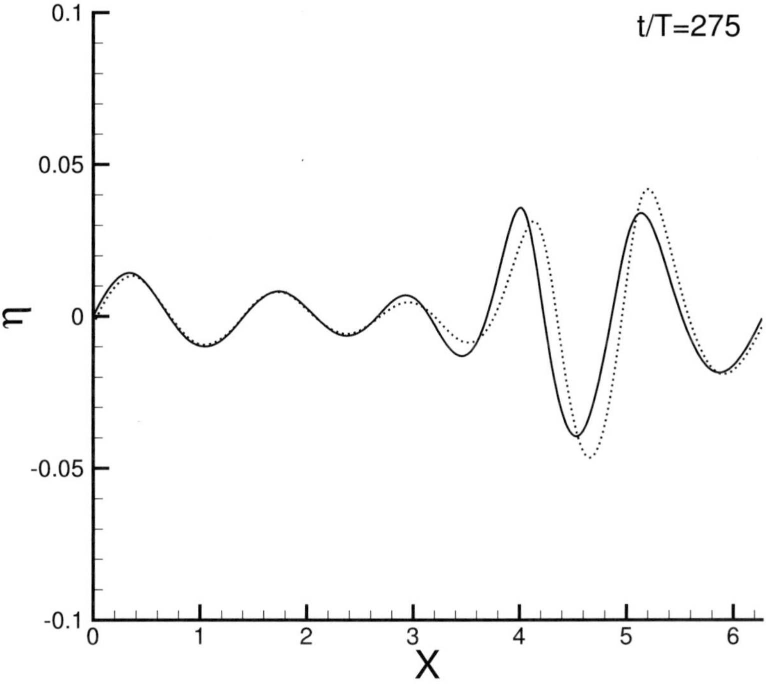

Fig. 14. Surface wave profile at $t = 275T$: without wind (solid line) and with wind (dotted line).

is dominant. At this stage a very steep wave occurs in the group as it can be seen in Figure 11.

Numerical simulation with wind action:

Figure 9 is similar to Figure 8, except that now water waves evolve under wind action. Wind forcing is applied over crests of slopes larger than $(\partial\eta/\partial x)_c = 0.405$. This condition is satisfied for $256T < t < 270T$, that is during the maximum of modulation which corresponds to the formation of the extreme wave event. When the values of the wind velocity are too high numerical simulations fail during the formation of the rogue wave event, due to breaking. During breaking wave process the slope of the surface becomes infinite, leading numerically to a spread of energy into high wavenumbers. This local steepening is characterized by a numerical blow-up (for methods dealing with an Eulerian description of the flow). In order to avoid a breaking wave too early, the wind velocity is fixed $U \approx 1.75c$. Owing to the weak effect of the wind on the phase velocity of the crests on which it acts, the phase velocity, c, is computed without wind. The effect of the wind reduces significantly the demodulation cycle and thus sustains the rogue wave event. This feature is clearly shown in Figure 10. The amplification factor is stronger in presence of wind and the rogue wave criterion given by equation (6), $A > 2$, is satisfied during a longer period of time. Figures 11, 12, 13 and 14 display water wave profiles at different instant of time in the vicinity of the maximum of modulation with and without wind. The solid lines correspond to waves propagating without wind while the dotted lines represent the wave profiles under wind action. These figures show that the wind does not modify the phase velocity of the very steep waves while it increases their height.

To summarize the results of this section, we can claim that extreme wave events generated by modulational instability in presence of wind behave similarly to those due to dispersive spatio-temporal focusing discussed in previous section 4 and subsection 5.1.

6 Conclusion

Two main mechanisms yielding to rogue wave events have been investigated experimentally and numerically. Within the framework of extreme wave events due to spatio-temporal focusing a good qualitative agreement is found between experiments and numerical simulations. It is shown that the wind amplifies the height of the steep waves and increases their duration. A second series of numerical simulations have been performed within the framework of rogue waves due to modulational instability . For this case no experiments have been conducted. Nevertheless the simulations have confirmed the results found for extreme wave events due to the spatio-temporal focusing phenomenon.

References

[1] M. I. Banner and W.K. Melville (1976) On the separation of air flow over water waves, J. Fluid Mech. 77, 825–842.

[2] T.B. Benjamin and J.E. Feir 1967 The desintegration of wave trains on deep water. Part 1. Theory., J. Fluid Mech. 27, 417–430.

[3] D. Clamond and J. Grue (2002) Interaction between envelope solitons as a model for freak wave formation. Pt 1: Long-time interaction, C.R. Mecanique 330, 575–580.

[4] D. Clamond, M. Francius, J. Grue and C. Kharif (2006) Strong interaction between envelope solitary surface gravity waves, Eur. J. Mech. B/Fluids, 25(5), 536-553.

[5] F.Dias and C. Kharif (1999) Nonlinear gravity and capillary-gravity waves, Annu. Rev. Fluid Mech. 31, 301–346.

[6] D.G. Dommermuth and D.K.P. Yue (1987) A high-order spectral method for the study of nonlinear gravity waves, J. Fluid Mech. 184, 267–288.

[7] K.B. Dysthe (2001) Modelling a "Rogue Wave" - Speculations or a realistic possibility?, Rogues Waves 2000 edited by M. Olagnon and G.A. Athanassoulis (Brest, France), 255–264, 2001.

[8] J.P. Giovanangeli, C. Kharif and E. Pelinovsky (2004) Experimental study of the wind effect on the focusing of transient wave groups, Proc. of abtracts of Rogue waves 2004,Brest.

[9] H. Jeffreys (1925) On the formation of wave by wind , Proc. Roy. Soc. A 107, 189–206.

[10] C. Kharif and A. Ramamonjiarisoa (1988) Deep water gravity wave instabilities at large steepness, Phys. Fluids 31, 1286–1288.

[11] C. Kharif and E. Pelinovsky (2003) Physical mechanisms of the rogue wave phenomenon, Eur. J. Mech. B/Fluids 22, 603–634.

[12] I.V. Lavranov (1998) The wave energy concentration at the Agulhas current off South Africa, Natural Hazards 17, 117–127.

[13] G. Lawton (2001) Monsters of the deep (The perfect wave), New Scientist 170 (2297), 28–32.

[14] M.S. Longuet-Higgins (1985) Bifurcation in gravity waves, J. Fluid Mech. 151, 457–475.

[15] S.R. Massel (1996) Ocean surface waves: Their physics and prediction, Word Scientific, Singapour.

[16] J.W. McLean (1982) Instabilities of finite-amplitude water waves, J. Fluid Mech. 114, 315–330.

[17] J.W. McLean, Y.C. Ma, D.U. Martin, P.G. Saffman and H.C. Yuen (1981) Three-dimensional instability of finite-amplitude water waves, Phys. Rev. Lett. 46, 817–820.

[18] E. Pelinovsky, T. Talipova and C. Kharif (2000) Nonlinear dispersive mechanism of the freak wave formation in shallow water, Physica D 147(1-2), 83–94.

[19] P. Peterson, T. Soomere, J. Engelbrecht and E. van Groesen (2003) Soliton interaction as a possible model for extreme waves, Nonlinear Processes in Geophysics 10, 503–510.

[20] T. Soomere and J. Engelbrecht (2005) Extreme elevations and slopes of interacting solitons in shallow water, Wave Motion 41, 179–192.

[21] C. Skandrani, C. Kharif and J. Poitevin (1996) Nonlinear evolution of water surface waves: The frequency downshifting phenomenon, Contemp. Math. 200, 157–171.
[22] A. Slunyaev, C. Kharif, E. Pelinovsky and T. Talipova (2002) Nonlinear wave focusing on water of finite depth, Physica D 173(1-2),77–96.
[23] R. Smith (1976) Giant waves, J. Fluid Mech. 77, 417–431.
[24] H.U. Sverdrup and W.H. Munk (1947) Wind, sea, and swell; theory of relations for forecasting, U.S. Navy Hydrographic Office, H.O. 601.
[25] J. Touboul, J.P. Giovanangeli, C. Kharif and E. Pelinovsky (2006) Freak waves under the action of wind: Experiments and simulations, Eur. J. Mech. B/Fluids 25, 662-676.
[26] B.S. White and B. Fornberg (1998) On the chance of freak waves at the sea, J. Fluid Mech. 255, 113–138.
[27] G.B. Whitham (1965) A general approach to linear and non-linear dispersive waves using a Lagrangian, J. Fluid Mech. 22, 273–283.
[28] G.B. Whitham (1967) Nonlinear dispersion of water waves, J. Fluid Mech. 27, 399–412.
[29] C.H. Wu and A. Yao (2004) Laboratory measurements of limiting freak waves on currents, J. Geophys. Res. 109, C12002, 1–18.

Numerical Verification of the Hasselmann equation

Alexander O. Korotkevich[1], Andrei N. Pushkarev[2,3], Don Resio[4], and Vladimir E. Zakharov[5,2,1,3]

[1] Landau Institute for Theoretical Physics RAS, 2, Kosygin Str., Moscow, 119334, Russian Federation kao@landau.ac.ru
[2] Lebedev Physical Institute RAS,53, Leninsky Prosp., GSP-1 Moscow, 119991, Russian Federation
[3] Waves and Solitons LLC, 918 W. Windsong Dr., Phoenix, AZ 85045, USA andrei@cox.net
[4] Coastal and Hydraulics Laboratory, U.S. Army Engineer Research and Development Center, Halls Ferry Rd., Vicksburg, MS 39180, USA
[5] Department of Mathematics, University of Arizona, Tucson, AZ 85721, USA zakharov@math.arizona.edu

Summary. The purpose of this article is numerical verification of the thory of weak turbulence . We performed numerical simulation of an ensemble of nonlinearly interacting free gravity waves (swell) by two different methods: solution of primordial dynamical equations describing potential flow of the ideal fluid with a free surface and, solution of the kinetic Hasselmann equation, describing the wave ensemble in the framework of the theory of weak turbulence . Comparison of the results demonstrates pretty good applicability of the weak turbulent approach. In both cases we observed effects predicted by this theory: frequency downshift, angular spreading as well as formation of Zakharov-Filonenko spectrum $I_\omega \sim \omega^{-4}$. To achieve quantitative coincidence of the results obtained by different methods one has to accomplish the Hasselmann kinetic equation by an empirical dissipation term S_{diss} modeling the coherent effects of white-capping. Adding of the standard dissipation terms used in the industrial wave predicting model (WAM) leads to significant improvement but not resolve the discrepancy completely, leaving the question about optimal choice of S_{diss} open.

Numerical modeling of swell evolution in the framework of the dynamical equations is affected by the side effect of resonances sparsity taking place due to finite size of the modeling domain. We mostly overcame this effect using fine integration grid of 512×4096 modes. The initial spectrum peak was located at the wave number $k = 300$. Similar conditions can be hardly realized in the laboratory wave tanks. One of the results of our article consists in the fact that physical processes in finite size laboratory wave tanks and in the ocean are quite different, and the results of such laboratory experiments can be applied to modeling of the ocean phenomena with extra care. We also present the estimate on the minimum size of the laboratory installation, allowing to model open ocean surface wave dynamics.

1 Introduction

The theory of weak turbulence is designed for statistical description of weakly-nonlinear wave ensembles in dispersive media. The main tool of weak turbulence is kinetic equation for squared wave amplitudes, or a system of such equations. Since the discovery of the kinetic equation for bosons by Nordheim (1) (see also paper by Peierls (2)) in the context of solid state physics, this quantum-mechanical tool was applied to wide variety of classical problems, including wave turbulence in hydrodynamics, plasmas, liquid helium, nonlinear optics, etc. (see monograph by Zakharov, Falkovich and L'vov (3)). Such kinetic equations have rich families of exact solutions describing weak-turbulent Kolmogorov spectra. Also, kinetic equations for waves have self-similar solutions describing temporal or spatial evolution of weak –turbulent spectra.

However, the most remarkable example of weak turbulence is wind-driven sea. The kinetic equation describing statistically the gravity waves on the surface of ideal liquid was derived by Hasselmann (4). Since this time the Hasselmann equation is widely used in physical oceanography as foundation for development of wave-prediction models: *WAM, SWAN* and *WAVEWATCH* – it is quite special case between other applications of the theory of weak turbulence due to the strength of industrial impact.

In spite of tremendous popularity of the Hasselmann equation, its validity and applicability for description of real wind-driven sea has never been completely proven. It was criticized by many respected authors, not only in the context of oceanography. There are at least two reasons why the weak–turbulent theory could fail, or at least be incomplete.

The first reason is presence of the coherent structures. The weak-turbulent theory describes only weakly-nonlinear resonant processes. Such processes are spatially extended, they provide weak phase and amplitude correlation on the distances significantly exceeding the wave length. However, nonlinearity also causes another phenomena, much stronger localized in space. Such phenomena – solitons, quasi-solitons and wave collapses are strongly nonlinear and cannot be described by the kinetic equations. Meanwhile, they could compete with weakly-nonlinear resonant processes and make comparable or even dominating contribution in the energy, momentum and wave-action balance. For gravity waves on the fluid surface the most important coherent structures are white-cappings (or wave-breakings), responsible for essential dissipation of wave energy.

The second reason limiting the applicability of the weak-turbulent theory is finite size of any real physical system. The kinetic equations are derived only for infinite media, where the wave vector runs continuous D-dimensional Fourier space. Situation is different for the wave systems with boundaries, e.g. boxes with periodical or reflective boundary conditions. The Fourier space of such systems is infinite lattice of discrete eigen-modes. If the spacing of the lattice is not small enough, or the level of Fourier modes is not big enough, the whole physics of nonlinear interaction becomes completely different from the continuous case.

For these two reasons verification of the weak turbulent theory is an urgent problem, important for the whole physics of nonlinear waves. The verification can be done by direct numerical simulation of the primitive dynamical equations describing wave turbulence in nonlinear medium.

So far,the numerical experimentalists tried to check some predictions of the weak-turbulent theory, such as weak-turbulent Kolmogorov spectra. For the gravity wave turbulence the most important is Zakharov-Filonenko spectrum $F_\omega \sim \omega^{-4}$ (5). At the moment, this spectrum was observed in numerous numerical experiments (6)-(19).

The attempts of verification of weak turbulent theory through numerical simulation of primordial dynamical equations has been started with numerical simulation of 2D optical turbulence (20), which demonstrated, in particular, co–existence of weak – turbulent and coherent events.

Numerical simulation of 2D turbulence of capillary waves was done in (6), (7), and (8). The main results of the simulation consisted in observation of classical regime of weak turbulence with spectrum $F_\omega \sim \omega^{-19/4}$, and discovery of non-classical regime of "frozen turbulence", characterized by absence of energy transfer from low to high wave-numbers. The classical regime of turbulence was observed on the grid of 256×256 points at relatively high levels of excitation, while the "frozen" regime was realized at lower levels of excitation, or more coarse grids. The effect of "frozen" turbulence is explained by sparsity of 3-wave resonance, both exact and approximate. The classical regime of turbulence becomes possible due to nonlinear shift of the linear frequencies caused by enhanced level of excitation. Conclusion has been made that in the reality the turbulence of waves in limited systems is practically always the mixture of classical and "frozen" regimes.

In fact, the "frozen" turbulence is close to KAM regime, when the dynamics of turbulence is close to the behavior of integrable system (8).

The first attempt to perform modeling of the system of nonlinear waves (swell on the surface of deep ocean), solving simultaneously kinetic equation and primordial dynamic equations, has been done in the article (15). The results of this simulation again confirmed ubiquity of the weak-turbulent Zakharov-Filonenko asymptotic ω^{-4} and shown existence of the inverse cascade, but presented essentially different scenario of the spectral peak evolution. Detailed analysis shown, that even on the grids as fine as 256×2048 modes, the energy transport is realized mostly by the network of few selected modes – "oligarchs" – posed in the optimal resonant condition. This regime, transitional between weak turbulence and "frozen" turbulence, should be typical for wave turbulence in the systems of medium size. It was called "mesoscopic turbulence". Similar type of turbulence was observed in (17), (18).

In this article we present the results of new numerical experiments on modeling of swell propagation in the framework of both dynamical and kinetic equations, using fine grid containing, corresponding to 512×4096 Fourier modes. We think that our results can be considered as first in the world literature direct verification of wave kinetic equation.

One important point should be mentioned. In our experiments we observed not only weak turbulence, but also additional nonlinear dissipation of the wave energy, which could be identified as the dissipation due to white-capping. To reach agreement with dynamic experiments, we had to accomplish the kinetic equation by a phenomenological dissipation term S_{diss}. In this article we examined dissipation terms used in the industrial wave-prediction models *WAM Cycle 3* and WAM cycle 4. Dissipation term used in *WAM Cycle 3* works fairly, while S_{diss} used in WAM Cycle 4 certainly overestimate nonlinear dissipation. This fact means that for getting better agreement between dynamic and kinetic computations, we need to take into consideration more sophisticated dissipation term.

2 Deterministic and statistic models

In the "dynamical" part of our experiment the fluid was described by two functions of horizontal variables x, y and time t: surface elevation $\eta(x, y, t)$ and velocity potential on the surface $\psi(x, y, t)$. They satisfy the canonical equations (23)

$$\frac{\partial \eta}{\partial t} = \frac{\delta H}{\delta \psi}, \quad \frac{\partial \psi}{\partial t} = -\frac{\delta H}{\delta \eta}, \tag{1}$$

Hamiltonian H is presented by the first three terms in expansion on powers of nonlinearity $\nabla \eta$

$$\begin{aligned} H &= H_0 + H_1 + H_2 + ..., \\ H_0 &= \frac{1}{2} \int \left(g\eta^2 + \psi \hat{k} \psi \right) dxdy, \\ H_1 &= \frac{1}{2} \int \eta \left[|\nabla \psi|^2 - (\hat{k}\psi)^2 \right] dxdy, \\ H_2 &= \frac{1}{2} \int \eta (\hat{k}\psi) \left[\hat{k}(\eta(\hat{k}\psi)) + \eta \nabla^2 \psi \right] dxdy. \end{aligned} \tag{2}$$

Here $\hat{k}$ is the linear integral operator $\hat{k} = \sqrt{-\nabla^2}$, defined in Fourier space as

$$\hat{k}\psi_{\mathbf{r}} = \frac{1}{2\pi} \int |k| \psi_{\mathbf{k}} e^{-\mathrm{i}\mathbf{k}\mathbf{r}} \mathrm{d}\mathbf{k}, \; |k| = \sqrt{k_x^2 + k_y^2}. \tag{3}$$

Using Hamiltonian (2) and equations (1) one can get the dynamical equations (6):

$$\begin{aligned} \dot{\eta} &= \hat{k}\psi - (\nabla(\eta \nabla \psi)) - \hat{k}[\eta \hat{k} \psi] + \\ &\quad + \hat{k}(\eta \hat{k}[\eta \hat{k} \psi]) + \tfrac{1}{2} \nabla^2 [\eta^2 \hat{k} \psi] + \\ &\quad \tfrac{1}{2} \hat{k} [\eta^2 \nabla^2 \psi] + \hat{F}^{-1} [\gamma_k \eta_k], \\ \dot{\psi} &= -g\eta - \tfrac{1}{2} \left[(\nabla \psi)^2 - (\hat{k}\psi)^2 \right] - \\ &\quad - [\hat{k}\psi] \hat{k} [\eta \hat{k} \psi] - [\eta \hat{k} \psi] \nabla^2 \psi + \hat{F}^{-1} [\gamma_k \psi_k]. \end{aligned} \tag{4}$$

Here $\hat{F}^{-1}$ corresponds to inverse Fourier transform. We introduced artificial dissipative terms $\hat{F}^{-1}[\gamma_k \psi_k]$, corresponding to pseudo-viscous high frequency damping.

It is important to stress that we added dissipation terms in both equations. In fact, equation for $\dot{\eta}$ is just kinematic boundary condition, and adding a smoothing term to this equation has no any physical sense. Nevertheless, adding of this term is necessary for stability of the numerical scheme.

The model (1)-(4) was used in the numerical experiments (6) – (8), (12), (13), (15), (17), (18).

Introduction of the complex normal variables $a_{\mathbf{k}}$

$$a_{\mathbf{k}} = \sqrt{\frac{\omega_k}{2k}} \eta_{\mathbf{k}} + \mathrm{i} \sqrt{\frac{k}{2\omega_k}} \psi_{\mathbf{k}}, \tag{5}$$

where $\omega_k = \sqrt{gk}$, transforms equations (1) into

$$\frac{\partial a_{\mathbf{k}}}{\partial t} = -\mathrm{i} \frac{\delta H}{\delta a_{\mathbf{k}}^*}. \tag{6}$$

To proceed with statistical description of the wave ensemble, first, one should perform the canonical transformation $a_{\mathbf{k}} \to b_{\mathbf{k}}$, which excludes the cubical terms in the Hamiltonian. The details of this transformation can be found in the paper by Zakharov (1999) (24). After the transformation the Hamiltonian takes the forms

$$H = \int \omega_{\mathbf{k}} b_{\mathbf{k}} b_{\mathbf{k}}^* + \frac{1}{4} \int T_{\mathbf{k}\mathbf{k}_1\mathbf{k}_2\mathbf{k}_3} b_{\mathbf{k}}^* b_{\mathbf{k}_1}^* b_{\mathbf{k}_2} b_{\mathbf{k}_3} \times \\ \times \delta_{\mathbf{k}+\mathbf{k}_1-\mathbf{k}_2-\mathbf{k}_3} d\mathbf{k}_1 d\mathbf{k}_2 d\mathbf{k}_3. \tag{7}$$

where T is a homogeneous function of the third order:

$$T(\varepsilon\mathbf{k}, \varepsilon\mathbf{k}_1, \varepsilon\mathbf{k}_2, \varepsilon\mathbf{k}_3) = \varepsilon^3 T(\mathbf{k}, \mathbf{k}_1, \mathbf{k}_2, \mathbf{k}_3). \tag{8}$$

Connection between $a_{\mathbf{k}}$ and $b_{\mathbf{k}}$ together with explicit expression for $T_{\mathbf{k}\mathbf{k}_1\mathbf{k}_2\mathbf{k}_3}$ can be found, for example, in (24).

Let us introduce the pair correlation function

$$< a_{\mathbf{k}} a_{\mathbf{k}'}^* >= g N_{\mathbf{k}} \delta(\mathbf{k} - \mathbf{k}'), \tag{9}$$

where $N_{\mathbf{k}}$ is the spectral density of the wave function. This definition of the wave action is common in oceanography.

We also introduce the correlation function for transformed normal variables

$$< b_{\mathbf{k}} b_{\mathbf{k}'}^* >= g n_{\mathbf{k}} \delta(\mathbf{k} - \mathbf{k}') \tag{10}$$

Functions $n_{\mathbf{k}}$ and $N_{\mathbf{k}}$ can be expressed through each other in terms of cumbersome power series (24). On deep water their relative difference is of the order of μ^2 (μ is the characteristic steepness) and can be neglected (in most cases of swell evolution (or wave evolution) experimental results shows $\mu \simeq 0.1$).

Spectrum $n_{\mathbf{k}}$ satisfies Hasselmann (kinetic) equation (4)

$$\begin{aligned} &\frac{\partial n_{\mathbf{k}}}{\partial t} = S_{nl}[n] + S_{diss} + 2\gamma_k n_{\mathbf{k}}, \\ &S_{nl}[n] = 2\pi g^2 \int |T_{\mathbf{k},\mathbf{k}_1,\mathbf{k}_2,\mathbf{k}_3}|^2 (n_{\mathbf{k}_1} n_{\mathbf{k}_2} n_{\mathbf{k}_3} + \\ &+ n_{\mathbf{k}} n_{\mathbf{k}_2} n_{\mathbf{k}_3} - n_{\mathbf{k}} n_{\mathbf{k}_1} n_{\mathbf{k}_2} - n_{\mathbf{k}} n_{\mathbf{k}_1} n_{\mathbf{k}_3}) \times \\ &\times \delta(\omega_k + \omega_{k_1} - \omega_{k_2} - \omega_{k_3}) \times \\ &\times \delta(\mathbf{k} + \mathbf{k}_1 - \mathbf{k}_2 - \mathbf{k}_3)\, d\mathbf{k}_1 d\mathbf{k}_2 d\mathbf{k}_3. \end{aligned} \tag{11}$$

Here S_{diss} is an empiric dissipative term, corresponding to white-capping.

Stationary conservative kinetic equation

$$S_{nl} = 0 \tag{12}$$

has the rich family of Kolmogorov-type (25) exact solutions. Among them is Zakharov-Filonenko spectrum (5) for the direct cascade of energy

$$n_k \sim \frac{1}{k^4}, \tag{13}$$

and Zakharov-Zaslavsky (26), (27) spectra for the inverse cascade of wave action

$$n_k \sim \frac{1}{k^{23/6}}, \tag{14}$$

3 Deterministic Numerical Experiment

3.1 Problem Setup

The dynamical equations (4) have been solved in the real-space domain $2\pi \times 2\pi$ on the grid 512×4096 with the gravity acceleration set to $g = 1$. The solution has been performed by the spectral code, developed in (21) and previously used in (22),(12),(13),(15). We have to stress that in the current computations the resolution in Y-direction (long axis) is better than the resolution in X-direction by the factor of 8.

This approach is reasonable if the swell is essentially anisotropic, almost one-dimensional. This assumption will be validated by the proper choice of the initial data for computation. As the initial condition, we used the Gaussian-shaped distribution in Fourier space (see Fig. 1):

$$\begin{cases} |a_{\mathbf{k}}| = A_i \exp\left(-\frac{1}{2}\frac{|\mathbf{k}-\mathbf{k}_0|^2}{D_i^2}\right), |\mathbf{k}-\mathbf{k}_0| \le 2D_i, \\ |a_{\mathbf{k}}| = 10^{-12}, |\mathbf{k}-\mathbf{k}_0| > 2D_i, \end{cases}$$
$$A_i = 0.92 \times 10^{-6}, D_i = 60, \mathbf{k}_0 = (0; 300), \omega_0 = \sqrt{gk_0}. \tag{15}$$

The initial phases of all harmonics were random. The average steepness of this initial condition was $\mu \simeq 0.167$.

To realize similar experiment in the laboratory wave tank, one has to to generate the waves with wave-length 300 times less than the length of the tank. The width of the tank would not be less than 1/8 of its length. The minimal wave length of the gravitational wave in absence of capillary effects can be estimated as $\lambda_{min} \simeq 3cm$. The leading wavelength should be higher by the order of magnitude $\lambda \simeq 30cm$.

In such big tank of 200×25 meters experimentators can observe the evolution of the swell until approximately $700T_0$ – still less than in our experiments. In the tanks of smaller size, the effects of discreetness the Fourier space will be dominating, and experimentalists will observe either "frozen", or "mesoscopic" wave turbulence, qualitatively different from the wave turbulence in the ocean.

To stabilize high-frequency numerical instability, the damping function has been chosen as

$$\gamma_k = \begin{cases} 0, k < k_d, \\ -\gamma(k-k_d)^2, k \ge k_d, \end{cases}$$
$$k_d = 1024, \gamma = 5.65 \times 10^{-3}. \tag{16}$$

The simulation was performed until $t = 336$, which is equivalent to $926T_0$, where T_0 is the period of the wave, corresponding to the maximum of the initial spectral distribution.

3.2 Zakharov-Filonenko spectra

Like in the previous papers (10),(12),(13) and (15), we observed fast formation of the spectral tail, described by Zakharov-Filonenko law for the direct cascade $n_k \sim k^{-4}$ (5) (see Fig.2). In the agreement with (15), the spectral maximum slowly down-shifts to the large scales region, which corresponds to the inverse cascade (26),(27).

Also, the direct measurement of energy spectrum has been performed during the final stage of the simulation, when the spectral down shift was slow enough.

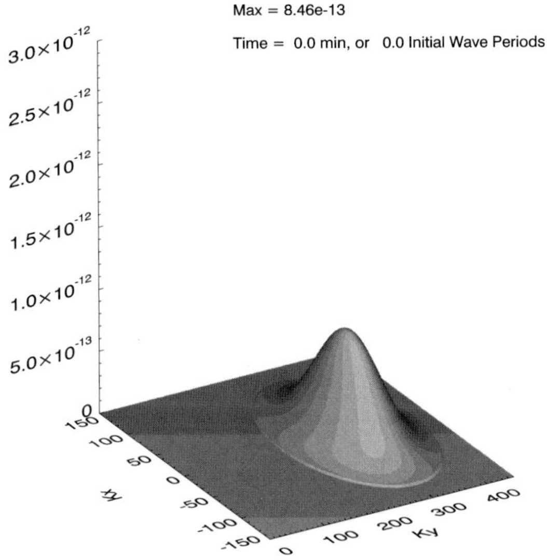

Fig. 1. Initial distribution of $|a_{\mathbf{k}}|^2$ on **k**-plane.

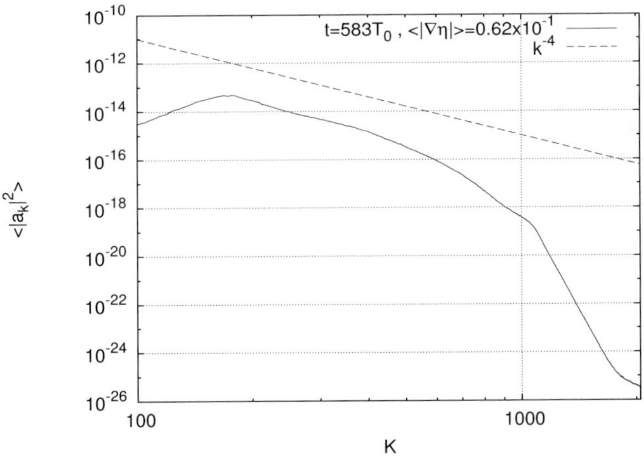

Fig. 2. Angle-averaged spectrum $n_k =< |a_{\mathbf{k}}|^2 >$ in a double logarithmic scale. The tail of distribution fits to Zakharov-Filonenko spectrum.

This experiment can be interpreted as the ocean buoy record – the time series of the surface elevations has been recorded at one point of the surface during $T_{buoy} \simeq 300T_0$. The Fourier transform of the autocorrelation function

$$E(\omega) = \frac{1}{2\pi} \int\limits_{-T_{buoy}/2}^{T_{buoy}/2} < \eta(t+\tau)\eta(\tau) > e^{i\omega t} \mathrm{d}\tau \mathrm{d}t. \tag{17}$$

allows to detect the direct cascade spectrum tail proportional to ω^{-4} (see Fig.3), well known from experimental observations (28),(29),(30).

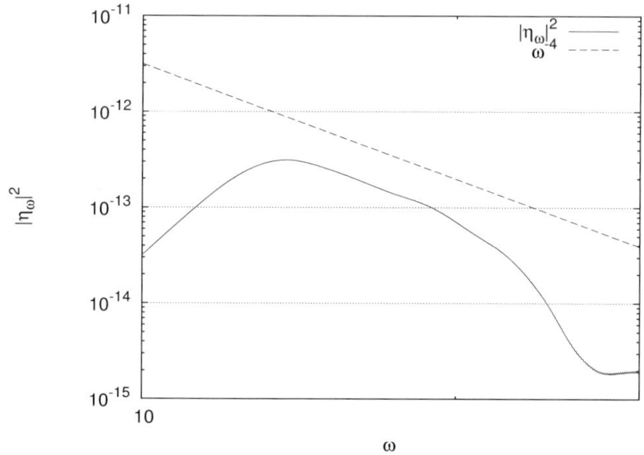

Fig. 3. Energy spectrum in a double logarithmic scale. The tail of distribution fits to asymptotics ω^{-4}.

3.3 Is the weak-turbulent scenario realized

Presence of Kolmogorov asymptotics in spectral tails, however, is not enough to validate applicability of the weak-turbulent scenario for description of wave ensemble. We have also to be sure that statistical properties of this ensemble correspond to weak-turbulent theory assumptions.

We should stress that in our experiments at thebeginning $|a_{\mathbf{k}}|^2$ is a smooth function of $\mathbf{k}$. Only phases of individual waves are random. As shows numerical simulation, the initial condition (15) (see Fig.1) does not preserve its smoothness – it becomes rough almost immediately (see Fig.4). The picture of this roughness is remarkably preserved in many details, even as the spectrum down-shifts as a whole. This roughness does not contradict the weak-turbulent theory. According to this theory, the wave ensemble is almost Gaussian, and both real and imaginary parts of each separate harmonics are not-correlated. However, according to the weak-turbulent theory, the spectra must become smooth after averaging over long enough time of more than $1/\mu^2$ periods. Earlier we observed such restoring of smoothness in the numerical experiments of the MMT model (see (45),(46), (47) and (48)). However, in the experiments discussed in the article, the roughness still persists and the averaging does not suppresses it completely. It can be explained by sparsity of the resonances.

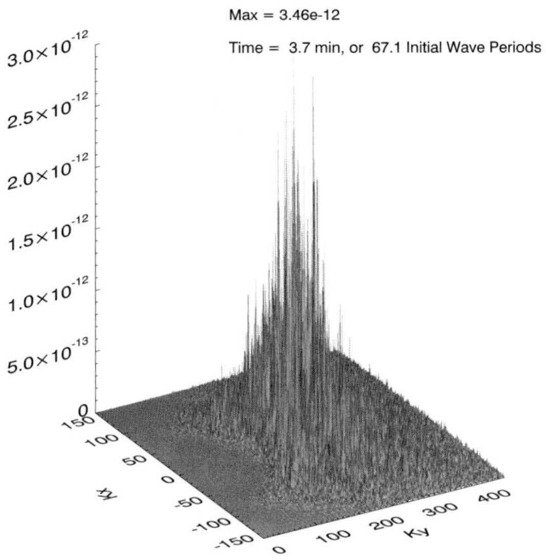

Fig. 4. Surface $|a_{\mathbf{k}}|^2$ at the moment of time $t \simeq 67T_0$.

Resonant conditions are defined by the system of equations:

$$\begin{aligned} \omega_k + \omega_{k_1} &= \omega_{k_2} + \omega_{k_3}, \\ \mathbf{k} + \mathbf{k}_1 &= \mathbf{k}_2 + \mathbf{k}_3, \end{aligned} \tag{18}$$

These resonant conditions define five-dimensional hyper-surface in six-dimensional space $\mathbf{k}, \mathbf{k}_1, \mathbf{k}_2$. In any finite system, (18) turns into Diophantine equation. Some solutions of this equation are known (31), (17). In reality, however, the energy transport is realized not by exact, but "approximate" resonances, posed in a layer near the resonant surface and defined by

$$|\omega_k + \omega_{k_1} - \omega_{k_2} - \omega_{k+k_1-k_2}| \leq \Gamma, \tag{19}$$

where Γ is a characteristic inverse time of nonlinear interaction.

In the finite systems $\mathbf{k}, \mathbf{k}_1, \mathbf{k}_2$ take values on the nodes of the discrete grid. The weak turbulent approach is valid, if the density of discrete approximate resonances inside the layer (19) is high enough. In our case the lattice constant is $\Delta k = 1$, and typical relative deviation from the resonance surface

$$\frac{\Delta\omega}{\omega} \simeq \frac{\omega'_k}{\omega}\Delta k = \frac{\omega'_k}{\omega} \simeq \frac{1}{600} \simeq 2 \times 10^{-3}. \tag{20}$$

Inverse time of the interaction Γ can be estimated from our numerical experiments: wave amplitudes change essentially during 30 periods, and one can assume: $\Gamma/\omega \simeq 10^{-2} \gg \frac{\delta\omega}{\omega}$. It means that the condition for the applicability of weak turbulent theory is typically satisfied, but the "reserve" for their validity is rather modest.

As a result, some particular harmonics, posed in certain "privileged" point of k-plane could form a "network" of almost resonant quadruplets and realize significant part of energy transport. Amplitudes of these harmonics exceed the average level essentially. This effect was described in the article (15), where such "selected few" harmonics were called "oligarchs". If "oligarchs" realize most part of the energy flux, the turbulence is "mesoscopic", not weak.

3.4 Statistics of the harmonics

According to the weak-turbulent scenario, statistics of the $a_{\mathbf{k}}(t)$ in any given k should be close to Gaussian. It presumes that the PDF for the squared amplitudes is

$$P(|a_{\mathbf{k}}|^2) \simeq \frac{1}{D} e^{-|a_{\mathbf{k}}|^2/D}, \tag{21}$$

here $D =< |a_{\mathbf{k}}|^2 >$ — mean square amplitude. To check equation (21) we need to find a way for calculation of $D(\mathbf{k})$. If the ensemble is stationary in time, $D(\mathbf{k})$ could be found for any given k by averaging in time. In our case, the process is non-stationary, and we have a problem with determination of $D(\mathbf{k})$.

To resolve this problem, we used low-pass filtering instead of time averaging. The low-pass filter was chosen in the form

$$f(\mathbf{n}) = e^{-(|\mathbf{n}|/\Delta)^3}, \Delta = 0.25Nx/2, Nx = 4096. \tag{22}$$

This choice of the low-pass filter preserves the values of total energy, wave action and the total momentum within three percent accuracy, see Fig.5. Then it is possible to average the PDF over different areas in k-space. The results for two different moments of time $t \simeq 70T_0$ and $t \simeq 933T_0$ are presented in Fig.6 and Fig.7. The thin line gives PDF after averaging over dissipation region harmonics, while bold line presents averaging over the non-dissipative area $|\mathbf{k}| < k_d = 1024$. One can see that statistics in the last case is quite close to the Gaussian, while in the dissipation region it deviates from Gaussian. However, deviation from the Guassianity in the dissipation region doesn't create any problems, since the "dissipative" harmonics do not contain any essential amount of the total energy, wave action and momentum.

One should remember, that the bold lines in the Fig.6 and Fig.7 are the results of averaging over a million of harmonics. Among them there is a population of "selected few", or "oligarchs", with the amplitudes exceeding the average value by the factor of more than ten times. The "oligarchs" exist because our grid is still not fine enough.

In our case "oligarchs" do exist, but their contribution in the total wave action is not more 4%. Ten leading "oligarchs" at the end of the experiment are presented in the Appendix A.

3.5 Two-stage evolution of the swell

Fig. 8-11 demonstrate time evolution of main characteristics of the wave field: wave action, energy, characteristic slope and mean frequency.

Fig.10 should be specially commented. Here and further we define the characteristic slope as follows

$$\mu = \sqrt{2}\left[< (\nabla\eta)^2 >\right]^{1/2}. \tag{23}$$

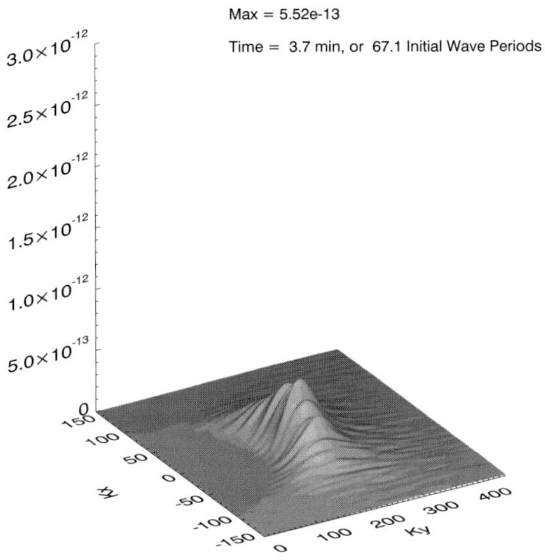

Fig. 5. Low-pass filtered surface $|a_{\mathbf{k}}|^2$ at $t \simeq 67T_0$.

Following this definition for the Stokes wave of small amplitude

$$\eta = a\cos(kx),$$
$$\mu = ak.$$

According to this definition of steepness for the classical Pierson-Moscowitz spectrum $\mu = 0.095$. Our initial steepness $\mu \simeq 0.167$ exceeds this value essentially.

Evolution of the spectrum can be conventionally separated in two phases. On the first stage we observe fast drop of wave action, slope and especially energy. Then the drop is stabilized, and we observe slow down-shift of mean frequency together with angular spreading. Level lines of smoothed spectra in the first and in the last moments of time are shown in Fig.12-13

Presence of two stages can be explained by study of the PDFs for elevation of the surface. In the initial moment of time PDF is Gaussian (Fig.14). However, very soon intensive super-Gaussian tails appear (Fig.15). Then they decrease slowly, and in the last moment of simulation, when characteristics of the sea are close to Peirson-Moscowitz, statistic is close to Gaussian again (Fig.16). Moderate tails do exist and, what is interesting, the PDF is not symmetric — elevations are more probable troughs. PDF for η_y — longitudinal gradients in the first moments of time is Gaussian (Fig.17). Then in a very short period of time strong non-Gaussian tails appear and reach their maximum at $t \simeq 14T_0$ (Fig.18). Here $T_0 = 2\pi/\sqrt{k_0}$ — period of initial leading wave. Since this moment the non-Gaussian tails decrease. In the last moment of simulation they are essentially reduced(Fig.19).

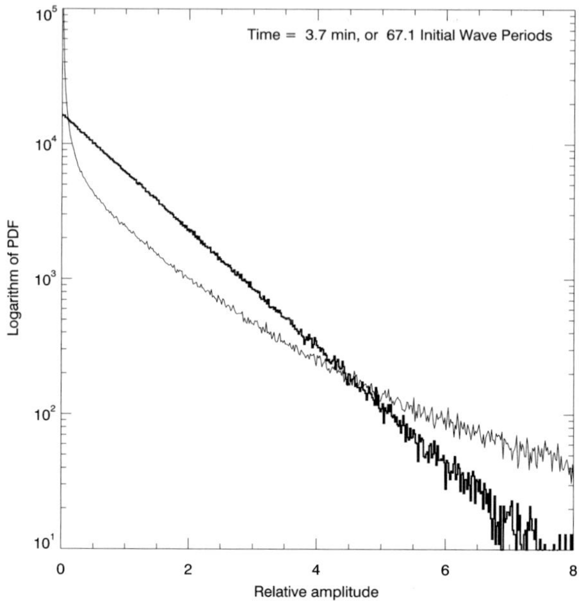

Fig. 6. Probability distribution function (PDF) for relative squared amplitudes $|a_k|^2/ < |a_k|^2 >$. $t \simeq 67T_0$.

Fast growing of non-Gaussian tails can be explained by fast formation of coherent harmonics. Indeed, $14T_0 \simeq 2\pi/(\omega_0\mu)$ is a characteristic time of nonlinear processes due to quadratic nonlinearity. Note that the fourth harmonic in our system is fast decaying, Hence we cannot see "real" white caps.

Figures 20-22 present PDFs for gradients in the orthogonal direction.

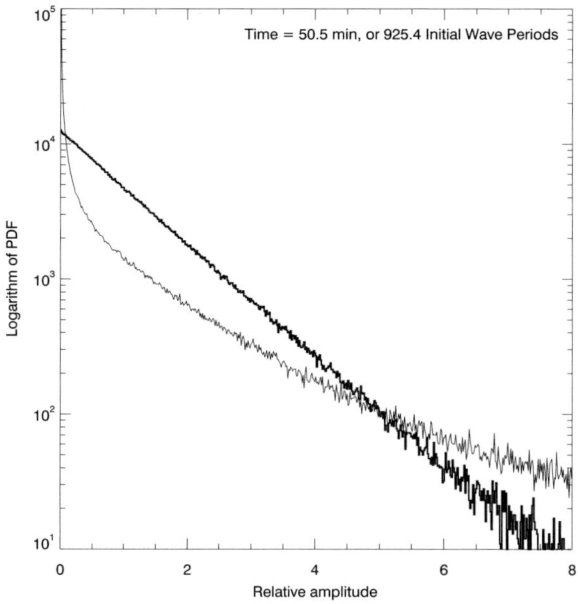

Fig. 7. Probability distribution function (PDF) for relative squared amplitudes $|a_k|^2/<|a_k|^2>$. $t \simeq 925T_0$.

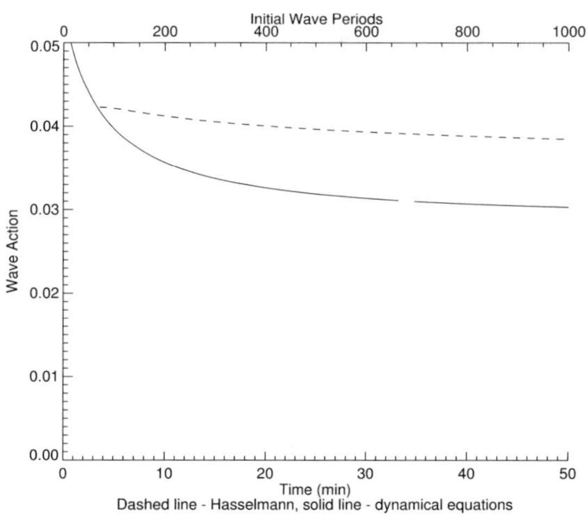

Fig. 8. Total wave action as a function of time for the artificial viscosity case.

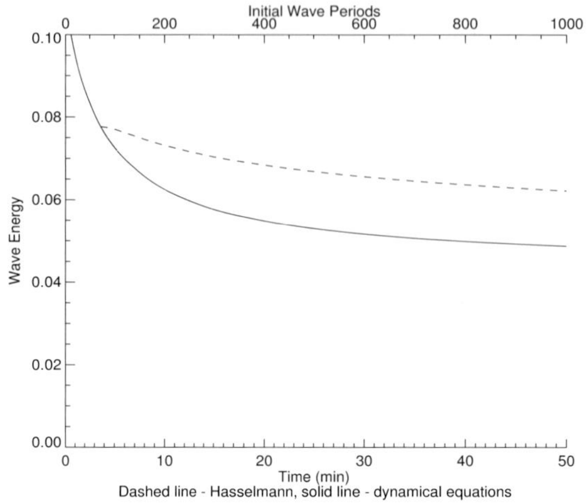

Fig. 9. Total wave energy as a function of time for the artificial viscosity case

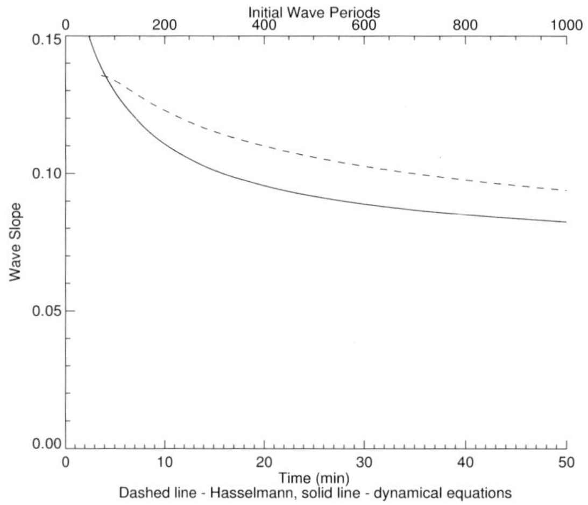

Fig. 10. Average wave slope as a function of time for the artificial viscosity case.

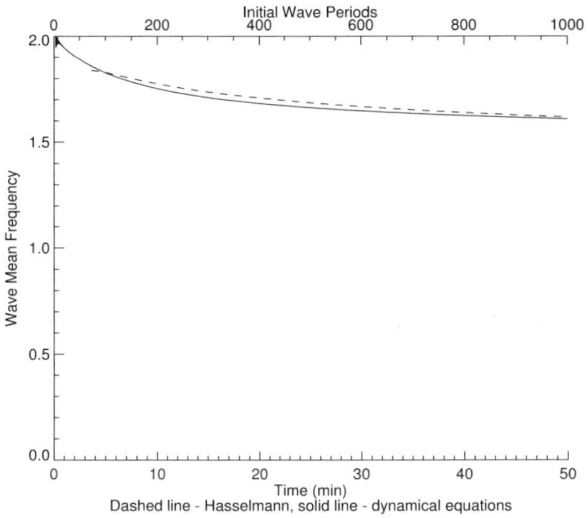

Fig. 11. Mean wave frequency as a function of time for the artificial viscosity case.

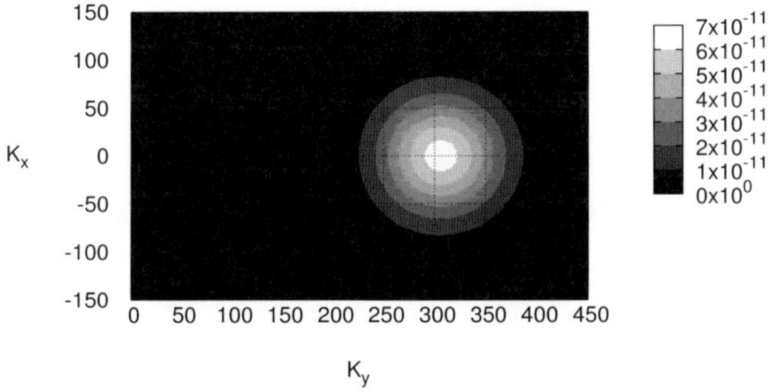

Fig. 12. Initial spectrum $|a_{\mathbf{k}}|^2$. $t = 0$.

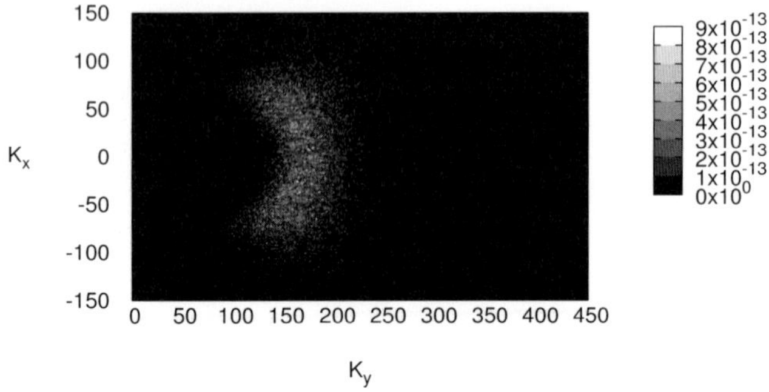

Fig. 13. Final spectrum $|a_{\mathbf{k}}|^2$. $t \simeq 933T_0$.

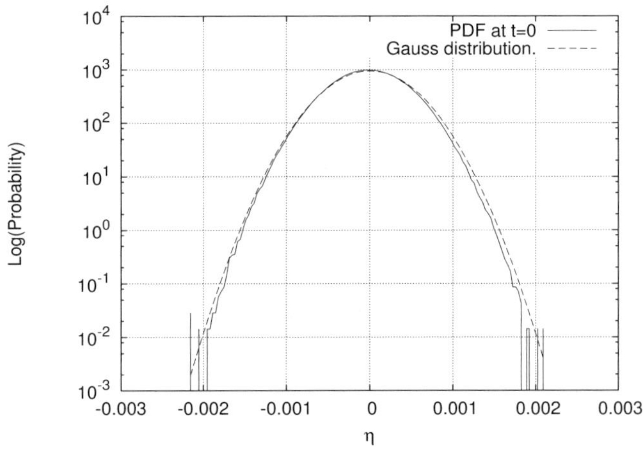

Fig. 14. PDF for the surface elevation η at the initial moment of time. $t = 0$.

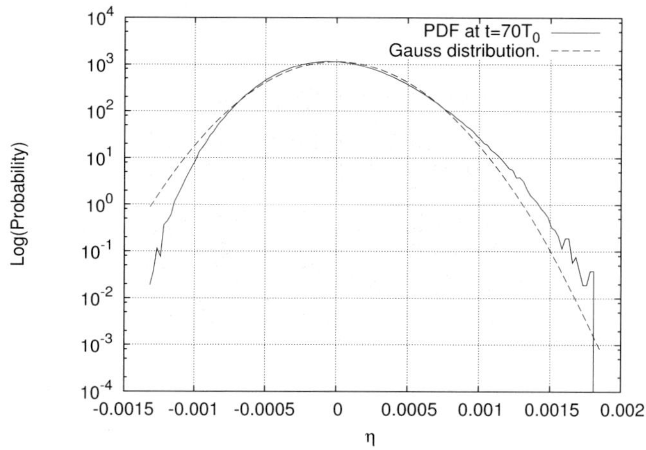

Fig. 15. PDF for the surface elevation η at some middle moment of time. $t \simeq 70T_0$.

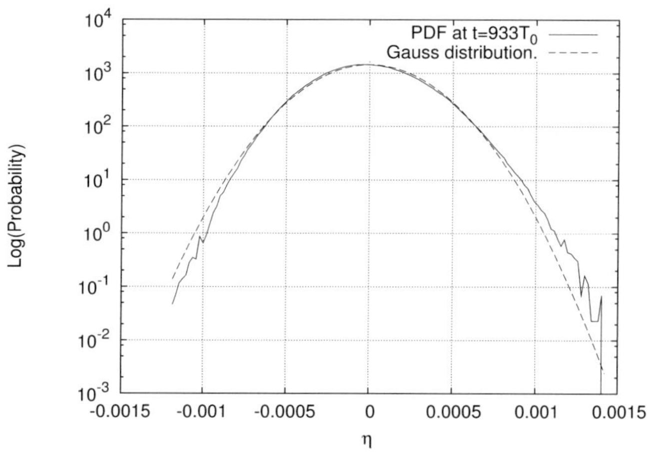

Fig. 16. PDF for the surface elevation η at the final moment of time. $t \simeq 933T_0$.

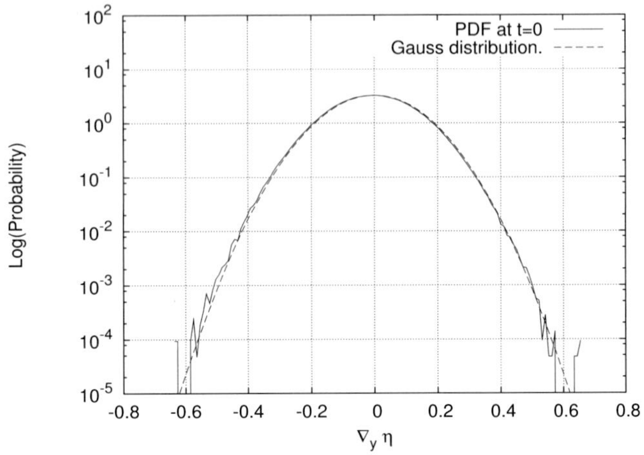

Fig. 17. PDF for $(\nabla\eta)_y$ at the initial moment of time. $t = 0$.

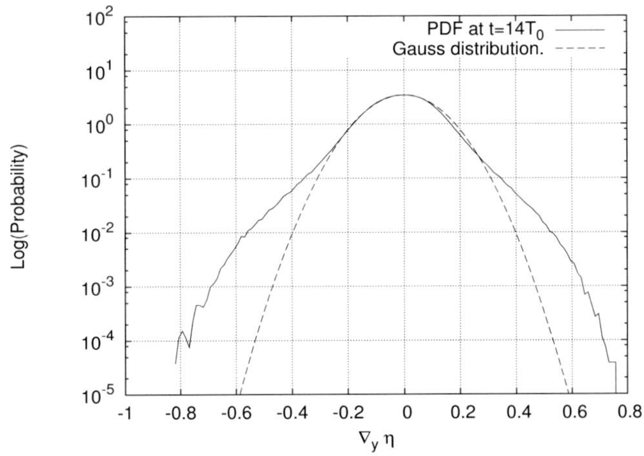

Fig. 18. PDF for $(\nabla\eta)_y$ at some middle moment of time. $t \simeq 14T_0$.

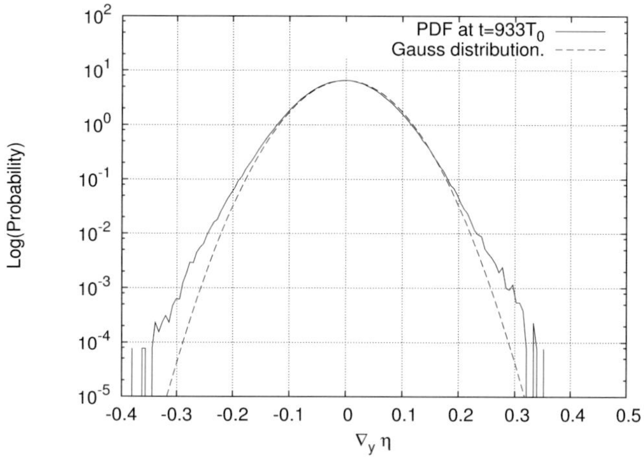

Fig. 19. PDF for $(\nabla\eta)_y$ at the final moment of time. $t \simeq 933T_0$.

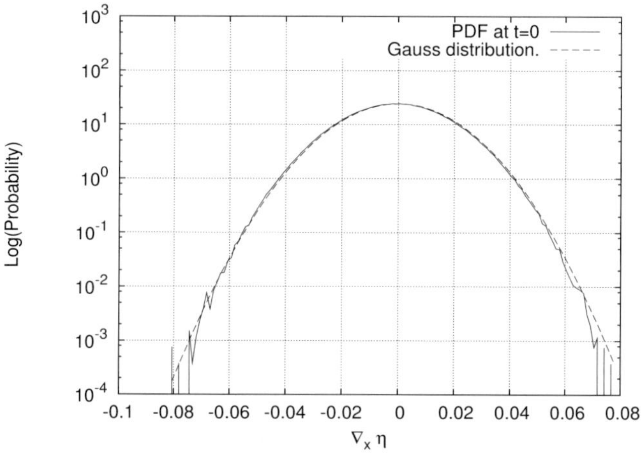

Fig. 20. PDF for $(\nabla\eta)_x$ at the initial moment of time. $t = 0$.

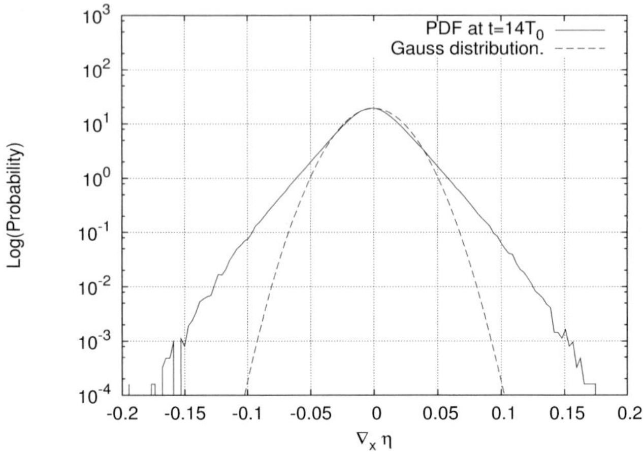

Fig. 21. PDF for $(\nabla\eta)_x$ at some middle moment of time. $t \simeq 14T_0$.

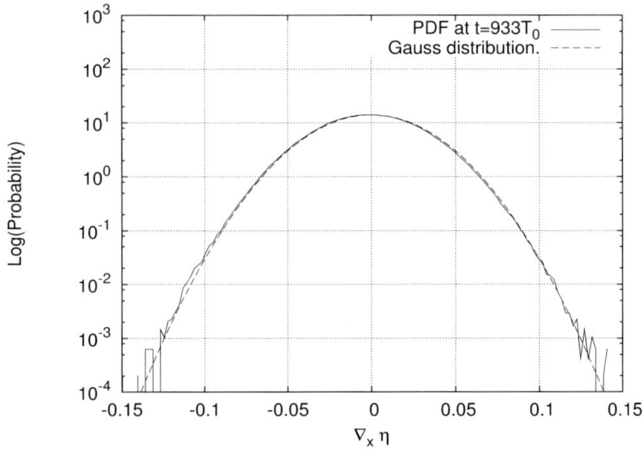

Fig. 22. PDF for $(\nabla\eta)_x$ at the final moment of time. $t \simeq 933T_0$.

Figures 23,24 present snapshots of the surface in the initial and final moments of simulation. Fig.25 is a snapshot of the surface in the moment of maximal roughness $T = 4.94 \simeq 14T_0$.

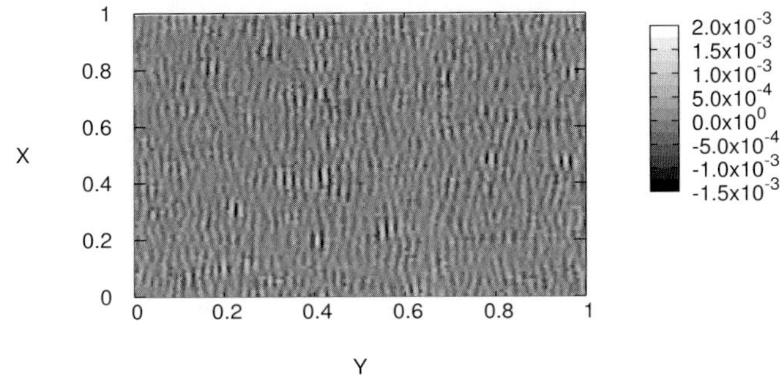

Fig. 23. Surface elevation at the initial moment of time. $t = 0$.

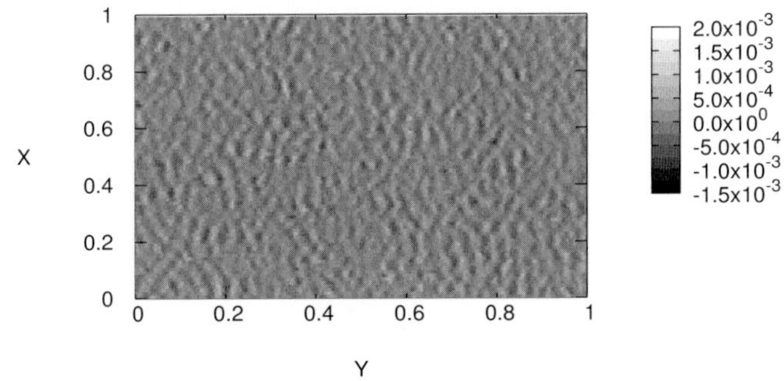

Fig. 24. Surface elevation at the final moment of time. $t \simeq 933T_0$.

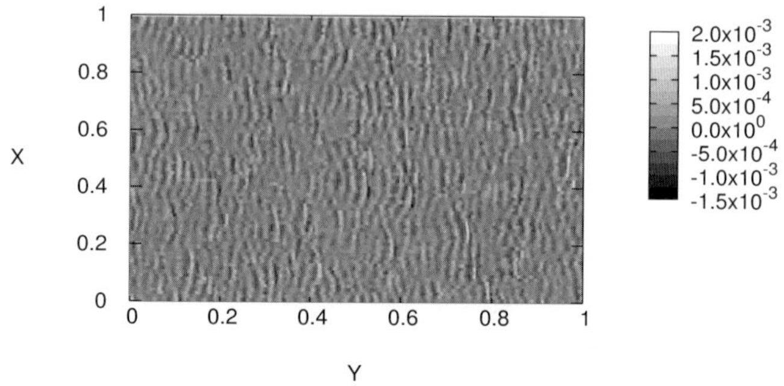

Fig. 25. Surface elevation at the moment of maximum roughness. $t \simeq 14T_0$. Gradients are more conspicuous.

4 Statistical numerical experiment

4.1 Numerical model for Hasselmann Equation

Numerical integration of kinetic equation for gravity waves on deep water (Hasselmann equation) was the subject of considerable efforts for last three decades. The "ultimate goal" of the effort – creation of the operational wave model for wave forecast based on direct solution of the Hasselmann equation – happened to be an extremely difficult computational problem due to mathematical complexity of the S_{nl} term, which requires calculation of a three-dimensional integral at every advance in time.

Historically, numerical methods of integration of kinetic equation for gravity waves exist in two "flavors".

The first one is associated with works of (32), (33), (34), (35), (36) and (37), and is based on transformation of 6-fold into 3-fold integrals using δ-functions. Such transformation leads to appearance of integrable singularities, which creates additional difficulties in calculations of the S_{nl} term.

The second type of models has been developed in works of (38) and (39), (40) and is currently known as Resio-Tracy model. It uses direct calculation of resonant quadruplet contribution into S_{nl} integral, based on the following property: given two fixed vectors $\mathbf{k}, \mathbf{k}_1$, another two $\mathbf{k}_2, \mathbf{k}_3$ are uniquely defined by the point "moving" along the resonant curve – locus.

Numerical simulation in the current work was performed with the help of modified version of the second type algorithm. Calculations were made on the grid 71×36 points in the frequency-angle domain $[\omega, \theta]$ with exponential distribution of points in the frequency domain and uniform distribution of points in the angle direction.

To date, Resio-Tracy model suffered rigorous testing and is currently used with high degree of trustworthiness: it was tested with respect to motion integrals conservation in the "clean" tests, wave action conservation in wave spectrum down-shift, realization of self – similar solution in "pure swell" and "wind forced" regimes (see (42), (41), (43)).

Description of scaling procedure from dynamical equations to Hasselman kinetic equation variables is presented in Appedix B.

4.2 Statistical model setup

The numerical model used for solution of the Hasselmann equation has been supplied with the damping term in three different forms:

1. Pseudo-viscous high frequency damping (16) used in dynamical equations;
2. $WAM1$ viscous term;
3. $WAM2$ viscous term;

Two last viscous terms referred as $WAM1$ and $WAM2$ are the "white-capping" terms, describing energy dissipation by surface waves due to white-capping, as used in $SWAN$ and WAM wave forecasting models, see (44):

$$\gamma_{\mathbf{k}} = C_{ds}\tilde{\omega}\frac{k}{\tilde{k}}\left((1-\delta) + \ delta\frac{k}{\tilde{k}}\right)\left(\frac{\tilde{S}}{\tilde{S}_{pm}}\right)^p \tag{24}$$

where k and ω are wave number and frequency, tilde denotes mean value; C_{ds}, δ and p are tunable coefficients; $S = \tilde{k}\sqrt{H}$ is the overall steepness; $\tilde{S}_{PM} = (3.02\times 10^{-3})^{1/2}$ is the value of $\tilde{S}$ for the Pierson-Moscowitz spectrum (note that the characteristic steepness $\mu = \sqrt{2}S$).

Values of tunable coefficients for $WAM1$ case (corresponding to *WAM cycle 3* dissipation) are:

$$C_{ds} = 2.36\times 10^{-5},\ \ \delta = 0,\ \ p = 4 \tag{25}$$

and for $WAM2$ case (corresponding to *WAM cycle 4* dissipation) are:

$$C_{ds} = 4.10\times 10^{-5},\ \ \delta = 0.5,\ \ p = 4 \tag{26}$$

In all three cases we used as initial data smoothed (filtered) spectra (Fig.5) obtained in the dynamical run at the time $T_* = 3.65min = 24.3 \simeq 70T_0$, which can be considered as a moment of the end of the fist "fast" stage of spectral evolution.

The natural question stemming in this point, is why calculation of the dynamical and Hasselmann model cannot be started from the initial conditions (15) simultaneously?

There are good reasons for that:

As it was mentioned before, the time evolution of the initial conditions (15) in presence of the damping term can be separated in two stages: relatively fast total energy drop in the beginning of the evolution and succeeding relatively slow total energy decrease as a function of time, see Fig.9. We explain this phenomenon by existence of the effective channel of the energy dissipation due to strong nonlinear effects, which can be associated with the white-capping.

We have started with relatively steep waves $\mu \simeq 0.167$. As we see, at that steepness white-capping is the leading effect. This fact is confirmed by numerous field and laboratory experiments. From the mathematical view-point the white-capping is formation of coherent structures – strongly correlated multiple harmonics. The spectral peak is posed in our experiments initially at $k \simeq 300$, while the edge of the damping area $k_d \simeq 1024$. Hence, only the second and the third harmonic can be developed, while hire harmonics are suppressed by the strong dissipation. Anyway, even formation of the second and the third harmonic is enough to create intensive non-Gaussian tail of the PDF for longitudinal gradients. This process is very fast. In the moment of time $T = 14T_0$ we see fully developed tails. Relatively sharp gradients mimic formation of white caps. Certainly, the "pure" Hasselmann equation is not applicable on this early stage of spectral evolution, when energy intensively dissipates.

As steepness decreases and spectral maximum of the swell down-shifts, an efficiency of such mechanism of energy absorption becomes less important when the steepness value drops down to $\mu \simeq 0.1$the white-capping becomes a negligibly small effect. It happens at $T \simeq 280T_0$. We decided to start comparison between deterministic and statistical modeling in some intermediate moment of time $T \simeq 70T_0$.

5 Comparison of deterministic and statistical experiments

5.1 Statistical experiment with pseudo-viscous damping term

First simulation has been performed with pseudo-viscous damping term, equivalent to (16).

Fig.8 – 11 show total wave action, total energy, mean wave slope and mean wave frequency as the functions of time.

Fig.32 shows the time evolution of angle-averaged wave action spectra as the functions of frequency for dynamical and Hasselmann equations.

Temporal behavior of angle-averaged spectrum is presented on Fig.32. We see the down-shift of the spectral maximum both in dynamic and Hasselmann equations. The correspondence of the spectral maxima is not good at all.

It is obvious that the influence of the artificial viscosity is not strong enough to reach the correspondence of two models.

5.2 Statistical experiments with $WAM1$ damping term

Fig.33 – 36 show total wave action, total energy, mean wave slope and mean wave frequency as the functions of time.

The temporal behavior of total wave action, energy and average wave slope is much better than in the artificial viscosity term, and for 50 min duration of the experiment we observe decent correspondence between dynamical and Hasselmann equations. However for longer time the $WAM1$ model deviates from the exact calculations significantly.

It is important to note that the curves of temporal behavior of the total wave action, energy and average wave slope diverge at the end of simulation time with different derivatives, and the correspondence cannot be expected to be that good outside of the simulation time interval.

Fig.37 shows the time evolution of the angle-averaged wave action spectra as the functions of frequency for dynamical and Hasselmann equations. As in the artificial viscosity case, we observe distinct down-shift of the spectral maxima. Correspondence of the time evolution of the amplitudes of the spectral maxima is much better then in artificial viscosity case.

5.3 Statistical experiments with $WAM2$ damping term

Fig.38 – 41 shows the temporal evolution of the total wave action, total energy, mean wave slope and mean wave frequency, which are divergent in time.

Fig.42 show time evolution of angle-averaged wave action spectra as the functions of frequency for dynamical and Hasselmann equations. While as in the artificial viscosity and $WAM1$ cases we also observe distinct down-shift of the spectral maxima, the correspondence of the time evolution of the amplitudes of the spectral maxima is worse than in $WAM1$ case.

Despite the fact that historically $WAM2$ appeared as an improvement of $WAM1$ damping term, it does not improve the correspondence of two models, observed in $WAM1$ case, and is presumably too strong for description of the reality.

6 Down-shift and angular spreading

The major process of time-evolution of the swell is frequency down-shift. During $T = 933T_0$ the mean frequency has been decreased from $\omega_0 = 2$ to $\omega_1 = .6$. On the last stage of the process, the mean frequency slowly decays as

$$< \omega > \simeq t^{-0.067} \simeq t^{-1/15} \tag{27}$$

The Hasselmann equation has self-similar solution, describing the evolution of the swell $n(\mathbf{k}, t) \simeq t^{4/11} F\left(\frac{\mathbf{k}}{t^{2/11}}\right)$ (see (41), (43)). For this solution

$$< \omega > \simeq t^{-1/11} \tag{28}$$

The difference between (27) and (28) can be explained as follows. What we observed, is not a self-similar behavior. Indeed, a self-silmilarity presumes that the angular structure of the solution is constant in time. Meanwhile, we observed intensive angular spreading of the initially narrow in angle, almost one-dimensional wave spectrum.

Level lines of the spectra after low-pass filtering, obtained in dynamical equations simulation, for two moments of time are presented on Fig. 26-27. Level lines of the spectra in the same moments of time, obtained by solution of the Hasselmann equation are presented on Fig. 28-29. One can see good correspondance between results of both experiments. Comparison of time-evolution of the mean angular spreading calculated from action and energy spectra are presented on Fig. 30-31.

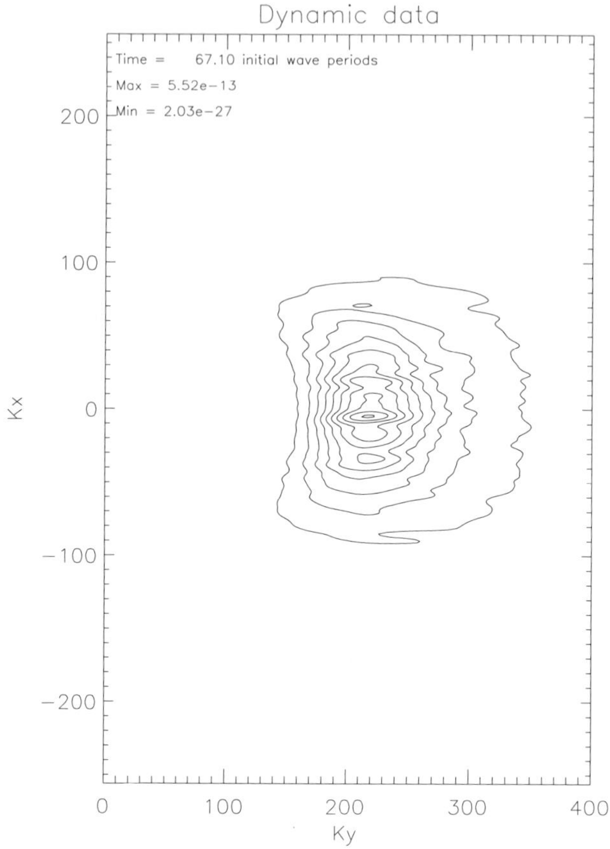

Fig. 26. Level lines of the spectra at $t = 67T_0$. Dynamical equations.

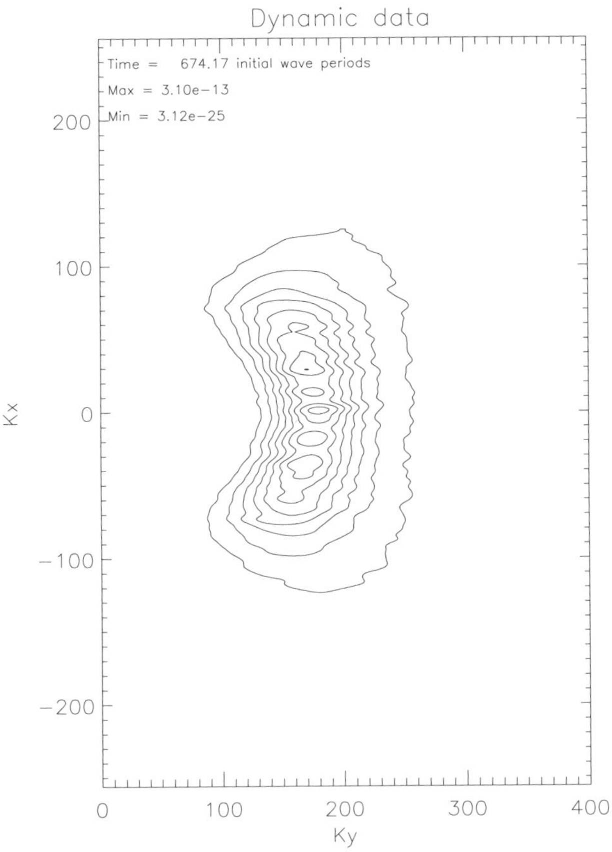

Fig. 27. Level lines of the spectra at $t = 674T_0$. Dynamical equations.

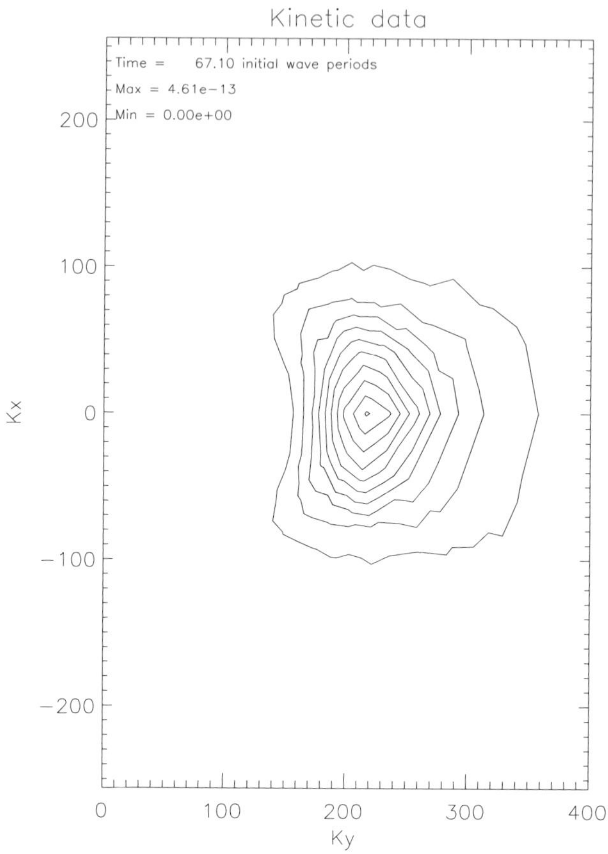

Fig. 28. Level lines of the spectra at $t = 67T_0$. Hasselmann equation.

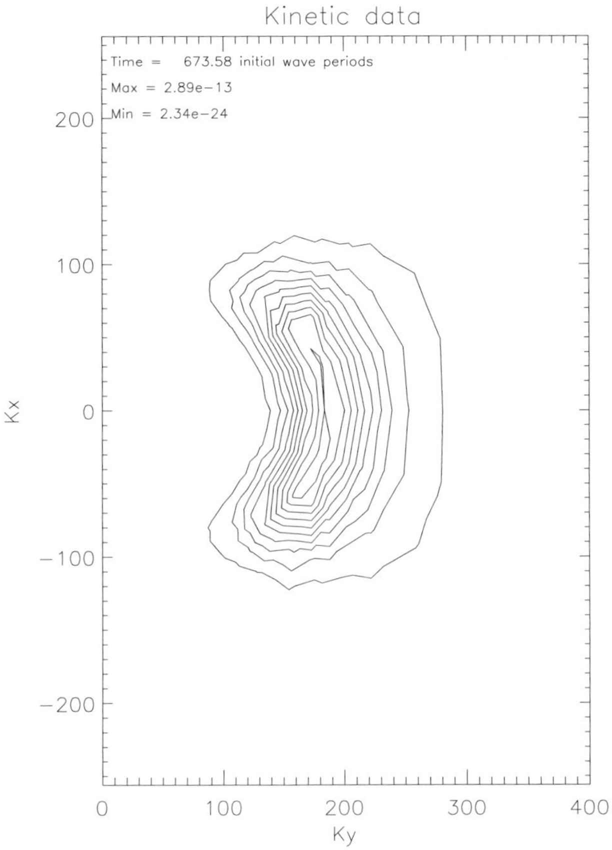

Fig. 29. Level lines of the spectra at $t = 674T_0$. Hasselmann equation.

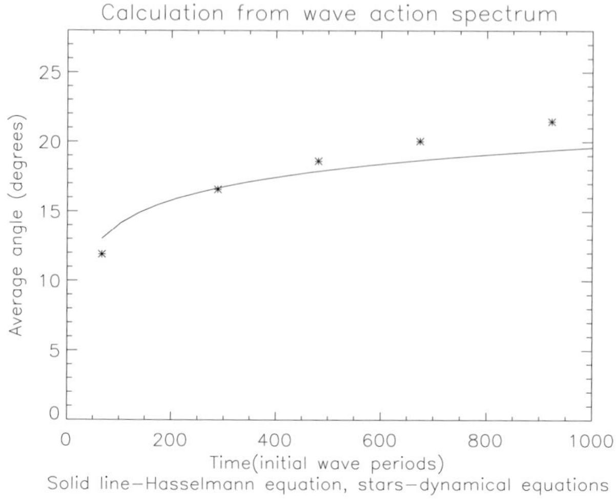

Fig. 30. Comparison of time-evolution of the mean angular spreading $\left(\int |\theta| n(\mathbf{k}) \mathrm{d}\mathbf{k}\right) / \left(\int n(\mathbf{k}) \mathrm{d}\mathbf{k}\right)$ calculated through wave action spectra.

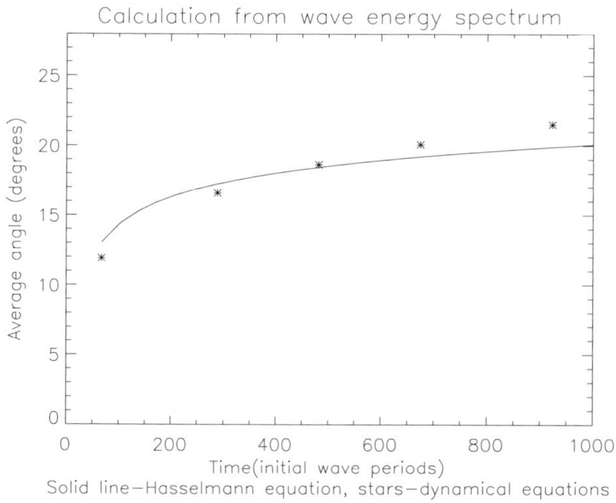

Fig. 31. Comparison of time-evolution of the mean angular spreading $\left(\int |\theta| \omega n(\mathbf{k}) \mathrm{d}\mathbf{k}\right) / \left(\int \omega n(\mathbf{k}) \mathrm{d}\mathbf{k}\right)$ calculated through wave energy spectra.

One has to expect that the angular spreading will be arrested at later times, and the spectra will take a universal self-similar shape.

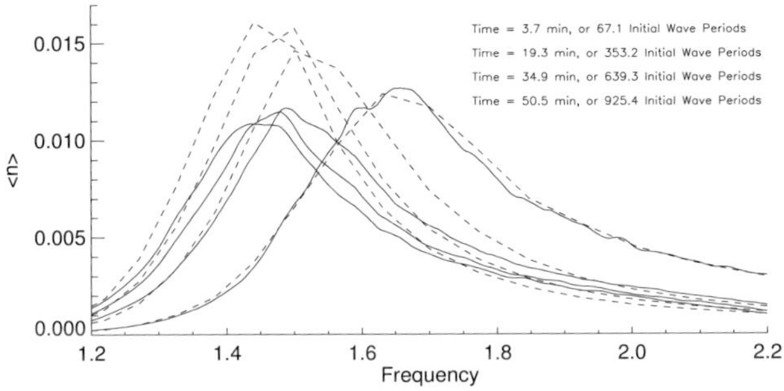

Fig. 32. Angle-averaged spectrum as a function of time for dynamical and Hasselmann equations for artificial viscosity case.

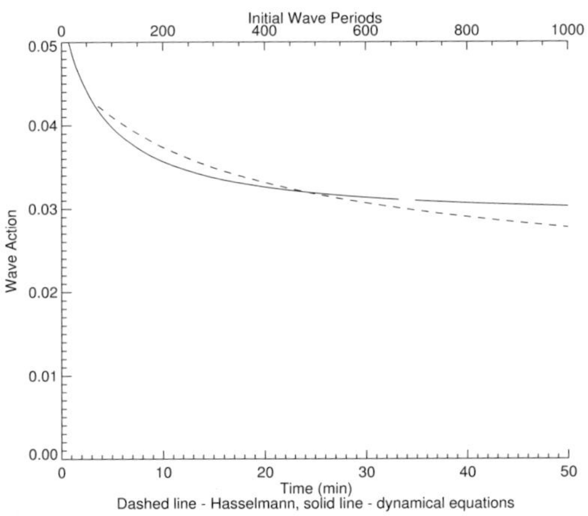

Fig. 33. Total wave action as a function of time for $WAM1$ case.

spectral down-shift of the spectral maximum wave-number and Zakharov-Filonenko spectral tails.

weak, while $WAM2$ damping term is obviously too strong. The quality of $WAM2$ damping can be significantly improved via development of new damping term to reach better correspondence on longer time evolution interval.

spectral modes. This observation modifies earlier definition of the weak turbulence as the wave ensemble with amplitudes slowly varying in time.

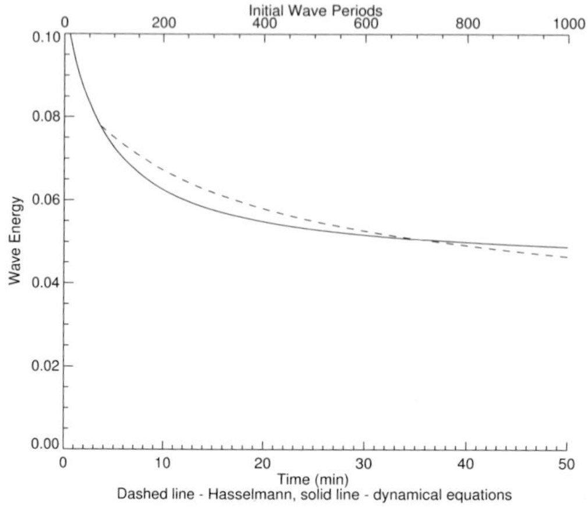

Fig. 34. Total wave energy as a function of time for $WAM1$ case

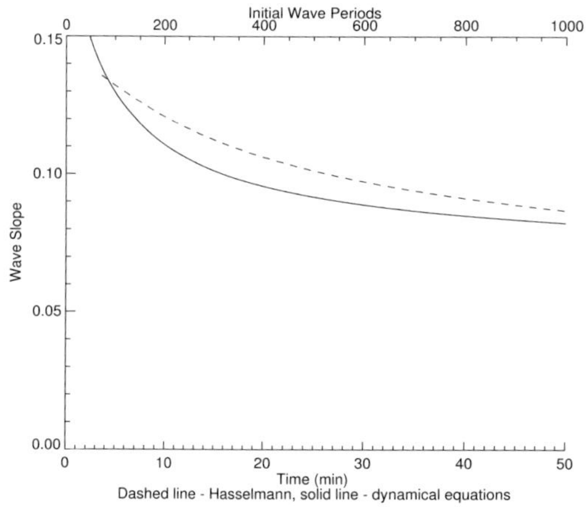

Fig. 35. Average wave slope as a function of time for $WAM1$ case.

7 Conclusion

1. We started our experiment with characteristic steepness $\mu \simeq 0.167$. This is three times less than steepness of the Stokes wave of limiting amplitude, but still it is a large steepness typical for young waves. For waves of such steepness white-capping effect could be essential. However, in our experiments we cannot observe such effects due to the strong pseudo-viscosity. Indeed, third harmonics of our initial leading wave is situated near the edge of damping area, while fourth and higher harmonics are far in the damping area. This circumstanse provides an intensive energy dissipation , which is not described by the Hasselmann equation.

Anyway, on the first stage of the process we observe intensive generation of coherent higher harmonics which reveal itself in tails of PDF for longitudinal gradients.

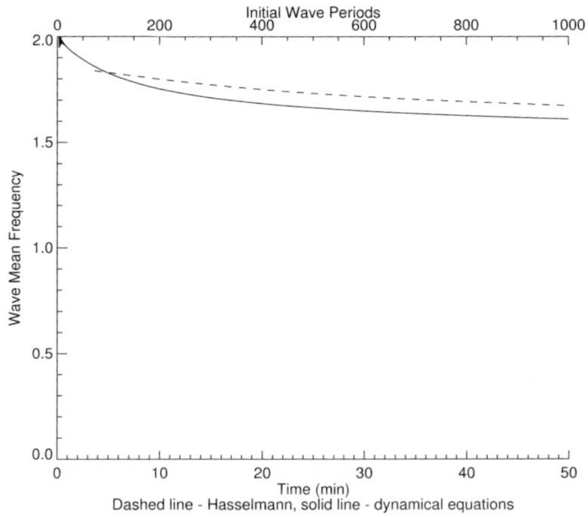

Fig. 36. Mean wave frequency as a function of time for $WAM1$ case.

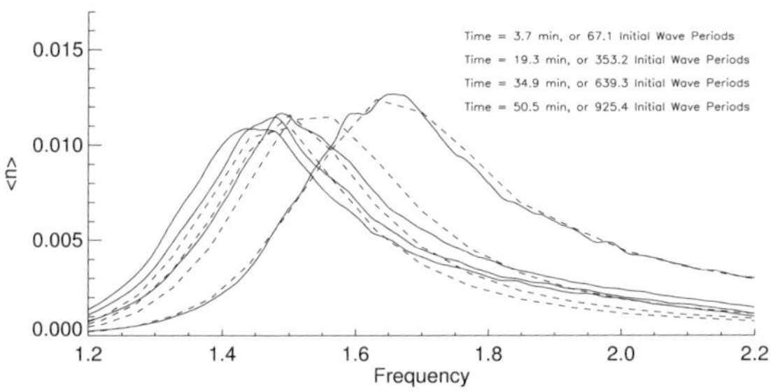

Fig. 37. Angle-averaged spectrum as a function of time for dynamical and Hasselmann equations a function of time for $WAM1$ case.

If our damping region would be shifted further to higher wave numbers, we could observe sharp crests formation.

2. We ended up with steepness $\mu \simeq 0.09$. This is close to mature waves, typically observed in the ocean and described by Hasselmann equation pretty well. We observed characteristic effects predicted by the weak-turbulent theory — down-shift of mean frequency formation, Zakharov-Filonenko weak turbulent spectrum ω^{-4} and strong angular spreading. Comparison of time-derivatives of the average quantities shows that for this steepness wave-breaking (white-capping) become not essential at $\mu \simeq 0.09$.

In general, our experiments validate Hasselmann equation. However, it has to be accomplished by a proper dissipation term.

3. The dissipative term used in the $WAM1$ model fairly describe damping due to white capping on the initial stage of evolution. It overestimate damping, however, for moderate steepness $\mu \simeq 0.09$

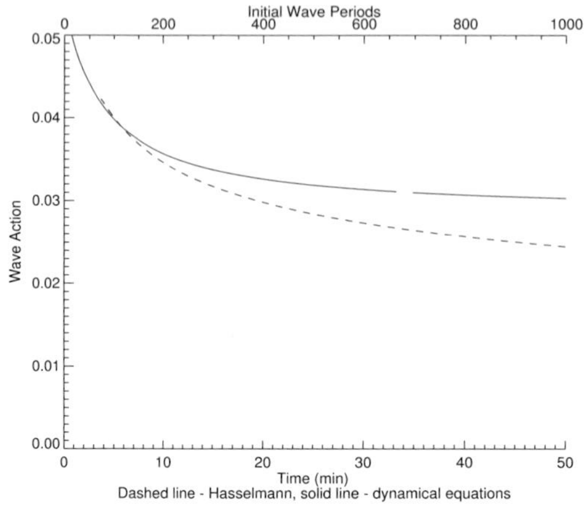

Fig. 38. Total wave action as a function of time for $WAM2$ case.

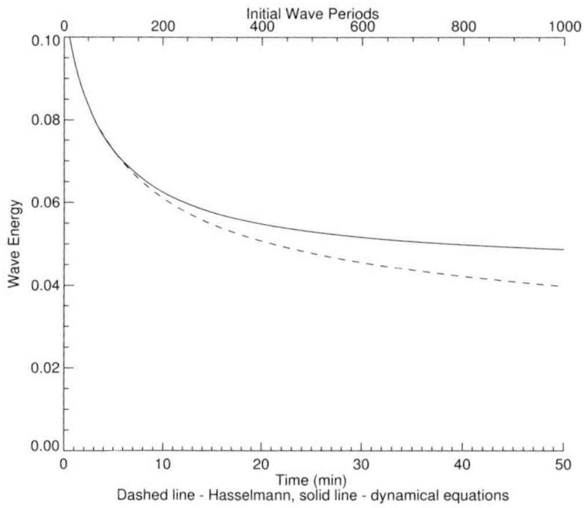

Fig. 39. Total wave energy as a function of time for $WAM2$ case

The dissipative term, used in the $WAM2$ model is not good. It overestimates damping essentially.

4. Presence of abnormally intensive harmonics, so called "oligarchs" show that, in spite of using a very fine grid, we did not eliminated effects of discreteness completely. More accurate verification of the Hasselmann equation should be made on the grid containing more than 10^7 modes. This is quite realistic task for modern supercomputers, and we hope to realize such an experiment.

Another conclusion is more pessimistic. Our results show that it is very difficult to reproduce real ocean conditions in any laboratory wave tank. Even a tank of size 200×200 meters is not large enough to model ocean due to the presence of wave numbers grid discreteness.

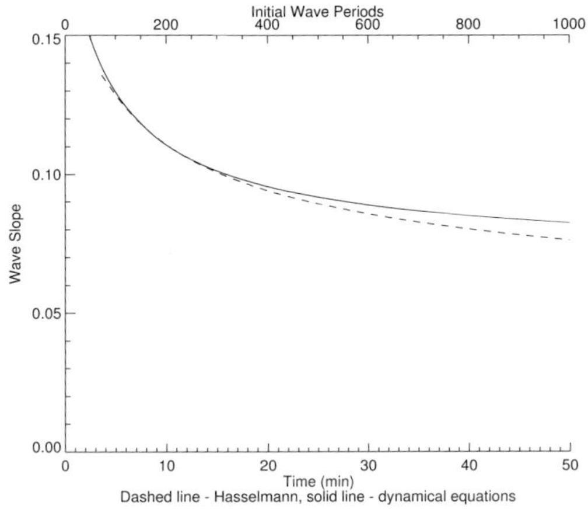

Fig. 40. Average wave slope as a function of time for $WAM2$ case.

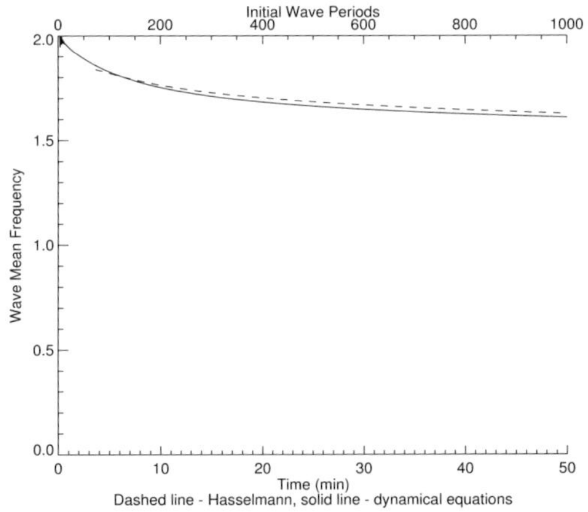

Fig. 41. Mean wave frequency as a function of time for $WAM2$ case.

Acknowledgements

This work was supported by ONR grant N00014-03-1-0648, RFBR grant 06-01-00665-a, INTAS grant 00-292, the Programme "Nonlinear dynamics and solitons" from the RAS Presidium and "Leading Scientific Schools of Russia" grant, also by US Army Corps of Engineers Grant DACW 42-03-C-0019 and by NSF Grant NDMS0072803.

A.O. Korotkevich was supported by Russian President grant for young scientist MK-1055.2005.2.

Also authors want to thank the creators of the open-source fast Fourier transform library FFTW (49) for this fast, portable and completely free piece of software.

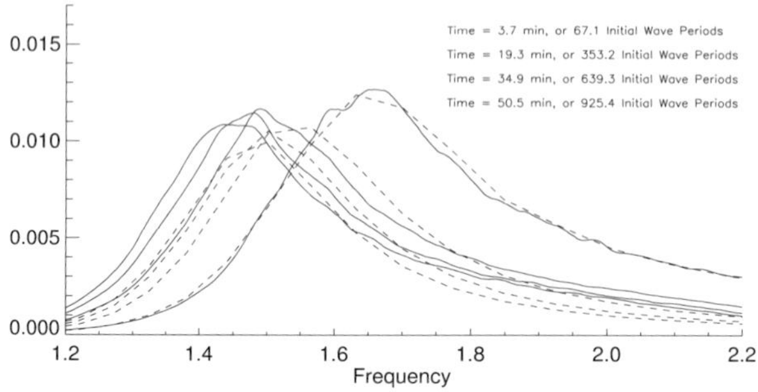

Fig. 42. Angle-averaged spectrum as a function of time for dynamical and Hasselmann equations a function of time for $WAM2$ case.

References

[1] Nordheim LW (1928) Proc.R.Soc. A119:689
[2] Peierls R (1929) Ann. Phys. (Leipzig) 3:1055
[3] Zakharov VE, Falkovich G, Lvov VS (1992) Kolmogorov Spectra of Turbulence I. Springer-Verlag, Berlin
[4] Hasselmann K (1962) J.Fluid Mech. 12:1
[5] Zakharov VE, Filonenko NN (1966) Doklady Acad. Nauk SSSR 160:1292
[6] Pushkarev AN, Zakharov VE (1996) Phys. Rev. Lett. 76:3320
[7] Pushkarev AN (1999) European Journ. of Mech. B/Fluids 18:345
[8] Pushkarev AN, Zakharov VE (2000) Physica D 135:98
[9] Tanaka M (2001) Fluid Dyn. Res. 28:41
[10] Onorato M, Osborne AR, Serio M, at al. (2002) Phys. Rev. Lett. 89:144501 arXiv:nlin.CD/0201017
[11] Dysthe KB, Trulsen K, Krogstad HE, Socquet-Juglard H (2003) J. Fluid Mech. 478:1–10
[12] Dyachenko AI, Korotkevich AO, Zakharov VE (2003) JETP Lett. 77:546 arXiv:physics/0308101
[13] Dyachenko AI, Korotkevich AO, Zakharov VE (2004) Phys. Rev. Lett. 92:134501 arXiv:physics/0308099
[14] Yokoyama N (2004) J. Fluid Mech. 501:169
[15] Zakharov VE, Korotkevich AO, Pushkarev AN, Dyachenko AI (2005) JETP Lett. 82:487 arXiv:physics/0508155.
[16] Dysthe K, Socquet-Juglard H, Trulsen K, at al. (2005) In "Rogue waves", Proceedings of the 14th 'Aha Huliko'a Hawaiian Winter Workshop 91:
[17] Lvov Yu, Nazarenko SV, Pokorni B (2006) Physica D 218:24 arXiv:math-ph/0507054
[18] Nazarenko SV (2006) J. Stat. Mech. L02002 arXiv:nlin.CD/0510054
[19] Annenkov SYu, Shrira VI (2006) Phys. Rev. Lett. 96:204501
[20] Dyachenko AI, Newell AC, Pushkarev AN, Zakharov VE (1992) Physica D 57:96

[21] Korotkevich AO (2003) Numerical Simulation of Weak Turbulence of Surface Waves. PhD thesis, L.D. Landau Institute for Theoretical Physics RAS, Moscow, Russia
[22] Dyachenko AI, Korotkevich AO, Zakharov VE (2003) JETP Lett. 77:477 arXiv:physics/0308100
[23] Zakharov VE (1968) J. Appl. Mech. Tech. Phys. 2:190
[24] Zakharov VE (1999) Eur. J. Mech. B/Fluids 18:327
[25] Kolmogorov A (1941) Dokl. Akad. Nauk SSSR 30:9 [Proc. R. Soc. London **A434**, 9 (1991)].
[26] V. E. Zakharov (1967) PhD thesis, Budker Institute for Nuclear Physics, Novosibirsk, USSR
[27] Zakharov VE, Zaslavskii MM (1982) Izv. Atm. Ocean. Phys. 18:747
[28] Toba Y (1973) J. Oceanogr. Soc. Jpn. 29:209
[29] Donelan MA, Hamilton J, Hui WH (1985) Phil. Trans. R. Soc. London A315:509
[30] Hwang PA, at al. (2000) J. Phys. Oceanogr 30:2753
[31] Dyachenko AI, Zakharov VE (1994) Phys. Lett., A190:144
[32] Hasselmann S, Hasselmann K, Barnett TP (1985) J. Phys. Oceanogr. 15:1378
[33] Dungey JC, Hui WH (1985) Proc. R. Soc. A368:239
[34] Masuda A (1981) J. Phys. Oceanogr. 10:2082
[35] Masuda A (1986) in Phillips OM, Hasselmann K, Waves Dynamics and Radio Probing of the Ocean Surface. Plenum Press, New York
[36] Lavrenov IV (1998) Mathematical modeling of wind waves at non-uniform ocean. Gidrometeoizdat, St.Petersburg, Russia
[37] Polnikov VG (2001) Wave Motion 1008:1
[38] Webb DJ (1978) Deep-Sea Res. 25:279
[39] Resio D., Tracy B (1982) Theory and calculation of the nonlinear energy transfer between sea waves in deep water. Hydraulics Laboratory, US Army Engineer Waterways Experiment Station, WIS Report 11
[40] Resio D, Perrie W (1991) J.Fluid Mech. 223:603
[41] Pushkarev A, Resio D, Zakharov VE (2003) Physica D 184:29
[42] Pushkarev A, Zakharov VE (2000) 6th International Workshop on Wave Hindcasting and Forecasting, November 6-10, Monterey, California, USA), 456 (published by Meteorological Service of Canada)
[43] Badulin SI, Pushkarev A, Resio D, Zakharov VE (2005) Nonlinear Processes in Geophysics
[44] SWAN Cycle III user manual, http://fluidmechanics.tudelft.nl/swan/index.htm
[45] Zakharov VE, Guyenne P, Dias F (2001) Wave turbulence in one-dimensional models. Physica D, 152-153:573
[46] Dias F, Pushkarev A, Zakharov VE (2004) One-Dimensional Wave Turbulence. Physics Reports, 398:1 (2004).
[47] Dias F, Guyenne P, Pushkarev A, Zakharov VE (2000) Wave turbulence in one-dimensional models. Preprint N2000-4, Centre de Methematiques et del leur Appl., E.N.S de CACHAN, 1
[48] Guyenne P, Zakharov VE, Dias F (2001) Turbulence of one-dimensional weakly nonlinear dispersive waves. Contemporary Mathematics, 283:107
[49] Frigo M, Johnson SG (2005) Proc. IEEE 93:216 http://fftw.org.

Appendix A: Forbes list of 15 largest harmonics

Here one can find 15 largest harmonics at the end of calculations in the framework of dynamical equations. Average square of amplitudes in dissipation-less region was taken from smoothed spectrum, while all these harmonics exceed level $|a_{\mathbf{k}}|^2 = 1.4 \times 10^{-12}$.

K_x	K_y	$\lvert a_{\mathbf{k}}\rvert^2$	$< \lvert a_{\mathbf{k}}\rvert^2 >_{filter}$	$\lvert a_{\mathbf{k}}\rvert^2 / < \lvert a_{\mathbf{k}}\rvert^2 >$
-59	155	1.563e-12	0.746e-13	2.095e+1
-37	166	1.903e-12	1.201e-13	1.585e+1
-37	185	1.569e-12	2.288e-13	0.686e+1
-36	162	1.477e-12	0.992e-13	1.489e+1
-33	157	1.442e-12	0.713e-13	2.022e+1
-26	164	3.351e-12	0.847e-13	3.956e+1
-17	189	1.463e-12	2.789e-13	0.525e+1
-14	173	1.408e-12	1.459e-13	0.965e+1
-2	176	1.533e-12	1.697e-13	0.903e+1
0	177	2.066e-12	1.741e-13	1.187e+1
10	179	1.427e-12	1.893e-13	0.754e+1
27	163	1.483e-12	0.832e-13	1.782e+1
31	174	1.431e-12	1.342e-13	1.066e+1
37	173	1.578e-12	1.581e-13	0.998e+1
60	133	1.565e-12	0.345e-13	4.536e+1

Appendix B: From Dynamical Equations to Hasselmann Equation

Standard setup for numerical simulation of the dynamical equations (4), implies $2\pi \times 2\pi$ domain in real space and gravity acceleration $g = 1$. Usage of the domain size equal 2π is convenient because in this case wave numbers are integers.

In the contrary to dynamical equations, the kinetic equation (11) is formulated in terms of real physical variables and it is necessary to describe the transformation from the "dynamical" variables into to the "physical" ones.

Eq.4 are invariant with respect to "stretching" transformation from "dynamical" to "real" variables:

$$\eta_{\mathbf{r}} = \alpha \eta'_{\mathbf{r}'}, \quad \mathbf{k} = \frac{1}{\alpha}\mathbf{k}', \quad \mathbf{r} = \alpha \mathbf{r}', \quad g = \nu g', \tag{29}$$

$$t = \sqrt{\frac{\alpha}{\nu}} t', \quad L_x = \alpha L'_x, \quad L_y = \alpha L'_y \tag{30}$$

where prime denotes variables corresponding to dynamical equations.

In current simulation we used the stretching coefficient $\alpha = 800.00$, which allows to reformulate the statement of the problem in terms of real physics: we considered $5026\,m \times 5026\,m$ periodic boundary conditions domain of statistically uniform ocean with the same resolution in both directions and characteristic wave length of the initial condition around $22\,m$. In oceanographic terms, this statement corresponds to the "duration-limited experiment".

Part II

Source & Run up

Runup of nonlinear asymmetric waves on a plane beach

Irina Didenkulova[1], Efim Pelinovsky[2], Tarmo Soomere[3] and Narcisse Zahibo[4]

[1] Institute of Applied Physics, Nizhny Novgorod, Russia
dii@hydro.appl.sci-nnov.ru
[2] Institute of Applied Physics, Nizhny Novgorod, Russia
pelinovsky@hydro.appl.sci-nnov.ru
[3] Institute of Cybernetics, Tallinn, Estonia soomere@cs.ioc.ee
[4] University of Antilles and Guyane, Guadeloupe, France
narcisse.zahibo@univ-ag.fr

Summary. The problem of the long wave runup on a beach is discussed in the framework of the rigorous solutions of the nonlinear shallow-water theory . The key and novel moment here is the analysis of the runup of a certain class of asymmetric waves, the face slope steepness of which exceeds the back slope steepness. Shown is that the runup height increases when the relative face slope steepness increases whereas the rundown weakly depends on the steepness. The results partially explain why the tsunami waves with the steep front (as it was for the 2004 tsunami in the Indian Ocean) penetrate deeper into inland compared with symmetric waves of the same height and length.

1 Introduction

The reliable estimate of the extension of the flooding zone is a key problem of the tsunami prevention and mitigation. Since the characteristic length of a tsunami wave in the coastal zone is several kilometres, the nonlinear shallow water theory is an appropriate theoretical model to describe the process of the tsunami runup on the beach. Carrier and Greenspan (2) first obtained rigorous mathematical results for the runup problem. They solved the nonlinear shallow-water equations for the case of 1+1 dimensions and a plane beach of constant slope. They applied the hodograph (Legendre) transformation to reduce the initial nonlinear hyperbolic equations in the spatial domain with an unknown boundary (*resp.* moving shoreline) to the linear wave equation on a fixed semi-axis.

After this pioneering study, also Spielfogel (16), Pelinovsky and Mazova (14), Tinti and Tonini (23) have found some particular exact explicit analytical solutions to this problem for specific beach profiles and/or types of incoming waves. The main difficulty in this problem is the implicit form of the hodograph transformation. For that reason the detailed analysis of runup characteristics usually requires numerical methods. Various shapes of the periodic incident wave trains such as the sine wave

(6), cnoidal wave (19), and bi-harmonic wave (4) have been analyzed in literature. The relevant analysis has been also performed for a variety of solitary waves and single pulses such as soliton (12); (18), sine pulse (10), Lorentz pulse (14), gaussian pulse ((3)), and N-waves (20). In particular, antisymmetric disturbances such as N-waves are considered now as the realistic initial conditions of the earthquake-forced tsunami (21), (23).

It is well known that nonlinear long wave evolution in shallow water even of constant depth results in the deformation of the wave profile and, finally, to the wave breaking (see, for instance, (17), (27), (26), (22), (8)). A tsunami wave is not an exception. It usually propagates over a long distance and, even if originally perfectly symmetric and linear entity, its shape is eventually modified due to nonlinearity. The increase of the steepness of the tsunami wave front is predicted theoretically in (11), (13) and is reproduced in the numerical simulation of the tsunami over long distances (25). There are a lot of observations of the wave breaking and its transformation into the undular bore made during the huge tsunami in the Indian Ocean on 26th December 2004. Analogous processes are commonly observed when tsunami waves enter an estuary or a river mouth (13), (24), or penetrate into straits or channels (15), (28), (1).

The main goal of this paper is to demonstrate the significant increase of the runup height in the particular case when the incoming wave has a steep front compared with the symmetric waves with the same parameters. This effect will be studied using the exact solutions of the nonlinear shallow-water equations. We also present a simple algorithm of calculating the conditions of the wave breaking (so-called gradient catastrophe) without using the Jacobian of the hodograph transformation . The paper is organized as follows. The method of solution of the runup problem in the framework of the nonlinear shallow-water theory based on the hodograph transformation is described in section 2. Matching of the runup zone with the shelf of constant depth, and the nonlinear transformation of the shallow-water wave above even bottom is considered in section 3. The runup of the nonlinear asymmetric waves is studied in section 4. Main results are summarized in section 5.

2 Mathematical Model and Hodograph Transformation

The classical nonlinear shallow water equations for 2D water waves in the ideal fluid with linearly slopping bottom (Fig. 1) are:

$$\frac{\partial \eta}{\partial t} + \frac{\partial}{\partial x}[(-\alpha x + \eta)u] = 0, \tag{1}$$

$$\frac{\partial u}{\partial t} + u\frac{\partial u}{\partial x} + g\frac{\partial \eta}{\partial x} = 0. \tag{2}$$

where η is the surface displacement, u is the depth-averaged velocity, g is the gravity acceleration, and α is the bottom slope. It is convenient to rewrite Eqs. (1, 2) through their Riemann invariants and to apply the hodograph transformation to the resulting equations (2). Doing so leads to the linear wave equation with respect to the wave function

$$\frac{\partial^2 \Phi}{\partial \lambda^2} - \frac{\partial^2 \Phi}{\partial \sigma^2} - \frac{1}{\sigma}\frac{\partial \Phi}{\partial \sigma} = 0, \tag{3}$$

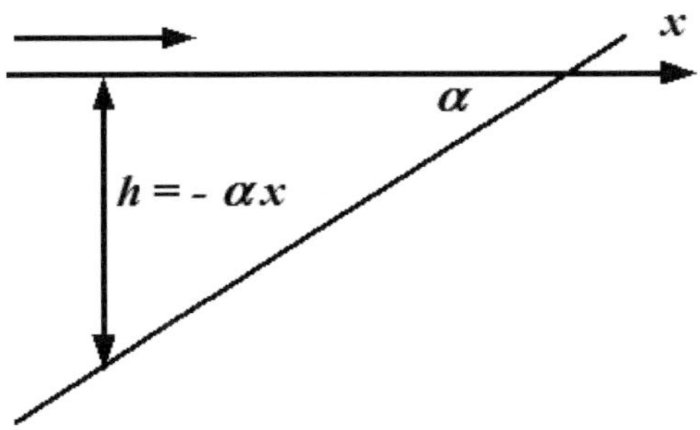

Fig. 1. Definition sketch for the wave runup problem

where the new coordinates λ and σ have been introduced and all variables can be expressed through the wave function $\Phi(\sigma, \lambda)$ as follows:

$$\eta = \frac{1}{2g}\left(\frac{\partial\Phi}{\partial\lambda} - u^2\right), \tag{4}$$

$$u = \frac{1}{\sigma}\frac{\partial\Phi}{\partial\sigma}, \tag{5}$$

$$t = \frac{1}{\alpha g}\left(\lambda - \frac{1}{\sigma}\frac{\partial\Phi}{\partial\sigma}\right), \tag{6}$$

$$x = \frac{1}{2\alpha g}\left(\frac{\partial\Phi}{\partial\lambda} - u^2 - \frac{\sigma^2}{2}\right). \tag{7}$$

Since

$$\sigma = 2\sqrt{g(-\alpha x + \eta)}, \tag{8}$$

and the point $\sigma = 0$ corresponds to the moving shoreline, it is sufficient to solve wave equation (3) on the semi-axis ($0 \leq \sigma < \infty$) with some initial or boundary conditions offshore. The dynamics of the moving shoreline is an extremely important feature of the flooding zone when tsunami waves approach to the coast. Its analysis for a class of asymmetric waves is the main goal of our study.

Similarly, linear equations of the shallow water theory

$$\frac{\partial\eta}{\partial t} + \frac{\partial}{\partial x}[-\alpha x u] = 0, \quad \frac{\partial u}{\partial t} + g\frac{\partial\eta}{\partial x} = 0 \tag{9}$$

can be also reduced to the linear wave equation (3), solution of which we call the "linear" wave function $\Phi_l(\sigma_l, \lambda_l)$ below. The difference compared with the above analysis consists in the use of the linear version of the hodograph transformation

$$\eta_l = \frac{1}{2g}\left(\frac{\partial\Phi_l}{\partial\lambda_l}\right), \quad u_l = \frac{1}{\sigma_l}\frac{\partial\Phi_l}{\partial\sigma_l}, \quad t_l = \frac{\lambda_l}{\alpha g}, \quad x_l = -\frac{\sigma_l^2}{4\alpha g}. \tag{10}$$

As above, Eq. (3) should be solved on a semi-axis. The point $\sigma_l = 0$ now corresponds to the unperturbed shoreline $x = 0$.

A long wave of small amplitude propagating in a deep open sea area is usually almost perfectly linear and can be described by linear theory with very high accuracy. For such an incident wave the boundary conditions for the "nonlinear" and "linear" wave equations coincide provided they are defined in a deep enough area. Consequently, the solutions of the nonlinear and linear problems also coincide, and $\Phi(\sigma, \lambda) = \Phi_l(\sigma_l, \lambda_l)$. Moreover, if the "linear" solution $\Phi_l(\sigma_l, \lambda_l)$ is known, the solution of the nonlinear problem (1, 2) can be directly found from expressions (4)-(7). In fact, it is difficult to do analytically but very easy numerically. In particular, description of properties of the moving shoreline $\sigma(x, t)$ is straightforward. From (5), (6) and (10) it follows that

$$u(\lambda) = \lambda - \lambda_l, \tag{11}$$

or, in an equivalent form,

$$u(\lambda) = u(\lambda_l + u), \tag{12}$$

which demonstrates that the speed of the shoreline displacement can be found through the Riemann transformation of time. As the functional forms of the "linear" and "nonlinear" solutions are identical, we may re-write (12) finally:

$$u(t) = U\left(t + \frac{u}{\alpha g}\right). \tag{13}$$

where $U(t)$ stands for the "linear" speed of the shoreline.

Thus, if the approaching wave is linear, a rigorous "two-step" method can be used to calculate the runup characteristics. Firstly, the wave properties on the unperturbed shoreline $x = 0$ such as the vertical displacement $Y(t)$ or the velocity of water tongue propagation

$$U(t) = \frac{1}{\alpha}\frac{dY(t)}{dt}, \tag{14}$$

are determined within the linear problem. Its solution can be found using traditional methods of the mathematical physics. Secondly, the properties of the solution to the nonlinear problem are found from expressions derived above. For example, the real "nonlinear" speed of the moving shoreline is found from (13), and finally, the vertical displacement of the water level and position of the shoreline at some time instant horizontal distance of the flooding (*resp.* the width of the flooded area) as

$$y(t) = \alpha x(t), \quad x(t) = \int u(t)dt. \tag{15}$$

Using (4), Eq. (15) can be re-written as

$$y(t) = \eta(t, \sigma = 0) = Y\left(t + \frac{u}{\alpha g}\right) - \frac{u^2}{2g}. \tag{16}$$

The important conclusion from expressions (13) and (16) is that the maxima of vertical displacements (equivalently, the runup or rundown height) and the velocity of the shoreline displacement in the linear and nonlinear theories coincide. Consequently, the linear theory adequately describes the runup height which is an extremely important characteristic of tsunami action on the shore. In fact, this conclusion was reached in many papers cited above for various shapes of the incident wave. The rigorous proof demonstrated here follows the work by Pelinovsky and Mazova (14).

There are no rigorous results in the nonlinear theory in the case of more complicated bottom profiles that cannot be approximated by the idealized beach of constant slope. Yet the linear theory can be used in some cases when the nearshore has such a slope alone. If the wave is nonlinear only at the runup stage, the linear theory frequently can be used to describe wave transformation in the ocean of variable depth and the resulting wave can be matched with the nonlinear solutions. This approach is quite popular, see (7) and references therein. Nonlinear effects in the transition zone (between the offshore and the runup zone) can be accounted for as the correction term to the boundary conditions far from the shoreline (9).

Another important outcome from proposed approach is the simple definition of the conditions of the first breaking of the waves on a beach. It is evident that long small-amplitude waves will not break at all and result in a slow rise of the water level resembling surge-like flooding. With increase of the wave amplitude, the breaking appears seawards from the runup maximum and, depending on the wave amplitude and the bottom slope, may occur relatively far offshore. The above approach allows determining the position of the first breaking from the expressions for the dynamics of the shoreline. The temporal derivative of the velocity of the moving shoreline, found from (13),

$$\frac{du}{dt} = \frac{dU/dt}{1 - \frac{dU/dt}{\alpha g}}, \tag{17}$$

tends to the infinity (equivalently, wave breaking occurs, (14)) when the denominator a the right-hand side of Eq. (17) approaches to zero. The condition of the first wave breaking therefore is

$$Br = \frac{\max(dU/dt)}{\alpha g} = \frac{\max(d^2Y/dt^2)}{\alpha^2 g} = 1. \tag{18}$$

This condition has a simple physical interpretation: the wave breaks if the maximal acceleration of the shoreline $Y''\alpha^{-1}$ along the slopping beach exceeds the along-beach gravity component (αg). This interpretation is figurative, because formally Y'' only presents the vertical acceleration of the shoreline in the linear theory and the "nonlinear" acceleration du/dt actually tends to infinity at the breaking moment.

The above-cited literature contains various examples of studies of long wave runup on the plane beach using the hodograph transformation . The effectiveness of this two-step approach can be demonstrated by considering the runup of a sine wave with frequency ω . The well-known bounded solution of the linear wave equation (3) is expressed in the Bessel functions

$$\eta(x,t) = RJ_0\left(\sqrt{\frac{4\omega^2|x|}{g\alpha}}\right)\cos(\omega t), \tag{19}$$

where R is the maximal wave amplitude at the unperturbed shoreline $x = 0$. As mentioned above, it is also the maximal runup height in the nonlinear theory. Far from the shoreline the wave field can be presented asymptotically as the superposition of two sine waves of equal amplitude propagating in the opposite directions

$$\eta(x,t) = A(x)\left[\sin\left(\omega(t-\tau)+\frac{\pi}{4}\right) + \sin\left(\omega(t+\tau)-\frac{\pi}{4}\right)\right], \tag{20}$$

where the instantaneous wave amplitude $A(x)$ is

$$A(x) = R\left(\frac{\alpha g}{\pi^2\omega^2|x|}\right)^{1/4}, \tag{21}$$

and the propagation time of this wave over some distance in a fluid of variable depth is

$$\tau(x) = \int \frac{dx}{\sqrt{gh(x)}}. \tag{22}$$

The maximum change of the amplitude of the approaching wave with the wavelength λ_0 (determines from dispersion relation, $\omega = ck$) and the initial amplitude A_0 at the fixed point $|x| = L$ is characterized by the amplification factor (equivalent to the shoaling coefficient in the linear surface wave theory), which can be found from (21):

$$\frac{R}{A_0} = \left(\frac{\pi^2 \omega^2 L}{g\alpha}\right)^{1/4} = 2\pi\sqrt{\frac{2L}{\lambda_0}}. \tag{23}$$

We emphasize that the amplification factor in (23) calculated in the framework of the linear theory is the same in the nonlinear theory. This feature allows to determine the extreme runup characteristics in both cases if the initial wave amplitude and length are known. Using (23), the extreme values for the velocity of the moving shoreline and the breaking criterion can be calculated as follows:

$$U_{ext} = \frac{\omega R}{\alpha}, \tag{24}$$

$$Br_{sin} = \frac{\omega^2 R}{g\alpha^2} = 1. \tag{25}$$

As a result, it is simple to predict the minimal value (threshold) of the runup height when the wave breaking appears from any given wave frequency (or period) and the bottom slope. The second step of solving the nonlinear runup problem consists in transformation of the "linear" expressions (36) and (38) for the water level displacement and velocity into "nonlinear" expressions for the moving shoreline with use of (13) and (15). Fig. 9 displays the "nonlinear" and "linear" time history of the water level and velocity of the moving shoreline (in dimensionless form) for a symmetrical sine incident wave ($s = s_0$). If the incident wave is strongly asymmetric wave ($s = 10s_0$, Fig. 11), the strong flow moves inland during a short time. The runup amplitude is higher than the rundown amplitude. Such intense flows can be distinguished on many images of the catastrophic 2004 tsunami in the Indian Ocean and eventually occurred in many sections of the affected coastline. In this case the incident wave is extremely steep and breaks rapidly, and there is almost no difference in the "linear" and "nonlinear" results.

3 Nonlinear Wave Deformation

An adequate theory of runup should take into account the potential asymmetry of the incoming waves. Such waves appear, for example, when the shelf offshore from the runup zone has a flat bottom (Fig. 2). In the water of constant depth, the exact one-wave solution of the nonlinear shallow water equations can be described by the following partial differential equation, which can be easily derived from (1)-(2):

$$\frac{\partial \eta}{\partial t} + V(\eta)\frac{\partial \eta}{\partial x} = 0.$$

$$V = \sqrt{gh} + \frac{3u}{2} = 3\sqrt{g(h+\eta)} - 2\sqrt{gh}, \quad u = 2(\sqrt{g(h+\eta)} - \sqrt{gh}). \tag{26}$$

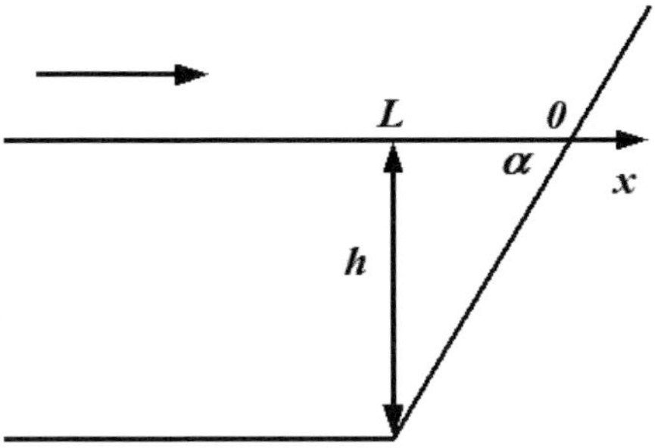

Fig. 2. Definition sketch of the coastal geometry

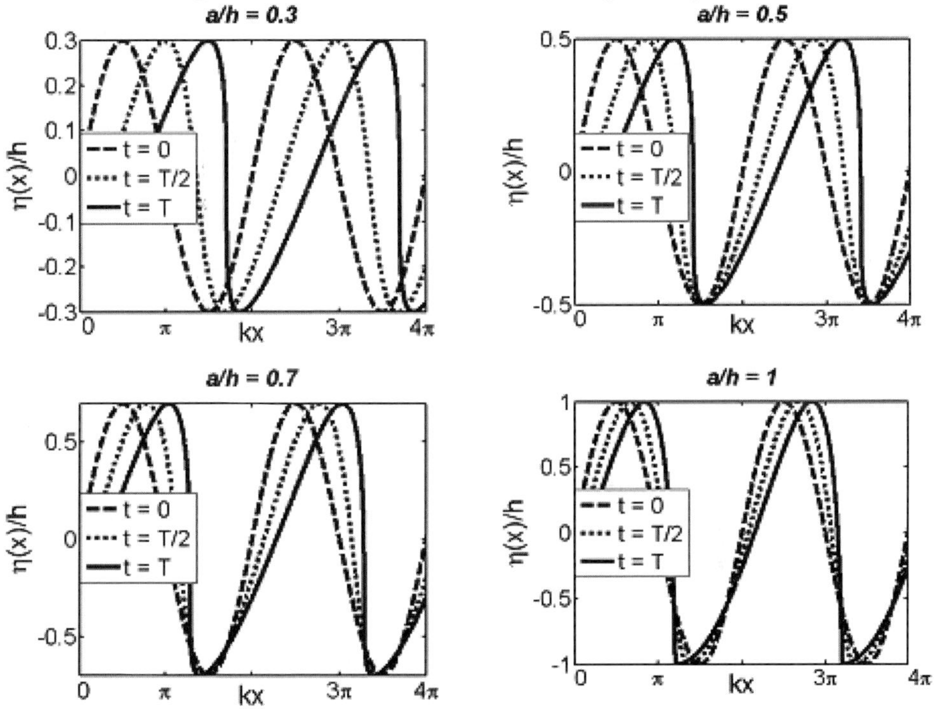

Fig. 3. Deformation of an initial sine wave (dashed line) in shallow water

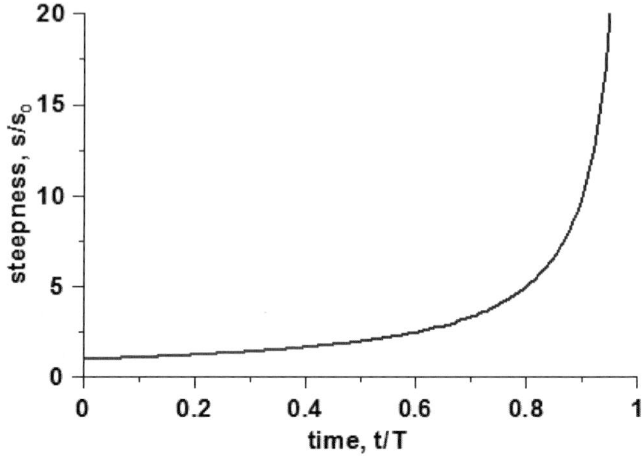

Fig. 4. Temporal evolution of the wave steepness

The solution of Eqs. (26) satisfying the initial condition $\eta(x,t) = \eta_0(x)$ is the Riemann wave

$$\eta(x,t) = \eta_0(x - Vt). \tag{27}$$

Its shape varies with distance and its steepness increases due to the difference in speed of the crest and trough. The instantaneous slope of the water surface at any point of the incoming wave is

$$\frac{\partial \eta}{\partial x} = \frac{\eta_0'}{1 + tV_0'}, \tag{28}$$

where the prime means $d/d\tilde{x}$, where $\tilde{x} = x - Vt$, and $V_0(x)$ is determined through the initial wave shape $\eta_0(x)$ with use of (27). On the face of the incident wave $\partial\eta/\partial x < 0$, $\partial V_0/\partial x < 0$ and the denominator at the right-hand side of (28) decreases with time; thus the wave steepness increases and becomes infinite at

$$t = T = \frac{1}{\max(-V_0')}. \tag{29}$$

As an example, we analyse the nonlinear deformation of the initial sine wave with an amplitude a and a wave number k propagating in water of constant depth. The temporal evolution of the wave shape is demonstrated in Fig. 3 for several initial dimensionless amplitudes a/h (17), (27), (13), (8), (5). The breaking time and the breaking distance X are

$$X = \sqrt{gh}T = \frac{1}{3k}\sqrt{\frac{2}{1 - \sqrt{1 - (a/h)^2}}}.\gamma \tag{30}$$

The breaking distance decreases when the wave amplitude increases. Large-amplitude waves break almost after their generation, but waves with small amplitudes may pass a long distance before breaking.

The following simplified formula for the maximum steepness of the face-slope (Fig. 4) can be derived from (29):

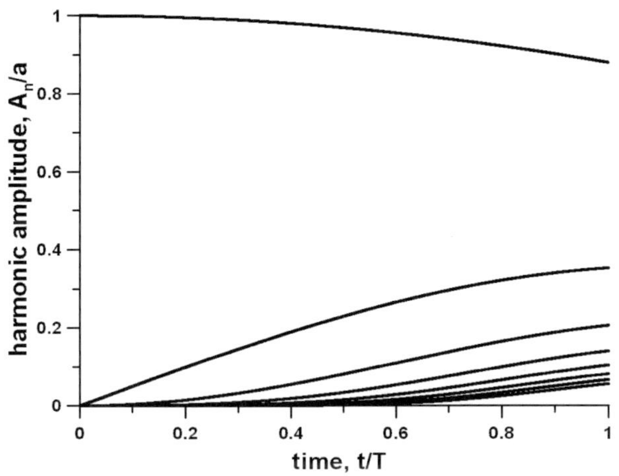

Fig. 5. Temporal evolution of the amplitude of the spectral harmonics. Curves are numbered from the top

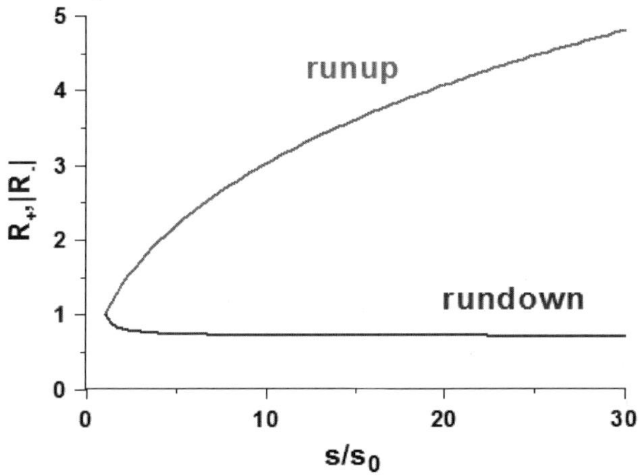

Fig. 6. Runup (R_+) and rundown (R_-) amplitudes versus the wave steepness

$$s = \max(\partial\eta/\partial x) = \frac{s_0}{1 - t/T}, \tag{31}$$

where $s_0 = ak$ is the initial wave steepness. At the breaking time, the steepness is infinite. For certain applications it is important to know the spectrum of the shallow-water wave. The spectral presentation of the Riemann wave in terms of sine harmonics can be presented explicitly (13), (5):

$$\eta(t, x) = \sum_{n=1}^{\infty} A_n(t) \sin(nk[x - \sqrt{gh}t]), \quad A_n(t) = 2a\frac{T}{nt} J_n\left(\frac{nt}{T}\right), \tag{32}$$

where J_n are the Bessel functions. The amplitudes of the higher harmonics increase with time whereas the amplitude of the basic harmonic corresponding to $n = 1$

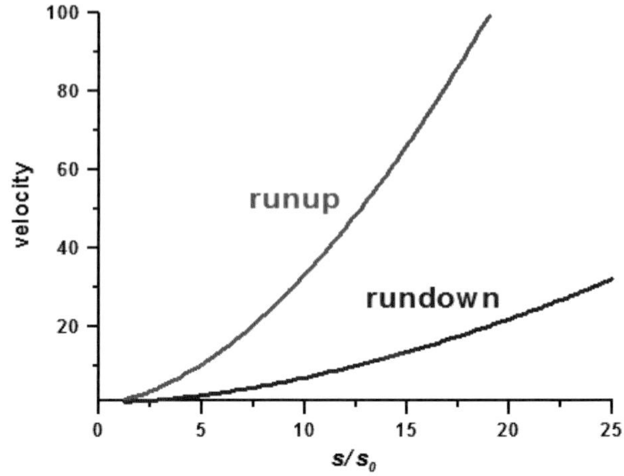

Fig. 7. Extreme runup and rundown velocities versus the wave steepness

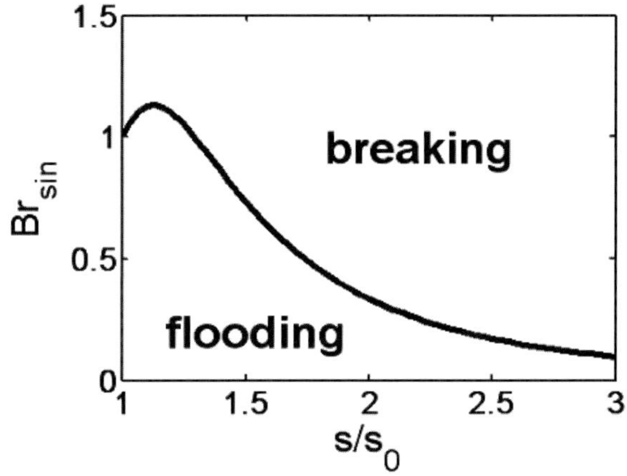

Fig. 8. Various scenarios of the wave runup on a beach

decreases (Fig. 5). The realistic tsunami wave evolution both in the open sea and in the coastal zone is extremely complicated due to effects of refraction, diffraction and resonance. The propagation time, used to characterize the wave properties in the simple example of nonlinear wave deformation considered above, technically may be used also in the general case. However, in the general case, the wave steepness is a more convenient measure of the wave shape than the propagation time. Using (31), the spectral amplitudes (32) can be expressed as

$$A_n(s) = \frac{2a}{n(1 - s_0/s)} J_n\left(n\left[1 - \frac{s_0}{s}\right]\right). \tag{33}$$

Since the spectral amplitudes with $n > 1$ increase with the steepness increasing, we may determine the relation between "local" or "current" wave characteristics without considering the wave history.

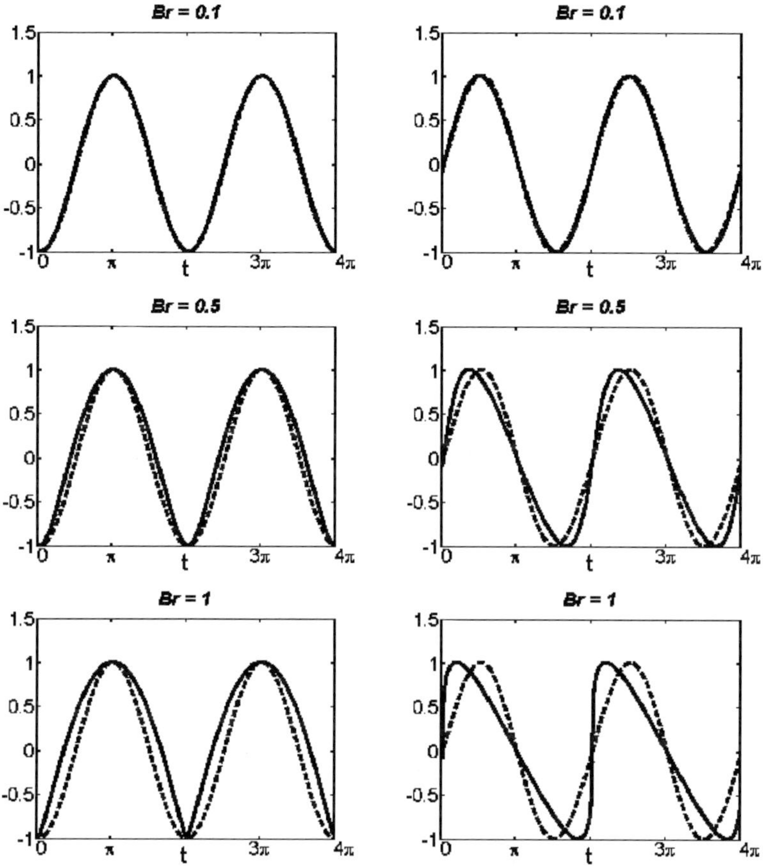

Fig. 9. Water level (left) and velocity (right) of the moving shoreline for various values of the breaking parameter and for a sinusoidal incident wave ($s = s_0$). Solid line corresponds to the "nonlinear" and dashed line to the "linear" solution.

4 Runup of Nonlinear Deformed Wave on a Plane Beach

The nonlinear long wave propagation in a large ocean of constant depth is thus always accompanied by a certain deformation of the wave shape. Such wave coming to the beach of constant slope (Fig. 2) has a front, much steeper, than its back. The runup of such asymmetric waves on the plane beach can be studied with the use of the model described in section 2.

The first step the two-step approach consists in solving the linear problem. For doing this we may use the linear superposition of elementary solutions (19) and match it with the Riemann wave (32) far from the shoreline. If the far-field wave in (20) is approximated by a (finite or infinite) superposition of harmonics, the use of expressions (21)-(23) leads to the following expressions for the incident wave on a

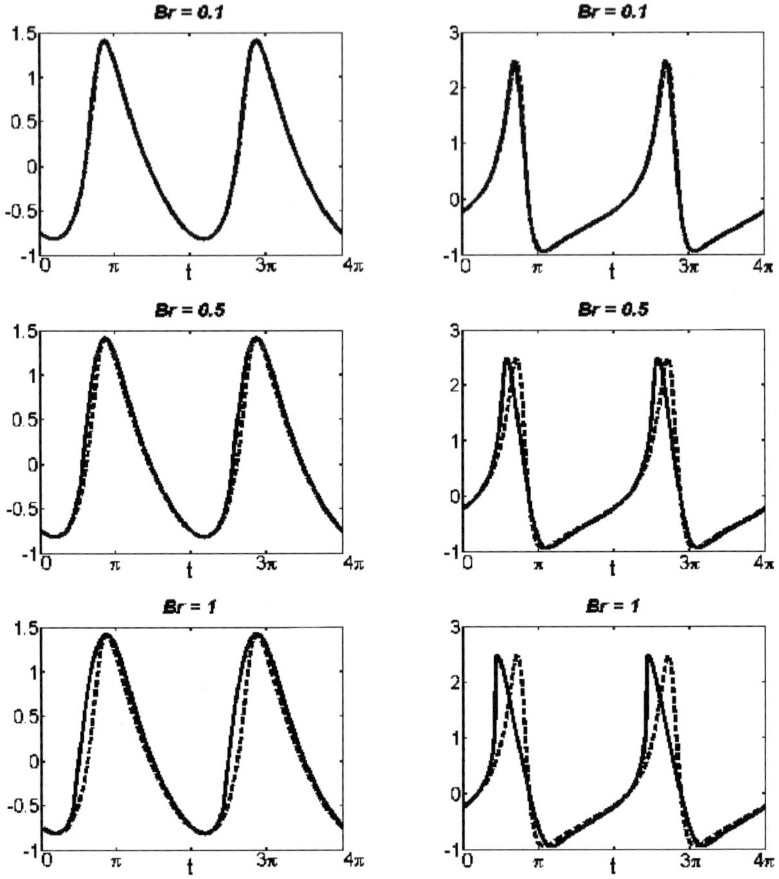

Fig. 10. Water level (left) and velocity (right) of the moving shoreline for various values of the breaking parameter and for a sinusoidal incident wave ($s = 2s_0$). Solid line corresponds to the "nonlinear" and dashed line to the "linear" solution.

distance L from the shoreline where the beach is matched with a shelf of constant depth and for the "linear" oscillations of the water level at the unperturbed shoreline:

$$\eta(t, x = -L) = \sum_{n=1}^{\infty} A_n(s) \sin(n\omega t), \quad A_n(s) = \frac{2a}{n(1 - s_0/s)} J_n\left(n\left[1 - \frac{s_0}{s}\right]\right), \quad (34)$$

$$Y(t) = \eta(t, x = 0) = \left(\frac{\pi\omega L}{gh}\right)^{1/2} \sum_{n=1}^{\infty} \sqrt{n} A_n(s) \sin\left[n\omega(t - \tau) + \frac{\pi}{4}\right], \quad (35)$$

where A_n is the amplitude of the n-th harmonic, $\omega = k\sqrt{gh}$ is the frequency of the initial sine wave, and τ is the time of wave propagation from $x = -L$ to $x = 0$. It is convenient to normalize water level oscillations at the shoreline against the runup height for the sine wave (23). The expression for the normalized water level

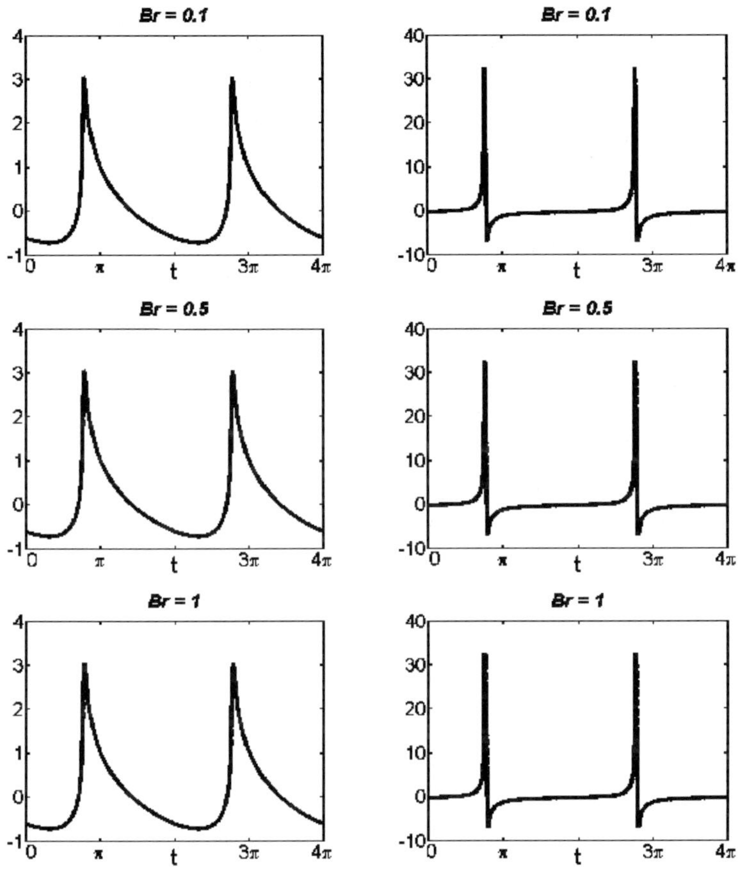

Fig. 11. Water level (left) and velocity (right) of the moving shoreline for various values of the breaking parameter and for a sinusoidal incident wave ($s = 10s_0$). Solid line corresponds to the "nonlinear" and dashed line to the "linear" solution.

oscillations at the shoreline is

$$Y^*(t^*) = \sum_{n=1}^{\infty} \sqrt{n}\frac{A_n}{a} \sin\left(nt^* + \frac{\pi}{4}\right). \tag{36}$$

where $t^* = \omega(t - \tau)$. For convenience the asterisk will be omitted in what follows. The above has shown that the extreme values of $Y(t)$ correspond to the maximal runup and rundown heights in the nonlinear theory, the dimensionless values of which (Fig. 6) are the functions of the wave steepness only. The rundown amplitude depends from the wave steepness weakly (no more then 30%), and we may use expression (23) to evaluate its approximate value. In the contrary, the runup height fast increases when the wave steepness increases. It tends to infinity for a shock wave that theoretically can be described by the model in question where the waves are assumed to be non-breaking. In realistic conditions, of course, the wave breaking will bound the runup height. The maximum runup height can be roughly approximated by (in dimension variables)

$$R_{\max} = 2\pi a \sqrt{\frac{2Ls}{\lambda_0 s_0}}. \tag{37}$$

Expression (37) shows that the wave steepness is the most significant parameter of the runup process. Further, expression (37) confirms that from all the waves of a fixed height and length from the class of waves in question, the wave with steepest front penetrates inland to the largest distance, and that all asymmetrical waves with the front steeper than the back create a larger flooding then a wave with a symmetrical shape. Many examples of extremely large penetration of tsunami waves to inland (including observations during the 2004 Indonesian tsunami) can be interpreted as the confirmation of the important role of the wave steepness.

Similar analysis can be performed for extreme properties of the shoreline velocity. The dimensionless expression for the "linear" velocity (normalized against the runup velocity for sine wave) is

$$U(t) = \sum_{n=1}^{\infty} n^{3/2} \frac{A_n}{a} \sin\left(nt + \frac{3\pi}{4}\right). \tag{38}$$

The extreme values of this function correspond to the "nonlinear" maximal values of runup and rundown velocity of the moving shoreline. The runup velocities exceed the rundown velocity (Fig. 7). The runup velocity can be approximated (in dimensional variables) by

$$U_{\max} = (2\pi)^2 a \sqrt{\frac{g}{h}} \left(\frac{Ls}{\lambda_0 s_0}\right)^{3/2}. \tag{39}$$

The linear theory of long wave runup allows also to estimate the parameters of the wave breaking that occurs when

$$Br = Br_{\sin} \max\left(\frac{d^2 Y}{dt^2}\right) = Br_{\sin} \max\left(\frac{dU}{dt}\right) = 1, \tag{40}$$

where $Br_{\sin}$ is the breaking parameter for sine wave (25), and all derivatives in (40) are calculated using dimensionless expressions (36) and (38). The curve defined by Eq. (40) on plane ($Br_{\sin}$, s/s_0) separates the surging and plunging scenarios of wave runup (Fig. 8). The breaking scenario is, as expected, more typical for asymmetric waves with a relatively steep front. If the wave is small ($Br << 1$), the position of the shoreline varies almost sinusoidally. An increase of the wave amplitude (equivalently, an increase of Br) in the nonlinear case is accompanied by forming of a region where the velocity changes very fast (equivalently, its graphical representation has a very steep front). The water surface displacement record tends to behave as a parabolic function; meanwhile the corresponding "linear" characteristics of the wave are sine functions. The first breaking appears at the stage of maximal rundown. The runup of the sine wave is described in (2), (14) and here reproduced for illustration.

The runup of an asymmetric wave is greatly different from the runup of the sine wave. The relevant results for the case $s = 2s_0$ are presented in Fig. 10. The temporal behaviour of both the position of the shoreline and the time record of velocity are asymmetric even when the wave amplitude is small. The runup amplitude and velocity of the shoreline displacement is higher than the rundown amplitude and the relevant velocity. The breaking point is located closer to the unperturbed shoreline than in the case of the sine wave runup .

5 Conclusion

The principal result of this study is the strong influence of the wave steepness on the runup characteristics of the long waves in the framework of the analytical theory of the nonlinear shallow-water waves. Among waves of a fixed amplitude and frequency (length), the steepest wave penetrates to inland to the largest distance and with largest speed. Consequently, the least dangerous is the symmetric sine wave.

This research is supported particularly by grants from INTAS (03-51-4286) and RFBR (05-05-64265) for ID and EP; University of Antilles and Guyane for EP and NZ, Marie Curie network SEAMOCS (MRTN-CT-2005-019374) for ID, and ESF grant 5762 for TS.

References

[1] Caputo, J-G, Stepanyants YA (2003) Bore formation, evolution and disintegration into solitons in shallow inhomogeneous channels. Nonlinear Processes in Geophysics 10: 407–424

[2] Carrier, GF, Greenspan HP (1958) Water waves of finite amplitude on a sloping beach. J. Fluid Mech. 4: 97 – 109

[3] Carrier, GF, Wu, TT, Yeh H (2003) Tsunami run-up and draw-down on a plane beach. J. Fluid Mech. 475: 79–99

[4] Didenkulova, II, Kharif Ch (20050) Runup of biharmonic long waves on a beach. Izvestiya, Russian Academy of Engineering Sciences 14: 91–97

[5] Didenkulova, I, Zahibo, N, Kurkin, A, Pelinovsky E (2006) Steepness and spectrum of nonlinear deformed shallow waves. Izvestiya, Atmospheric and Oceanic Physics (accepted)

[6] Kaistrenko, VM, Mazova, RKh, Pelinovsky, EN, Simonov KV (1991) Analytical theory for tsunami run up on a smooth slope. Int. J. Tsunami Soc. 9: 115 – 127

[7] Kanoglu U (2004) Nonlinear evolution and runup-rundown of long waves over a sloping beach. J. Fluid Mech. 513: 363–372

[8] Kapinski J (2006) On modeling of long waves in the Lagrangian and Eulerian description. Coastal Engineering, in press, available online through Science Direct.

[9] Li, Y, Raichlen F (2001) Solitary wave runup on plane slopes. J. Waterway, Port, Coastal and Ocean Engineering 127: 33–44.

[10] Mazova, RKh, Osipenko, NN, Pelinovsky EN (1991) Solitary wave climbing a beach without breaking. Rozprawy Hydrotechniczne 54: 71 – 80

[11] Murty T (1977) Seismic Sea Waves - Tsunamis. Canada.

[12] Pedersen G, Gjevik B (1983) Runup of solitary waves. J. Fluid Mech. 142: 283-299

[13] Pelinovsky EN (1982) Nonlinear Dynamics of Tsunami Waves. Applied Physics Institute Press, Gorky.
[14] Pelinovsky, E, Mazova R (1992) Exact analytical solutions of nonlinear problems of tsunami wave run-up on slopes with different profiles. Natural Hazards 6: 227 – 249
[15] Pelinovsky, EN, Troshina EN (1994) Propagation of long waves in straits. Phys. Oceanography 5: 43 – 48
[16] Spielfogel LO (1976) Run-up of single waves on a sloping beach. J. Fluid Mech. 74,: 685–694
[17] Stoker JJ (1957) Water waves. Willey Inter Science, NY.
[18] Synolakis CE (1987) The runup of solitary waves. J. Fluid Mech. 185: 523–545
[19] Synolakis CE (1991) Tsunami runup on steep slopes: How good linear theory really is. Natural Hazards 4: 221 – 234
[20] Tadepalli, S, Synolakis CE (1994) The Runup of N-waves. Proc. Royal Society London A445: 99–112.
[21] Tadepalli, S, Synolakis CE (1996) Model for the leading waves of tsunamis. Physical Review Letters 77: 2141–2145
[22] Tan WY (1992) Shallow water hydrodynamics. Elsevier, N.Y.
[23] Tinti, S, Tonini R (2005) Analytical evolution of tsunamis induced by near-shore earthquakes on a constant-slope ocean. J. Fluid Mech. 535: 33–64.
[24] Tsuji, Y, Yanuma, T, Murata, I, Fujiwara C (1991) Tsunami ascending in rivers as an undular bore. Natural Hazards 4: 257–266
[25] Zahibo, N, Pelinovsky, E, Talipova, T, Kozelkov, A, Kurkin A (2006) Analytical and numerical study of nonlinear effects at tsunami modelling. Applied Mathematics and Computation 174: 795–809
[26] Voltsinger, NE, Klevanny, KA, Pelinovsky EN (1989) Long wave dynamics of the coastal zone, Hydrometeoisdat, Leningrad.
[27] Whitham GB (1974) Linear and Nonlinear Waves. Wiley, N.Y.
[28] Wu, YH, Tian J-W (2000) Mathematical analysis of long-wave breaking on open channels with bottom friction. Ocean Engineering 26: 187–201.

Tsunami Runup in Lagrangian Description

Koji Fujima

Dept. of Civil and Environmental Eng., National Defense Academy. 1-10-20 Hashirimizu, Yokosuka, 239-8686 Japan. fujima@nda.ac.jp

1 Lagrangian Point of View

In general analyses of fluid dynamics, we fix attention on a particular point of space. On the other hand, in the Lagrangian point of view, we fix attention on a particular fluid particle and follow its motion. Independent variables are the initial position of the particle, (a, b, c), and the time, t. The position of the particle, $(\hat{x}, \hat{y}, \hat{z})$, and the pressure, $\hat{p}$, are functions of (a, b, c, t).

The Lagrangian description is not suitable for analyzing violent fluid motion with large deformation, because $(\hat{x}, \hat{y}, \hat{z})$ becomes a discontinuous function if the wave breaks. Further, it is not suitable for analyzing viscous fluid motion, because the viscous terms become complicated in the Lagrangian coordinates. However, it has been applied to research on long-wave runup in some cases ((1) to (6)). This is because the Lagrangian coordinates have the following advantages for analyzing wave runup:

- It is easy to follow up the wave front. If the shoreline is set at $a = 0$ and wave breaking is negligible, $\hat{x}(0, b, c, t)$ always expresses the position of the wave front.
- Boundary conditions at the bottom and free surface are satisfied at the wave front automatically.
- A nonlinear effect is included in the analysis to some extent even if the equations used are linear.

In this paper, a nonlinear shallow-water wave equation is derived in the Lagrangian description first, and the characteristics of a long wave are discussed using the Lagrangian theories.

2 Derivation of Nonlinear Shallow-water Wave Equation

2.1 Basic Equations

Let us consider a fluid particle with infinitesimal volume dV at time t. The position of the particle in cartesian coordinates is denoted as $(\hat{x}, \hat{y}, \hat{z})$, therefore we have

$dV = d\hat{x}d\hat{y}d\hat{z}$. The initial position and initial volume of the particle are written as (a, b, c) and dV_0, respectively, therefore $dV_0 = dadbdc$ and

$$d\hat{x}d\hat{y}d\hat{z} = Jdadbdc\,, \quad J = \frac{\partial(\hat{x}, \hat{y}, \hat{z})}{\partial(a, b, c)}\,, \tag{1}$$

where J is the Jacobian of the coordinates.

Because the equation of continuity implies $dV = dV_0$ for incompressible fluid, the following equation is obtained as the continuity equation in the Lagrangian description:

$$\frac{\partial(\hat{x}, \hat{y}, \hat{z})}{\partial(a, b, c)} = 1\,. \tag{2}$$

The equation of motion for inviscid incompressible fluid is given by

$$\begin{pmatrix} \hat{x}_a & \hat{y}_a & \hat{z}_a \\ \hat{x}_b & \hat{y}_b & \hat{z}_b \\ \hat{x}_c & \hat{y}_c & \hat{z}_c \end{pmatrix} \begin{pmatrix} \hat{x}_{tt} \\ \hat{y}_{tt} \\ \hat{z}_{tt} + g \end{pmatrix} + \frac{1}{\rho} \begin{pmatrix} \hat{p}_a \\ \hat{p}_b \\ \hat{p}_c \end{pmatrix} = 0\,, \tag{3}$$

where $\hat{x}$ and $\hat{y}$ are horizontal axes and the $\hat{z}$-axis is taken vertically upwards, the function with subscript of a, b, c and t denotes the derivative with respect to a, b, c and t, respectively, g is the acceleration due to gravity, and ρ the density. The above equation is transformed to

$$\frac{\partial^2 \hat{x}}{\partial t^2} + \frac{1}{\rho}\frac{\partial(\hat{p}, \hat{y}, \hat{z})}{\partial(a, b, c)} = 0\,, \tag{4}$$

$$\frac{\partial^2 \hat{y}}{\partial t^2} + \frac{1}{\rho}\frac{\partial(\hat{x}, \hat{p}, \hat{z})}{\partial(a, b, c)} = 0\,, \tag{5}$$

$$\frac{\partial^2 \hat{z}}{\partial t^2} + g + \frac{1}{\rho}\frac{\partial(\hat{x}, \hat{y}, \hat{p})}{\partial(a, b, c)} = 0\,. \tag{6}$$

Equation (3) and Eqs.(4)–(6) are equivalent; however the latter are convenient in the present discussion.

In addition to mass and momentum conservations, irrotation of the fluid element is assumed in the vertical plane, because the friction force is ignored. Figure 1 is a sketch of the fluid element in the $x - z$ plane.

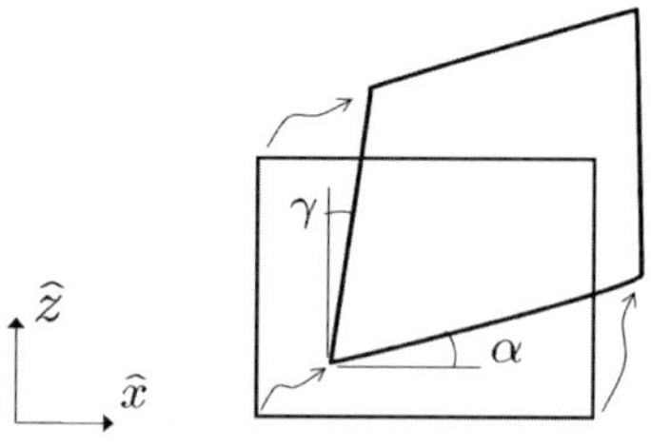

Fig. 1. Deformation of fluid element

In the figure, α and γ are evaluated as

$$\tan\alpha = \frac{\hat{z}_a}{\hat{x}_a}\,, \quad \tan\gamma = \frac{\hat{x}_c}{\hat{z}_c}\,,$$

and the fluid element is irrotational if $\tan\alpha = \tan\gamma$. Thus, the condition of irrotation of the fluid element is obtained by

$$\frac{\partial\hat{x}}{\partial a}\frac{\partial\hat{x}}{\partial c} = \frac{\partial\hat{z}}{\partial a}\frac{\partial\hat{z}}{\partial c}\,, \tag{7}$$

in the $x - z$ plane. Similarly, the irrotational condition in the $y - z$ plane yields

$$\frac{\partial\hat{y}}{\partial b}\frac{\partial\hat{y}}{\partial c} = \frac{\partial\hat{z}}{\partial b}\frac{\partial\hat{z}}{\partial c}\,. \tag{8}$$

The boundary conditions become simple in the Lagrangian description. The bottom boundary condition has the physical meaning that a fluid particle at the bottom surface should move along the bottom. This implies that a particle satisfying $\hat{z} = -h$ keeps satisfying $\hat{z} = -h$, where $-h(\hat{x}, \hat{y})$ is the shape of the bottom[1]. It is convenient for tsunami application that h is regarded to vary with time. Thus, the boundary condition at the bottom is described as

$$\hat{z} = -h(\hat{x}, \hat{y}, t) \text{ on } c = -h(a, b, t_0)\,. \tag{9}$$

Note that $-h(a, b, t_0)$ denotes the initial bottom shape[2].

The kinematic condition at the water surface suggests that a particle at the water surface keeps its position at the water surface. This implies that the water surface elevation is given by the $\hat{z}$-value of the particle which occupied the water surface at the initial time. Similarly, the dynamic condition simply implies that the pressure $\hat{p}$ is zero at the same position. If the initial water surface lies $c = 0$ and the water surface elevation is denoted by η, the kinematic and dynamic conditions at the water surface are written as follows:

$$\hat{z} = \eta \text{ on } c = 0\,, \tag{10}$$

$$\hat{p} = 0 \text{ on } \hat{z} = \eta \text{ (i.e. } c = 0)\,. \tag{11}$$

2.2 Long-wave Assumption and Scaling

Tsunamis have a long wavelength in the order of hundreds of kilometers, although the water depth is only in the order of about one kilometer. Thus, it is realistic that the horizontal length scale is assumed to be greater than the vertical length scale. This is called 'long-wave assumption'.

We introduce the scaling parameters $\mathcal{L}$ and $\mathcal{D}$, which are the horizontal and vertical length scales, respectively. The nondimensional forms of variables are introduced as follows:

$$A = \frac{a}{\mathcal{L}}\,, \qquad B = \frac{b}{\mathcal{L}}\,, \qquad C = \frac{c}{\mathcal{D}}\,, \qquad T = \frac{\sqrt{g\mathcal{D}}}{\mathcal{L}}t\,,$$

$$\hat{X} = \frac{\hat{x}}{\mathcal{L}}\,, \qquad \hat{Y} = \frac{\hat{y}}{\mathcal{L}}\,, \qquad \hat{Z} = \frac{\hat{z}}{\mathcal{D}}\,, \qquad \hat{P} = \frac{\hat{p}}{\rho g\mathcal{D}}\,.$$

---

[1] It is natural that h is defined as a function of cartesian coordinates.

[2] The 'initial' position (a, b, c) does not have to be the position at $t = 0$. For some cases, a static state before generating the wave is convenient for the 'initial' condition.

Nondimensional equations are obtained by substituting the above equations to the basic equations (2), (4)–(6) and (7)–(8). Taking $\kappa = (\mathcal{D}/\mathcal{L})^2$, nondimensional forms of governing equations are as follows:

$$\frac{\partial(\hat{X},\hat{Y},\hat{Z})}{\partial(A,B,C)} = 1\,, \tag{12}$$

$$\frac{\partial^2 \hat{X}}{\partial T^2} + \frac{\partial(\hat{P},\hat{Y},\hat{Z})}{\partial(A,B,C)} = 0\,, \tag{13}$$

$$\frac{\partial^2 \hat{Y}}{\partial T^2} + \frac{\partial(\hat{X},\hat{P},\hat{Z})}{\partial(A,B,C)} = 0\,, \tag{14}$$

$$\kappa\frac{\partial^2 \hat{Z}}{\partial T^2} + 1 + \frac{\partial(\hat{X},\hat{Y},\hat{P})}{\partial(A,B,C)} = 0\,, \tag{15}$$

$$\frac{\partial \hat{X}}{\partial A}\frac{\partial \hat{X}}{\partial C} = \kappa\frac{\partial \hat{Z}}{\partial A}\frac{\partial \hat{Z}}{\partial C}\,, \tag{16}$$

$$\frac{\partial \hat{Y}}{\partial B}\frac{\partial \hat{Y}}{\partial C} = \kappa\frac{\partial \hat{Z}}{\partial B}\frac{\partial \hat{Z}}{\partial C}\,. \tag{17}$$

Because we can expect $\kappa \ll 1$ through long wave assumption, the first term of Eq.(15) and the right hand side of Eqs.(16)–(17) are negligible. Thus, the equation of motion in the z-direction is approximated as

$$1 + \frac{\partial(\hat{X},\hat{Y},\hat{P})}{\partial(A,B,C)} = 0 \quad \text{or} \quad g + \frac{1}{\rho}\frac{\partial(\hat{x},\hat{y},\hat{p})}{\partial(a,b,c)} = 0\,. \tag{18}$$

Equation (16) becomes $\hat{X}_A\hat{X}_C = 0$; however, $\hat{X}_A = 0$ is physically unreasonable, because this implies that fluid particles which occupied a different point at the initial time occupy the same $\hat{X}$ at the next moment. Thus, we have $\hat{X}_C = 0$. After similar discussion in Eq.(17), the conditions of irrotation of a fluid element are written by

$$\frac{\partial \hat{X}}{\partial C} = 0 \quad \text{or} \quad \frac{\partial \hat{x}}{\partial c} = 0\,, \tag{19}$$

$$\frac{\partial \hat{Y}}{\partial C} = 0 \quad \text{or} \quad \frac{\partial \hat{y}}{\partial c} = 0\,. \tag{20}$$

The above equations suggest that the velocity distribution is uniform vertically.

2.3 Equation of Continuity

From Eq.(2) and Eqs.(19)–(20), the following relation is obtained:

$$\frac{\partial(\hat{x},\hat{y},\hat{z})}{\partial(a,b,c)} = \begin{vmatrix} \hat{x}_a & \hat{y}_a & \hat{z}_a \\ \hat{x}_b & \hat{y}_b & \hat{z}_b \\ \hat{x}_c & \hat{y}_c & \hat{z}_c \end{vmatrix} = \begin{vmatrix} \hat{x}_a & \hat{y}_a & \hat{z}_a \\ \hat{x}_b & \hat{y}_b & \hat{z}_b \\ 0 & 0 & \hat{z}_c \end{vmatrix} = \hat{z}_c\frac{\partial(\hat{x},\hat{y})}{\partial(a,b)} = 1$$

$$\therefore \frac{\partial \hat{z}}{\partial c} = \frac{1}{s}\,, \quad s = \frac{\partial(\hat{x},\hat{y})}{\partial(a,b)}\,. \tag{21}$$

Note that $\hat{x}$ and $\hat{y}$ are not a function of c, thus also $1/s$ is not a function of c. Thus, Eq.(21) can be easily integrated as follows:

$$\hat{z} = \frac{c}{s} + \text{constant}\,.$$

The integral constant of the above equation is determined by the bottom boundary condition, Eq.(9), and we obtain

$$\hat{z} = \frac{c + h(a,b,t_0)}{s} - h(\hat{x},\hat{y},t)\,. \tag{22}$$

From the kinematic boundary condition at the water surface, Eq.(10), the following equation is obtained as the equation of continuity for shallow-water wave theory:

$$\eta = \frac{h(a,b,t_0)}{s} - h(\hat{x},\hat{y},t)\,. \tag{23}$$

The equation of continuity, Eq.(23), has a clear physical interpretation based on mass conservation. Because of $\hat{x}_c = \hat{y}_c = 0$, a vertical water pile does not lean, but keeps vertical. If the above equation is written as $(h(\hat{x},\hat{y},t) + \eta)s\Delta a\Delta b = h(a,b,t_0)\Delta a\Delta b$, the right hand side implies the initial volume of the water pile, and the left hand side the volume of the water pile at the time t.

2.4 Equation of Motion

By a similar procedure to the derivation of Eq.(21), the following equation is obtained from Eq.(18):

$$\frac{\partial(\hat{x},\hat{y},\hat{p}/\rho g)}{\partial(a,b,c)} = \begin{vmatrix} \hat{x}_a & \hat{y}_a & \hat{p}_a/\rho g \\ \hat{x}_b & \hat{y}_b & \hat{p}_b/\rho g \\ \hat{x}_c & \hat{y}_c & \hat{p}_c/\rho g \end{vmatrix} = \begin{vmatrix} \hat{x}_a & \hat{y}_a & \hat{p}_a/\rho g \\ \hat{x}_b & \hat{y}_b & \hat{p}_b/\rho g \\ 0 & 0 & \hat{p}_c/\rho g \end{vmatrix} = \frac{\hat{p}_c}{\rho g}\frac{\partial(\hat{x},\hat{y})}{\partial(a,b)} = -1$$

$$\therefore \frac{\partial}{\partial c}\left(\frac{\hat{p}}{\rho g}\right) = -\frac{1}{s}\,. \tag{24}$$

From Eqs.(21) and (24), we have $\frac{\partial}{\partial c}(\hat{z} + \frac{\hat{p}}{\rho g}) = 0$. Thus, $(\hat{z} + \frac{\hat{p}}{\rho g})$ is a function of a, b and t. By considering the dynamic boundary condition at the water surface, Eq.(11), we obtain

$$\hat{z} + \frac{\hat{p}}{\rho g} = \eta\,, \tag{25}$$

which suggests that the pressure is hydrostatic.

In addition, through Eqs.(10) and (24), the following relation is obtained:

$$\frac{\hat{p}}{\rho g} = -\frac{c}{s}\,. \tag{26}$$

Through Eq.(20), (21), (24) and (25), the second term of Eq.(4) is transformed as follows:

$$\begin{aligned}
\frac{\partial(\hat{p}/\rho,\hat{y},\hat{z})}{\partial(a,b,c)} &= \frac{\hat{p}_c}{\rho}\frac{\partial(\hat{y},\hat{z})}{\partial(a,b)} - \hat{y}_c\frac{\partial(\hat{p}/\rho,\hat{z})}{\partial(a,b)} + \hat{z}_c\frac{\partial(\hat{p}/\rho,\hat{y})}{\partial(a,b)} \\
&= -\frac{g}{s}\frac{\partial(\hat{y},\hat{z})}{\partial(a,b)} + \frac{1}{s}\frac{\partial(\hat{p}/\rho,\hat{y})}{\partial(a,b)} \\
&= \frac{g}{s}\left\{\frac{\partial(\hat{z},\hat{y})}{\partial(a,b)} + \frac{\partial(\hat{p}/\rho g,\hat{y})}{\partial(a,b)}\right\} \\
&= \frac{g}{s}\frac{\partial(\eta,\hat{y})}{\partial(a,b)}\,.
\end{aligned}$$

Finally, the equation of motion in the x-direction is obtained as follows:

$$\frac{\partial^2 \hat{x}}{\partial t^2} + \frac{g}{s}\frac{\partial(\eta, \hat{y})}{\partial(a, b)} = 0\,. \tag{27}$$

Similarly, the equation of motion in the y-direction is obtained as follows:

$$\begin{aligned}
\frac{\partial(\hat{x}, \hat{p}/\rho, \hat{z})}{\partial(a, b, c)} &= \hat{x}_c \frac{\partial(\hat{p}/\rho, \hat{z})}{\partial(a, b)} - \frac{\hat{p}_c}{\rho}\frac{\partial(\hat{x}, \hat{z})}{\partial(a, b)} + \hat{z}_c \frac{\partial(\hat{x}, \hat{p}/\rho)}{\partial(a, b)} \\
&= \frac{g}{s}\frac{\partial(\hat{x}, \hat{z})}{\partial(a, b)} + \frac{1}{s}\frac{\partial(\hat{x}, \hat{p}/\rho)}{\partial(a, b)} \\
&= \frac{g}{s}\left\{\frac{\partial(\hat{x}, \hat{z})}{\partial(a, b)} + \frac{\partial(\hat{x}, \hat{p}/\rho g)}{\partial(a, b)}\right\} \\
&= \frac{g}{s}\frac{\partial(\hat{x}, \eta)}{\partial(a, b)}
\end{aligned}$$

$$\therefore\ \frac{\partial^2 \hat{y}}{\partial t^2} + \frac{g}{s}\frac{\partial(\hat{x}, \eta)}{\partial(a, b)} = 0\,. \tag{28}$$

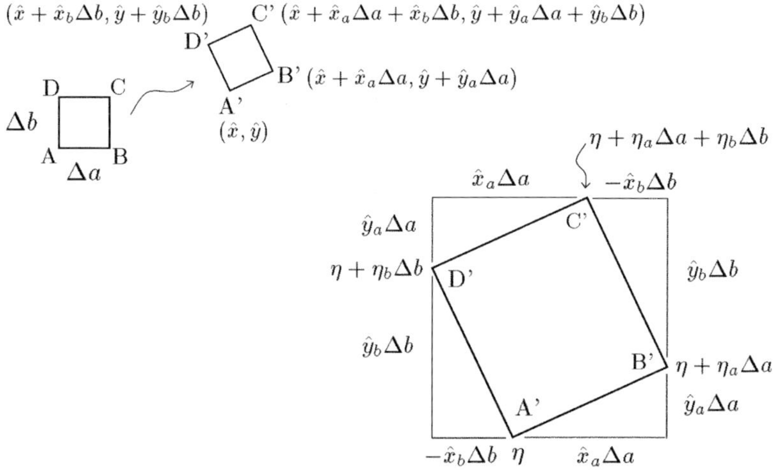

Fig. 2. Deformation of water pile; quadrilateral ABCD and A'B'C'D' shows a plane section of water pile

It is possible that we derive the equations of motion through a simple physical consideration. The mean water elevation along the side A'B' in Fig.2 is $\eta + \eta_a \Delta a/2$. Thus the pressure force acting to the side A'B' (from outside to inside of quadrilateral A'B'C'D') is estimated as $\rho g(\eta + \eta_a \Delta a/2)\sqrt{\hat{x}_a^2 + \hat{x}_b^2}\Delta a$; and the $\hat{x}$- and $\hat{y}$-direction component of pressure force is $\rho g(\eta + \eta_a \Delta a/2)\hat{y}_a \Delta a$ and $\rho g(\eta + \eta_a \Delta a/2)\hat{x}_a \Delta a$, respectively. The pressure force acting on the other sides can be estimated similarly. As result, the total force in the $\hat{x}$- and $\hat{y}$-direction is obtained as $-\frac{\partial(\eta, \hat{y})}{\partial(a,b)}\Delta a \Delta b$ and $-\frac{\partial(\hat{x}, \eta)}{\partial(a,b)}\Delta a \Delta b$, respectively. Because the area of quadrilateral A'B'C'D' is $s\Delta a \Delta b$, we obtain the same equations of motion.

2.5 Separation of Variables

The nonlinear shallow water wave equations are derived in the last section as follows:

$$\frac{\partial^2 \hat{x}}{\partial t^2} + \frac{g}{s}\frac{\partial(\eta, \hat{y})}{\partial(a,b)} = 0\,, \tag{29}$$

$$\frac{\partial^2 \hat{y}}{\partial t^2} + \frac{g}{s}\frac{\partial(\hat{x}, \eta)}{\partial(a,b)} = 0\,, \tag{30}$$

$$\eta = \frac{h(a,b,t_0)}{s} - h(\hat{x}, \hat{y}, t)\,, \tag{31}$$

$$s = \frac{\partial(\hat{x}, \hat{y})}{\partial(a,b)}\,. \tag{32}$$

For the linearization of equations, the following separation of variables is convenient:

$$\hat{x} = a + x\,, \tag{33}$$

$$\hat{y} = b + y\,. \tag{34}$$

Note that x and y denote the displacement of a fluid particle from its initial position. By substituting the above equations to the nonlinear shallow-water wave equations, the following equations are obtained:

$$\frac{\partial^2 x}{\partial t^2} + \frac{g}{s}\left\{\frac{\partial \eta}{\partial a} + \frac{\partial(\eta, y)}{\partial(a,b)}\right\} = 0\,, \tag{35}$$

$$\frac{\partial^2 y}{\partial t^2} + \frac{g}{s}\left\{\frac{\partial \eta}{\partial b} + \frac{\partial(x, \eta)}{\partial(a,b)}\right\} = 0\,, \tag{36}$$

$$\eta - \frac{h(a,b,t_0)}{s} + h(a+x, b+y, t) = 0\,, \tag{37}$$

$$s = 1 + \frac{\partial x}{\partial a} + \frac{\partial y}{\partial b} + \frac{\partial(x,y)}{\partial(a,b)}\,. \tag{38}$$

The above equations of motion are equivalent to those derived by Goto and Shuto (1980); however, the equation of continuity is slightly different. Some nonlinear terms are ignored in their theory, because they assumed that the displacement of the water particle was small.

If we assume that the displacement of the fluid particle is small, the higher order terms with respect to x, y and η are ignored and the linearized shallow-water wave equations are obtained as follows:

$$\frac{\partial^2 x}{\partial t^2} + g\frac{\partial \eta}{\partial a} = 0\,, \tag{39}$$

$$\frac{\partial^2 y}{\partial t^2} + g\frac{\partial \eta}{\partial b} = 0\,, \tag{40}$$

$$\eta - h(a,b,t_0)\left(1 - \frac{\partial x}{\partial a} - \frac{\partial y}{\partial b}\right) + h(a+x, b+y, t) = 0\,. \tag{41}$$

The above set of linear equations is the same as that derived by Shuto (1968).

3 One-dimensional Propagation

3.1 Horizontal Bottom Case

For one-dimensional problems, the governing equations (35), (37) and (38) are written as follows:

$$\frac{\partial^2 x}{\partial t^2} + \frac{g}{s}\frac{\partial \eta}{\partial a} = 0\,, \tag{42}$$

$$\eta - \frac{h(a,t_0)}{s} + h(a+x,t) = 0\,, \tag{43}$$

$$s = 1 + \frac{\partial x}{\partial a}\,. \tag{44}$$

At first, let us consider the case of a horizontal bottom. Because $h =$ constant, we can obtain $\eta_a = -hs_a/s^2$ and $s_a = x_{aa}$ from Eqs.(43) and (44), respectively. Substituting these relation to Eq.(42), we obtain the governing equation of x as folllows:

$$x_{tt} - \frac{gh}{s^3}x_{aa} = 0\,. \tag{45}$$

Similarly, the linear set of governing equations is written as follows:

$$\frac{\partial^2 x}{\partial t^2} + g\frac{\partial \eta}{\partial a} = 0\,, \tag{46}$$

$$\eta - h(a,t_0)\left(1 - \frac{\partial x}{\partial a}\right) + h(a+x,t) = 0\,. \tag{47}$$

The linear governing equation of x yields

$$x_{tt} - ghx_{aa} = 0\,. \tag{48}$$

The exact solution of Eq.(48) is obviously

$$x = Ae^{ik(a \pm c_0 t)}\,,\quad c_0 = \sqrt{gh}\,,$$

where A and k are coefficients. The wave profile is obtained by substituting the above solution into Eq.(47). The following set of equations is the progressive wave solution of the linear theory, where $\sigma = kc_0$:

$$\eta = \frac{H}{2}\cos(ka - \sigma t)\,, \tag{49}$$

$$\hat{x} = a - \frac{H}{2kh}\sin(ka - \sigma t)\,. \tag{50}$$

Figure 3 shows the wave profile estimated by Eqs.(49) and (50) for wave height $H = 6$ m, water depth $h = 20$ m, wavelength $L(= 2\pi/k) = 600$ m, and $t = 0$. The dotted line shown in the figure is the profile of Airy's theory with the form of $\cos(k\hat{x} - \sigma t)$. The Lagrangian theory clearly has a sharp crest, namely, a nonlinear effect is included in the Lagrangian analysis to some extent even if the equations used are linear. This is easily explained by transformation of Eq.(49) to cartesian coordinates as follows:

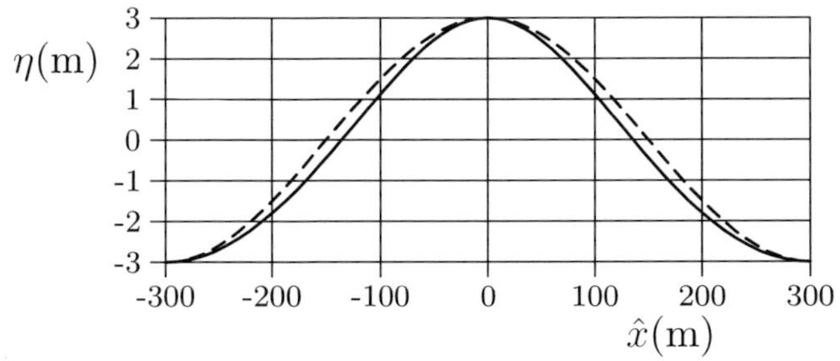

Fig. 3. Wave profile of progressive wave on a horizontal bottom obtained by Lagrangian linear theory ($H = 6$ m, $h = 20$ m, $L = 600$ m); —— : Lagrangian linear theory, - - - : Airy's theory

$$\begin{aligned}\eta &= \frac{H}{2}\cos[k\hat{x} + \frac{H}{2h}\sin(ka - \sigma t) - \sigma t] \\ &= \frac{H}{2}\left\{\cos(k\hat{x} - \sigma t)\cos\left[\frac{H}{2h}\sin(ka - \sigma t)\right]\right. \\ &\qquad \left. - \sin(k\hat{x} - \sigma t)\sin\left[\frac{H}{2h}\sin(ka - \sigma t)\right]\right\} \\ &\simeq \frac{H}{2}\left\{\cos(k\hat{x} - \sigma t) - \frac{H}{2h}\sin^2(k\hat{x} - \sigma t)\right\} \\ &= \frac{H}{2}\left\{\cos(k\hat{x} - \sigma t) + \frac{H}{2h}\frac{\cos 2(k\hat{x} - \sigma t) - 1}{2}\right\} .\end{aligned}$$

The linear solution of the Lagrangian equation essentially includes the half-wavelength component like Stokes wave. This is the reason why a nonlinear effect is partly included in the Lagrangian linear solution. However, the averaged water level does not become zero[3]. This inconvenience is eliminated by using a nonlinear equation. If Eq.(50) is substituted into the nonlinear continuity equation, Eq.(43), the wave profile is estimated as follows:

$$\begin{aligned}\eta &= \frac{h}{1 - \frac{H}{2h}\cos(ka - \sigma t)} - h \\ &\simeq h\left\{1 + \frac{H}{2h}\cos(ka - \sigma t) + \left[\frac{H}{2h}\cos(ka - \sigma t)\right]^2\right\} - h \\ &\simeq \frac{H}{2}\left\{\cos(k\hat{x} - \sigma t) - \frac{H}{2h}\sin^2(k\hat{x} - \sigma t)\right\} + \frac{H^2}{4h}\cos^2(k\hat{x} - \sigma t) \\ &= \frac{H}{2}\left\{\cos(k\hat{x} - \sigma t) + \frac{H}{2h}\cos 2(k\hat{x} - \sigma t)\right\} .\end{aligned}$$

Thus, the nonlinear continuity equation sets the averaged water level to zero. Note that the above equation does not suggest the existence of a sinusoidal stationary

[3] The averaged elevation of a particle at the water surface is zero; although the spatial average of the water level is not zero.

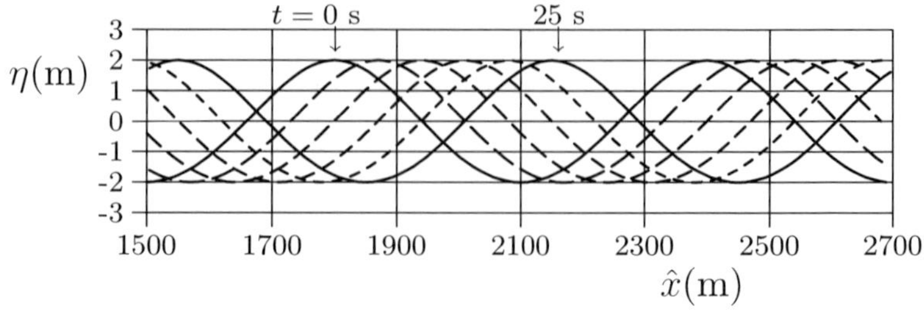

Fig. 4. Profile of sinusoidal wave propagating on a horizontal bottom evaluated by linear theory ($H = 4$ m, $h = 20$ m, $L = 600$ m); $t = 0$, 5, 10, 15, 20 and 25 s

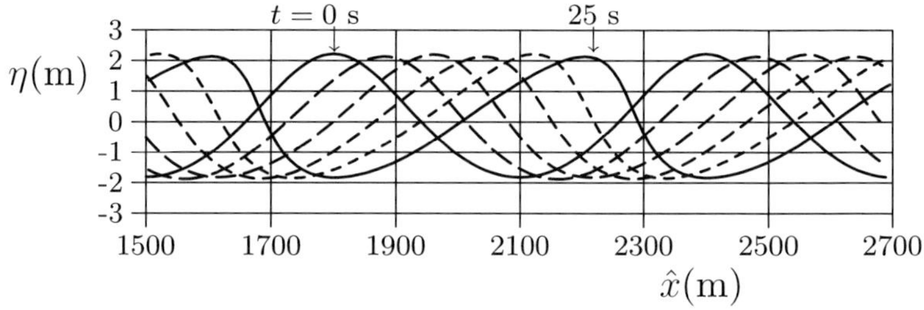

Fig. 5. Profile of sinusoidal wave propagating on a horizontal bottom evaluated by nonlinear theory ($H = 4$ m, $h = 20$ m, $L = 600$ m); $t = 0$, 5, 10, 15, 20 and 25 s

wave solution for a nonlinear shallow water wave on a horizontal bottom. It only provides the momentary wave profile when x is described as Eq.(50).

Through an analogy to the linear equation (48), the wave celerity of the nonlinear equation, Eq.(45), is estimated as follows:

$$c \approx \sqrt{\frac{gh}{s^3}} = \sqrt{gh}(1 + \frac{\eta}{h})^{3/2} \approx \sqrt{gh}(1 + \frac{3}{2}\frac{\eta}{h}).$$

The above relation is the same as that of a nonlinear shallow-water wave in cartesian coordinates. Because the phase speed of the crest is faster than that of other parts, the wave-crest moves forward from the center.

Figures 4 and 5 show comparisons of wave propagation of sinusoidal waves on a horizontal bottom estimated by linear and nonlinear theories. In both cases, the solutions were obtained numerically, and Eq.(50) is used as the initial condition. The linear solution clearly gives the progressive wave without transformation as described by the theoretical solution. However, in the nonlinear solution, the wave-crest propagates faster than other parts, and the wave-trough slower. Therefore, the wave crest moves forward and the wave trough relatively backward, and the water surface gradient of the fore-face becomes steep. In addition, the elevation of both the

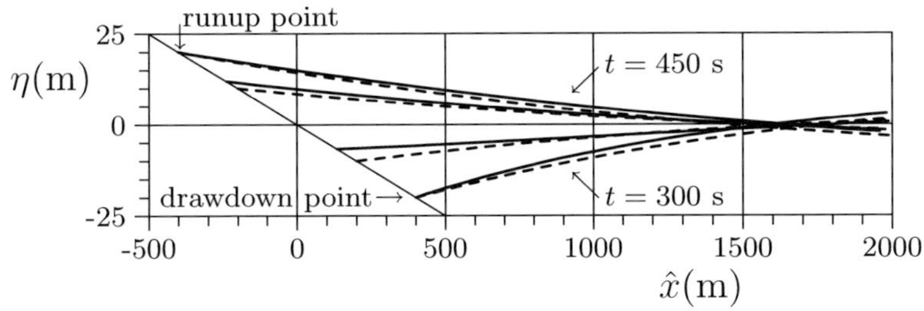

Fig. 6. Profile of standing wave on a sloping beach ($H = 20$ m, $\alpha = 1/20$, $T = 300$ s); $t = 300$, 350, 400, and 450 s; ——: nonlinear, - - -: linear

wave crest and wave trough becomes high in comparison to that of Fig.4, because the averaged elevation of the water surface becomes zero by the nonlinear term.

3.2 Uniform Sloping Bottom Case

In the case of a uniform sloping beach, $h = \alpha\hat{x}$, where α is the bottom slope. Thus, Eq.(43) is reduced to $\eta - \alpha a/s + \alpha(a + x) = 0$ and then $\eta_a = \alpha(1/s - a s_a/s^2 - s)$. Thus, the nonlinear governing equation of x is obtained as

$$x_{tt} + \alpha g(\frac{1}{s^2} - a\frac{s_a}{s^3} - 1) = 0\,. \tag{51}$$

The linearized equation is

$$x_{tt} - \alpha g(2x_a + a x_{aa}) = 0\,. \tag{52}$$

The solution of the above linear equation is given by

$$x = \frac{H}{\sigma}\sqrt{\frac{g}{\alpha a}} Z_1 \left(2\sigma\sqrt{\frac{a}{\alpha g}}\right) e^{i\sigma t}\,,$$

where Z_1 is the Bessel function of the first order. To obtain the progressive wave solution, we should adopt the combination of solution as $J_1(\cdot)\cos(\cdot) - N_1(\cdot)\sin(\cdot)$ that diverges at $a = 0$. The standing wave solution is obtained by $J_1(\cdot)\cos(\cdot)$. Shuto (1967) had discussed these solutions in detail.

The solution of η is obtained by substituting x into the linear continuity equation, Eq.(47). The combination of x and η of the standing wave solution is written as follows:

$$\eta = -H J_0 \left(2\sigma\sqrt{\frac{a}{\alpha g}}\right) \cos(\sigma t)\,, \tag{53}$$

$$\hat{x} = a + \frac{H}{\sigma}\sqrt{\frac{g}{\alpha a}} J_1 \left(2\sigma\sqrt{\frac{a}{\alpha g}}\right) \cos(\sigma t)\,. \tag{54}$$

Figure 6 shows a comparison of the linear theoretical solution and nonlinear numerical solution. To evaluate the numerical solution, Eq.(54) is used as the initial condition. The initial time of simulation is $t = 0$, and the wave profile at $t = T$ to

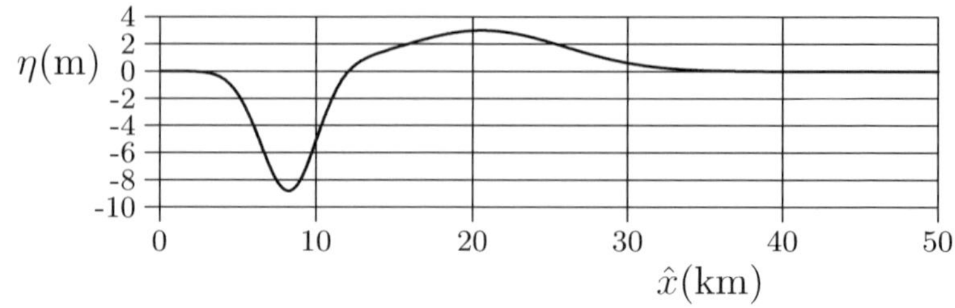

Fig. 7. Initial wave profile (Case d of reference (7))

$3T/2$ is drawn in the figure where $T(= 2\pi/\sigma) = 300$ s. Regarding the runup and drawdown heights, the linear and nonlinear solutions provide similar results. However, the runup process is slightly different. The wave front reaches the drawdown point at $t = 300$ s, and the wave front velocity is accelerated after $t = 300$ s. In the linear solution, the wave front velocity becomes the maximum when the wave front crosses $\hat{x} = 0$. In the nonlinear solution, the wave front velocity accelerates rapidly just after $t = 300$ s, and the maximum runup velocity appeares slightly before $\hat{x} = 0$. Thus, the wave front evaluated by the nonlinear theory goes forward in comparison to that evaluated by linear theory at $t = 350$ s.

Carrier, Wu and Yeh (2003) obtained solutions for the 1-D nonlinear shallow water wave equations on a uniform sloping beach, as an initial value problem. They used the distorted coordinates to transform the equations, although their basic equations are the nonlinear shallow water wave theory in the cartesian coordinates. It may be interesting to compare the Lagrangian solution with their solution. Figure 7 shows the initial wave profile of the test, Case d of Carrier et al. ((7)). The bottom slope α is 1/10. To evaluate the Lagrangian numerical solution, $x = x_t = 0$ at $t = 0$ is used as the initial condition, and the initial bottom shape is assumed $h = \alpha\hat{x} - \eta_0 (= \alpha a - \eta_0)$, where η_0 is the initial wave profile drawn in Fig.7. The bottom is deformed linearly into $h = \alpha\hat{x}$ from $t = 0$ to $t = 1$ s. At $t = 1$ s, the bottom becomes a uniform slope. In the exact meaning, the wave condition used in the present simulation is not the same as that used by Carrier et al. However, the water surface elevation at $t = 1$ s is almost equivalent to η_0.

Figure 8 shows the trajectory of shoreline, where $x_f = x(0, t)$, and Fig.9 the wave profiles. It is difficult for the Lagrangian nonlinear simulation to provide a stable solution in this case, because the wave condition might be close to the limitation of wave breaking and thus the nonlinear term $1/s$ often causes instability. Thus, it is possible that the numerical error is not negligible in the Lagrangian nonlinear simulation. However, the nonlinear simulation result agrees well with the result of Carrier et al. Regarding the runup and drawdown heights, the linear theory also reproduces the result of Carrier et al. satisfactorily. However, we can find a difference between the linear and nonlinear results in the runup process. The wave front reproduced by the nonlinear theory is accelerated rapidly after departing the drawdown point in comparison to that reproduced by the linear theory. This characteristic was already found in the standing wave analysis drawn in Fig.6.

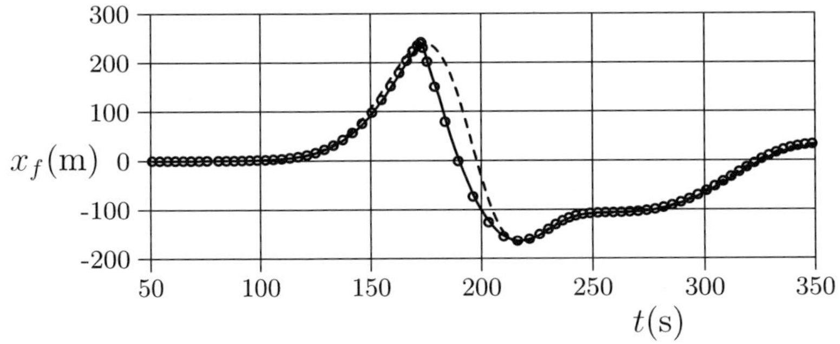

Fig. 8. Trajectory of shoreline; ○: Carrier et al., ——: nonlinear, - - -: linear

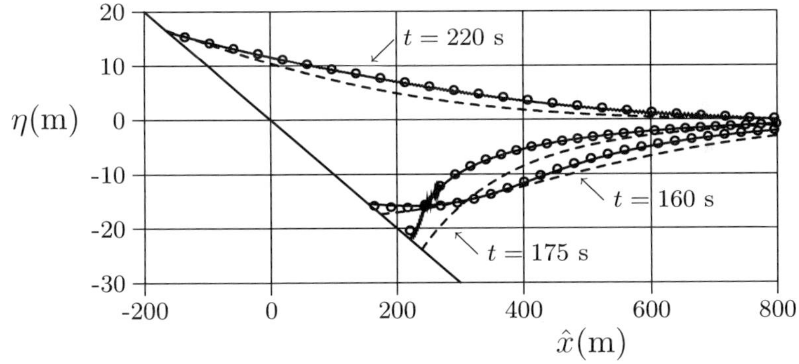

Fig. 9. Wave profiles at $t = 160$, 175 and 220 s; ○: Carrier et al., ——: nonlinear, - - -: linear

4 Two-dimenaional Propagation

Let us consider a two-dimensional linear theory for the case of a uniform plane beach, $h = \alpha\hat{x}$. In this case, the linear continuity equation, Eq.(41) is transformed as:

$$\eta + \alpha\left[(ax)_a + (ay)_b\right] = 0\,.$$

By substituting Eqs.(39) and (40) into the above equation, the following governing equation is obtained:

$$\eta_{tt} - \alpha g\left[(a\eta_a)_a + (a\eta_b)_b\right] = 0\,. \tag{55}$$

If we suppose the solution with the form of $\eta = f(a)e^{i(kb-\sigma t)}$, we obtain

$$f(a) = e^{-|k|a}L_n(2|k|a)\,, \tag{56}$$

$$\sigma = \pm\sigma_n\,,\quad \sigma_n = \sqrt{\alpha g|k|(2n+1)}\quad (n = 0, 1, 2, \cdots)\,, \tag{57}$$

where L_n denotes the Laguerre function, and Eq.(57) provides the dispersion relation. The above linear solution of an edge wave on a uniform sloping beach in Lagrangian coordinates was derived by Shuto(1968).

The combination of η, x and y is as follows:

$$\eta = \frac{H}{2} e^{-|k|a} L_n(2|k|a) e^{i(kb-\sigma t)}, \tag{58}$$

$$x = \frac{gH}{2\sigma^2} \frac{\partial}{\partial a} \left[e^{-|k|a} L_n(2|k|a) \right] e^{i(kb-\sigma t)}, \tag{59}$$

$$y = \frac{igkH}{2\sigma^2} e^{-|k|a} L_n(2|k|a) e^{i(kb-\sigma t)}. \tag{60}$$

At the shore ($a = 0$),

$$\eta(= -\alpha x) = \frac{H}{2} \cos(kb - \sigma t), \tag{61}$$

$$\hat{y} = b - \frac{gkH}{2\sigma^2} \sin(kb - \sigma t). \tag{62}$$

The above relation provides the profile of shoreline along the coast, and it is similar to the progressive wave profile on a horizontal bottom described by Eqs.(49) and (50). Thus, the shoreline profile of this solution has similar characteristics to the profile drawn in Fig.3. Namely, the Lagrangian solution shows a sharp crest in comparison to the solution in the cartesian coordinates. In addition, the averaged elevation of the shoreline is not zero in the Lagrangian linear theory. Yeh (1992) demonstrated that the averaged elevation of shoreline becomes zero by the higher order solution.

Because the fundamental solution has the same form as that in the cartesian coordinates, the solution for the arbitrary initial value problem is obtained by the procedure proposed by Fujima et al. (2000). The initial condition is $\eta = \eta_0(a, b)$ and $\eta_t = 0$ at $t = 0$, where η_0 is the given function. The solution is expressed as

$$\eta = \int_{-\infty}^{\infty} \sum_{n=0}^{\infty} A_{nk} e^{-|k|a} L_n(2|k|a) e^{ikb} (e^{-i\sigma_n t} + e^{i\sigma_n t}) dk, \tag{63}$$

$$x = \int_{-\infty}^{\infty} \sum_{n=0}^{\infty} \frac{g}{\sigma^2} A_{nk} \frac{\partial}{\partial a} \left[e^{-|k|a} L_n(2|k|a) \right] e^{ikb} (e^{-i\sigma_n t} + e^{i\sigma_n t}) dk, \tag{64}$$

$$y = \int_{-\infty}^{\infty} \sum_{n=0}^{\infty} \frac{igk}{\sigma^2} A_{nk} e^{-|k|a} L_n(2|k|a) e^{ikb} (e^{-i\sigma_n t} + e^{i\sigma_n t}) dk, \tag{65}$$

$$A_{nk} = |k| \int_0^{\infty} e^{-|k|a} L_n(2|k|a) c_k(a) da, \tag{66}$$

where c_k is the Fourier transform of η_0 in the b-direction. Namely,

$$\eta_0(a, b) = \int_{-\infty}^{\infty} c_k(a) e^{ikb} dk. \tag{67}$$

Of course the nonlinearity is ignored in this theory, however, linear analysis may reproduce the runup and drawdown heights accurately as discussed in the previous section. This theory is convenient for easily reproducing the edge wave behavior over a long time at a long distance.

Three test cases are conducted; in all cases, the bottom slope α is $1/100$ and the initial profile is given by the following equation:

Table 1. Conditions of source of Cases 1, 2 and 3

	H(m)	W(km)	L(km)	ℓ_a(km)	ℓ_b(km)
Case 1	5	60	100	70	-50
Case 2	5	60	100	0	-50
Case 3	5	100	60	-25	-30

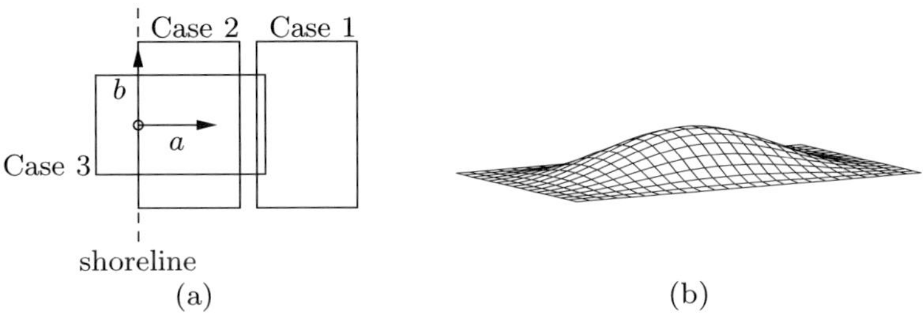

Fig. 10. Sketch of source area; (a) location of source area of Cases 1, 2 and 3, (b) bird's-eye view of initial profile η_0

$$\eta_0 = \begin{cases} H \sin^2 A_1 \sin^2 B_1 & (0 < A_1 < \pi, 0 < B_1 < \pi) \\ 0 & (\text{elsewhere}) \end{cases} , \tag{68}$$

$$A_1 = \pi(a - \ell_a)/W \ , \quad B_1 = \pi(b - \ell_b)/L \ . \tag{69}$$

The parameters are listed in Table 1. The source area (area where η_0 has a non-zero value) is drawn in Fig.10. The source area of Case 1 locates off the shoreline ($a = 0$), and the source area of Case 2 touches the shoreline. In Case 3, the source area is put across the shoreline, and the left quarter of the area is meaningless physically. However, the initial water elevation of the shoreline is not zero, namely, the shoreline is curved landward at $t = 0$.

Figure 11 shows the elevation of wave front in Cases 1, 2 and 3, where $\eta_f = \eta(0, b, t)$. In Fig.11, the top two figures show the results of Case 1, the middle figures the results of Case 2, and the bottom figures Case 3; the left side figures show the water elevation at $b = 0$, and the right side the elevation at $b = 400$ km. The source of Case 1 generates the highest tsunami height at $b = 0$. However, at $b = 400$ km, the sources of Case 2 and 3 generate a higher tsunami than the Case 1 source. This is caused by the generation of edge wave in Cases 2 and 3. In addition, the tsunami duration becomes long in Cases 2 and 3. For a community distant from a tsunami source, an earthquake near the shoreline is troublesome if an edge wave is generated. It will possibly provide a high tsunami height and long tsunami duration.

5 Summary

Characteristics of one- and two-dimensional propagation and runup of a long wave were discussed through the Lagrangian method. The Lagrangian method is not a tool for wide use. However, it has some advantages for analyzing long-wave motion.

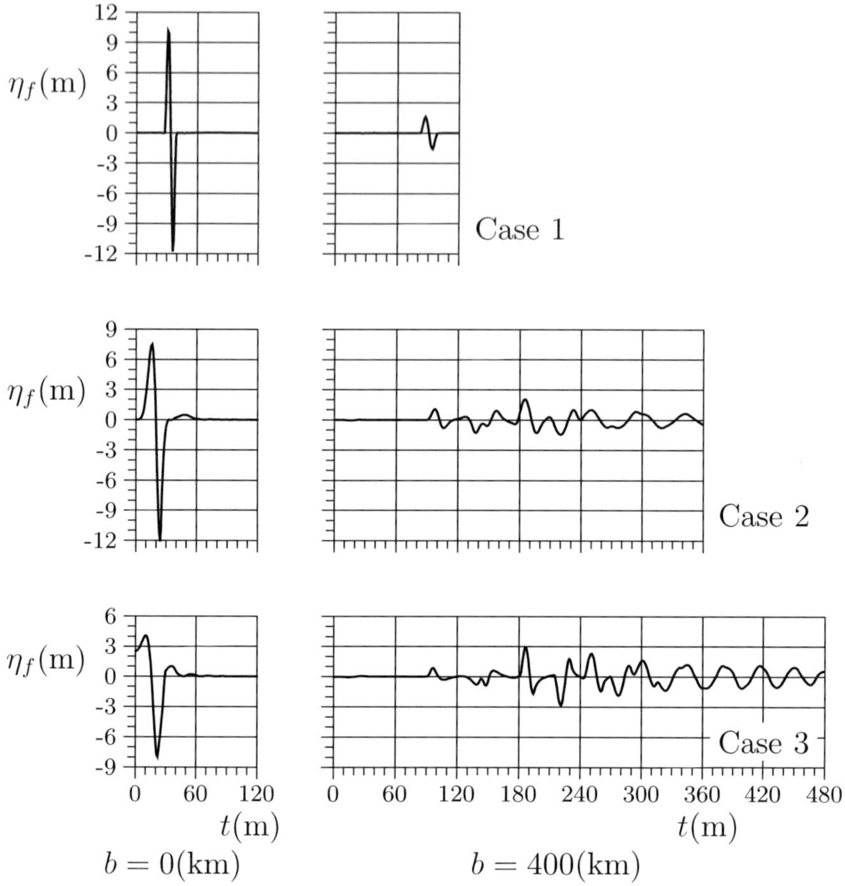

Fig. 11. Elevation of wave front at $b = 0$ and 400 km in Cases 1, 2 and 3

On the one-dimensional propagation on a horizontal bottom, the Lagrangian linear theory has a solution of the same form as a solution in the cartesian coordinates. However, the Lagrangian solution shows a sharp crest because it partly includes a nonlinear effect. In the nonlinear theory, the phase speed of the wave crest is faster than that of the other parts. Thus, the wave crest moves forward and the water surface gradient of the fore-face of the cest becomes steep.

On the one-dimensional runup on a uniform slope, linear and nonlinear theories give similar results for the runup and drawdown heights. A small difference is found in the runup process; in the nonlinear theory, the wave front is accelerated rapidly after departing the drawdown point.

On the two-dimensional propagation on a uniform plane beach, the shoreline profile evaluated by the Lagrangian linear theory has the same tendency as the wave profile of a progressive wave on a horizontal bottom. The procedure to evaluate the wave propagation from an arbitrary initial profile proposed by Fujima et al. (2000)

is adaptable to the Lagrangian theory, which enables estimation of the tsunami behavior over a long time and at a long distance from the source without large numerical simulation.

References

[1] Shuto N (1967) Runup of long waves on a sloping beach. Coastal Engineering in Japan, JSCE, 10:23-38

[2] Shuto N (1968) Three dimensional behavior of long waves on a sloping beach. Coastal Engineering in Japan, JSCE, 11:53-57

[3] Shuto N, Goto C (1978) Numerical simulation of tsunami runup. Coastal Engineering in Japan, JSCE, 21:13-20

[4] Goto C (1979) Nonlinear equation of long waves in the Lagrangian description. Coastal Engineering in Japan, JSCE, 22:1-9

[5] Goto C, Shuto N (1980) Runup of tsunamis by linear and nonlinear theories. Proc. 17th ICCE, ASCE:695-707

[6] Yeh H (1992) Shoreline profile of Stokes-mode edge waves. J. Waterway, Port, Coastal and Ocean Engineering, ASCE, 118(1):112-116

[7] Carrier GF, Wu TT, Yeh H (2003) Tsunami run-up and draw-down on a plane beach. J. Fluid Mech. 475:79-99

[8] Fujima K, Dozono R, Shigemura T (2000) Generation and propagation of tsunami accompanying edge waves on a uniform sloping beach. Coastal Engineering J., JSCE, 42(2):211-236

Analytical and numerical models for tsunami run-up

Per A. Madsen[1,2] and David R. Fuhrman[1,3]

[1] Technical University of Denmark, Mechanical Engineering Department, Nils Koppels Allé, Building 403, DK-2800 Kgs. Lyngby, Denmark
[2] prm@mek.dtu.dk
[3] drf@mek.dtu.dk

Summary. The classical analytical solution for the run-up of periodic long waves on an infinitely long slope is presented and discussed. This leads to simple expressions for the maximum run-up and the associated flow velocity in terms of the surf similarity parameter and the amplitude to depth ratio determined at some offshore location. We use these expressions to analyze the impact of tsunamis on beaches and relate the discussion to the recent Indian Ocean tsunami from December 26, 2004. An important conclusion is that the impact is very sensitive to the beach slope. Next, we present a numerical model based on a highly accurate Boussinesq-type formulation. This model incorporates nonlinear and dispersive effects, and is extended to include a moving shoreline. As a first step, the model is verified against the non-dispersive analytical run-up solution, demonstrating good quantitative accuracy. The model is then used to study an idealized three-dimensional nearshore-generated tsunami propagating over a hypothetical sound.

1 Introduction

This work is split into three parts: In the first part (§2), we present an analytical model for the run-up of tsunamis on beaches. The theoretical description is based on a solution to the linear and nonlinear shallow-water equations and follows previous work by Green (8), Lamb (19) and Carrier & Greenspan (2). The main result is a quantification of the maximum run-up and the associated flow velocity for periodic long waves on a plane beach. These results are given in terms of the surf similarity parameter and the amplitude to depth ratio determined at some offshore location. We use these expressions to discuss the impact of tsunamis on various beaches and relate the discussion to the recent Indian Ocean tsunami from December 26, 2004.

In the second part of this work (§3), we present a numerical model based on a highly accurate Boussinesq-type formulation. This formulation, which was originally derived by Madsen et al. (24; 25) for slowly varying bathymetries, has recently been extended by Madsen et al. (26) to allow rapidly varying bathymetries. We also introduce a moving boundary into the model, which allows for the simulation of run-up on beaches. Although the formulation accurately describes dispersive nonlinear wave motion, we concentrate on the run-up of long waves and compare the computations with the analytical expressions presented in §2.

In the third part of this work (§4), we use the numerical model to study an idealized three-dimensional tsunami, which is generated near the shore by a periodic point source, which then propagates over a hypothetical sound. This is a first and rather crude approximation to a three-dimensional submerged or subarial landslide, as previously studied e.g. by Lynett & Liu (22). The simulation includes dispersive effects, and we observe the formation of shore-trapped, edge wave modes. A linear refraction model is also presented and a comparison made with the numerical results.

2 An analytical model for tsunami run-up

2.1 Governing equations and basic assumptions

Tsunamis usually have relatively long wave periods, and it is often a good approximation to assume hydrostatic pressure and a uniform vertical distribution of the velocity field. This means that under most circumstances tsunamis can be approximated by the nonlinear shallow-water equations,

$$\frac{\partial \eta}{\partial t} + \frac{\partial}{\partial x}\left(U(h+\eta)\right) + \frac{\partial}{\partial y}\left(V(h+\eta)\right) = 0, \tag{1}$$

$$\frac{\partial U}{\partial t} + U\frac{\partial U}{\partial x} + V\frac{\partial U}{\partial y} + g\frac{\partial \eta}{\partial x} = 0, \tag{2}$$

$$\frac{\partial V}{\partial t} + U\frac{\partial V}{\partial x} + V\frac{\partial V}{\partial y} + g\frac{\partial \eta}{\partial y} = 0, \tag{3}$$

where η is the surface elevation, h the still water depth, g the acceleration of gravity, and (U,V) the velocity components in the (horizontal) x- and y-directions, respectively. One exception is the generation of tsunamis from subarial/submerged landslides where dispersive effects are important. Such events have been studied previously e.g. by (10; 9; 11; 35; 21; 22).

Typically, nonlinearity plays an important role, especially when the tsunami approaches coastal areas and starts to run up the beach. In this connection, wave breaking may occur, which effectively turns the tsunami into a moving bore. The details of such a bore are very similar to tidal bores, except that tidal bores cannot occur on beaches, but only in confined funnel-shaped bays. The propagation and run-up of bores has been treated analytically by e.g. Keller et al. (16), Ho & Meyer (13) and Shen & Meyer (31; 32), numerically by e.g. Hibberd & Peregrine (12) and Watson et al. (34), and experimentally by e.g. Yeh et al. (36).

In this section, we shall focus on analytical run-up expressions, which can be obtained by introducing the following additional assumptions: 1) The effects of wave breaking and bottom dissipation are neglected; 2) Full reflection from the shoreline is assumed; 3) Periodic wave motion is assumed i.e. the transient nature of the tsunami is ignored; and 4) The beach slope s is assumed to be constant. We will present and discuss the solutions previously derived by Green (8), Lamb (19), and Carrier & Greenspan (2).

We shall deal with one-dimensional motion only i.e. all waves are assumed perpendicular to the shoreline. Initially, we shall also assume that the linearized shallow-water equations are an adequate description of our problem. Hence, equations (1)-(3) are simplified to

$$\frac{\partial \eta}{\partial t} + \frac{\partial}{\partial x}(hU) = 0, \tag{4}$$

$$\frac{\partial U}{\partial t} + g\frac{\partial \eta}{\partial x} = 0. \tag{5}$$

By eliminating U we then obtain the linear shallow-water wave equation

$$\frac{\partial^2 \eta}{\partial t^2} - g\frac{\partial}{\partial x}\left(h\frac{\partial \eta}{\partial x}\right) = 0. \tag{6}$$

2.2 The linear far-field solution

A first description of waves approaching a sloping beach can be obtained by using the classical WKBJ expansion, assuming a mild beach slope. Green (8) derived this solution in his work on the motion of waves in a variable canal of small depth and width, see also Lamb (19), §185. We consider a slowly varying bathymetry $h(\delta x)$, where δ is an ordering parameter, with the x-axis starting at the shoreline and pointing towards the sea. A linear wave propagating towards the shore can now be described by

$$\eta(x,t) = A(\delta x)\cos\left(\omega t + \int k(\delta x)dx\right), \tag{7}$$

where ω is the cyclic frequency, $k(\delta x)$ is the wave number, and $A(\delta x)$ the wave amplitude, both slowly varying in space. We insert (7) into the wave equation (6) and collect terms in powers of δ. The zeroth-order problem now yields the dispersion relation

$$k = \frac{\omega}{\sqrt{gh}}, \tag{8}$$

while the first-order problem yields the linear shoaling equation

$$\frac{1}{h}\frac{dh}{dx} + \frac{1}{k}\frac{dk}{dx} + \frac{2}{A}\frac{dA}{dx} = 0. \tag{9}$$

We differentiate (8) with respect to x to obtain

$$\frac{1}{k}\frac{dk}{dx} = -\frac{1}{2h}\frac{dh}{dx}.$$

Inserting this result into (9) then yields

$$\frac{1}{A}\frac{dA}{dx} = -\frac{1}{4h}\frac{dh}{dx}.$$

This is the classical shoaling law by Green (8), which can also be expressed as

$$A(x) = a_0\left(\frac{h(x)}{h_0}\right)^{-1/4}, \tag{10}$$

where a_0 denotes the incoming wave amplitude at depth h_0.

Utilizing that the beach slope s is constant i.e. $h(x) = sx$, we obtain

$$k = \frac{\omega}{\sqrt{gsx}} \qquad \Rightarrow \qquad \int k dx = 2\omega\sqrt{\frac{x}{gs}}. \tag{11}$$

Hence, according to (11), (10) and (7) the incoming wave can be expressed by

$$\eta(x,t) = a_0 \left(\frac{sx}{h_0}\right)^{-1/4} \cos\left(\omega t + 2\omega\sqrt{\frac{x}{gs}}\right). \tag{12}$$

Notice that this expression is strictly not valid at the shoreline, as the amplitude goes to infinity for $x \to 0$. Nevertheless, we assume that the incoming wave is fully reflected from the shoreline and add an equivalent outgoing wave, to obtain the standing wave solution

$$\eta(x,t) = 2a_0 \left(\frac{sx}{h_0}\right)^{-1/4} \cos\left(2\omega\sqrt{\frac{x}{gs}} + \varphi\right) \cos\omega t \quad \text{for} \quad x > 0. \tag{13}$$

Here φ denotes an unknown phase shift, which will be determined in the following.

2.3 The linear near-field solution

Lamb (19), §186, presented a linear long wave solution, which is also valid in the vicinity of the shoreline. Again a constant slope is assumed, i.e. $h(x) = sx$, but the previous assumption of a mildly sloping beach is relaxed, and we look for solutions of the form

$$\eta(x,t) = A(x)\cos\omega t, \qquad U(x,t) = B(x)\sin\omega t. \tag{14}$$

By inserting (14) into (6) and (5) we obtain

$$x\frac{d^2A}{dx^2} + \frac{dA}{dx} + \frac{\omega^2}{gs}A = 0, \quad \text{and} \qquad B = -\frac{g}{\omega}\frac{dA}{dx}. \tag{15}$$

Next, we introduce the coordinate transformation

$$\sigma \equiv 2\omega\sqrt{\frac{x}{gs}}, \tag{16}$$

inspired by (12), by which (15) simplifies to

$$\frac{d^2A}{d\sigma^2} + \frac{1}{\sigma}\frac{dA}{d\sigma} + A = 0, \qquad \text{and} \qquad B = -\frac{2\omega}{\sigma s}\frac{dA}{d\sigma}. \tag{17}$$

This is the Bessel equation of the first kind, and the solution to this equation can be expressed by

$$A(\sigma) = RJ_0(\sigma), \quad \text{and} \quad B(\sigma) = \frac{2\omega R}{\sigma s}J_1(\sigma), \tag{18}$$

where J_n is the nth-order Bessel function of the first kind and R is an unknown amplitude. Consequently, the surface elevation and the particle velocity are determined by

$$\eta(x,t) = RJ_0(\sigma)\cos\omega t, \tag{19}$$

$$U(x,t) = \frac{2\omega R}{\sigma s}J_1(\sigma)\sin\omega t. \tag{20}$$

Approaching the shoreline, i.e. for $\sigma \to 0$, we have that $J_0(\sigma) \to 1$ and $J_1(\sigma) \to \sigma/2$ i.e.

$$\eta \to R \cos\omega t \quad \text{for} \quad \sigma \to 0, \qquad\qquad U \to \frac{\omega R}{s} \sin\omega t \quad \text{for} \quad \sigma \to 0. \tag{21}$$

Hence, R defines the surface elevation at the shoreline, which in this linear approach is identical to a vertical measure of the run-up.

2.4 The nonlinear near-field solution

Carrier & Greenspan (2) derived an analytical solution to the nonlinear shallow-water equations by using a number of elegant coordinate transformations. A summary of their method can be found in e.g. Mei et al. (28). First, they transform the independent coordinates from $(x,\, t)$ to the Riemann invariants $(\alpha,\, \beta)$ defined by

$$\alpha \equiv U + 2\Lambda - mt, \qquad\qquad \beta \equiv U - 2\Lambda - mt, \tag{22}$$

where $\Lambda \equiv \sqrt{g(h+\eta)}$ and $m \equiv gs$. Second, they introduce $(\rho,\, \lambda)$ as new independent variables defined by

$$\rho \equiv \alpha - \beta = 4\Lambda, \qquad\qquad \lambda \equiv \alpha + \beta = 2(U - mt). \tag{23}$$

Third, they introduce the dependent variable $\psi(\rho, \lambda)$ defined by

$$\frac{\partial \psi}{\partial \rho} = -\frac{-U}{m}\rho, \qquad\qquad \frac{\partial \psi}{\partial \lambda} = 4\left(x - \frac{U^2}{2m} - \frac{\rho^2}{16m}\right). \tag{24}$$

This simplifies the nonlinear shallow-water equations to the following linear equation

$$\frac{\partial}{\partial \rho}\left(\rho \frac{\partial \psi}{\partial \rho}\right) - \rho \frac{\partial^2 \psi}{\partial \lambda^2} = 0, \tag{25}$$

which provides analytical solutions for various initial or boundary conditions on ψ.

In this notation the moving shoreline is defined as $\rho = 0$. A solution to (25), which is finite for $\rho = 0$ reads

$$\psi(\rho, \lambda) = -\frac{8gR}{\omega} J_0(\gamma\rho)\sin(\gamma\lambda), \qquad\qquad \gamma \equiv \frac{\omega}{2m}, \tag{26}$$

where J_0 is the zeroth-order Bessel function of the first kind and R is an unknown amplitude. The remaining problem is to determine $\eta, U, x,$ and t in terms of $\psi, \rho,$ and λ. From (24) and (23) we obtain

$$U(\rho, \lambda) = -\frac{m}{\rho}\frac{\partial \psi}{\partial \rho}, \tag{27}$$

$$x(\rho, \lambda) = \frac{1}{4}\frac{\partial \psi}{\partial \lambda} + \frac{U^2}{2m} + \frac{\rho^2}{16m}, \tag{28}$$

$$\eta(\rho, \lambda) = s\left(\frac{\rho^2}{16m} - x\right) = -s\left(\frac{1}{4}\frac{\partial \psi}{\partial \lambda} + \frac{U^2}{2m}\right), \tag{29}$$

$$t(\rho, \lambda) = \frac{1}{m}\left(U - \frac{\lambda}{2}\right), \tag{30}$$

which define the nonlinear solution as a set of parametric curves in terms of ρ and λ.

While ψ is a single-valued function of ρ and λ, this is not necessarily the case in terms of x and t. Carrier & Greenspan (2) showed that the Jacobian

$$J = \frac{\partial(x,t)}{\partial(\rho,\lambda)},$$

vanishes for increasing values of R, and they established the breaking criterion

$$R \leq R_{\max} = \frac{gs^2}{\omega^2}, \tag{31}$$

beyond which the nonlinear solution is no longer valid. For $R \to R_{\max}$ the free surface elevation during draw-down will become vertical.

Far away from the shoreline, where wave conditions can be linearized, we obtain

$$\gamma\rho = \frac{\omega}{2m} 4\sqrt{g(h+\eta)} \to 2\omega\sqrt{\frac{x}{gs}} = \sigma,$$
$$\gamma\lambda = \frac{\omega}{2m} 2(U - mt) \to -\omega t,$$

which indicates that ρ represents the spatial coordinate, while λ represents the temporal coordinate. Furthermore, (29) linearizes to

$$\eta \to -s\left(\frac{1}{4}\frac{\partial\psi}{\partial\lambda}\right) = RJ_0\left(2\omega\sqrt{\frac{x}{gs}}\right)\cos\omega t,$$

which is in agreement with (19).

The nonlinear solutions at the shoreline can be obtained by inserting (26) in (27), (28), and (29), and taking the limit for $\rho \to 0$. This leads to

$$\lim_{\rho\to 0} U(\rho,\lambda) = -\frac{\omega R}{s}\sin(\gamma\lambda),$$
$$\lim_{\rho\to 0} x(\rho,\lambda) = -\frac{R}{s}\left(\cos(\gamma\lambda) - \frac{1}{2}\frac{R}{R_{\max}}\sin^2(\gamma\lambda)\right),$$
$$\lim_{\rho\to 0} \eta(\rho,\lambda) = R\left(\cos(\gamma\lambda) - \frac{1}{2}\frac{R}{R_{\max}}\sin^2(\gamma\lambda)\right).$$

As $R \leq R_{\max}$, the maximum run-up and drawn-down elevation will be identical to R, which is formally in agreement with the linear result given in (21). Furthermore, the maximum velocity will be identical to $\omega R/s$, also in agreement with (21). However, in contrast to the linear result, which has a fixed shoreline at $x = 0$, the nonlinear solution has a moving shoreline, where x varies between $-R/s$ and R/s. Because of this, the spatial and temporal variations of η and U will actually deviate significantly from the linear solution close to the shoreline.

2.5 Matching of near-field and far-field solutions

In order to determine the run-up elevation and shoreline velocity of a given incoming tsunami, it is necessary to match the near-field and far-field solutions. Far away from the shore, the asymptotic value of $J_0(\sigma)$ becomes

$$J_0(\sigma) \to \sqrt{\frac{2}{\pi}}\sigma^{-0.5}\cos\left(\sigma - \frac{\pi}{4}\right) \qquad \text{for} \quad \sigma \to \infty.$$

Hence, the far field expansion of the near-field solution (19) becomes

$$\eta(x,t) \to \frac{R}{\sqrt{\pi\omega}}\left(\frac{x}{gs}\right)^{-0.25}\cos\left(2\omega\sqrt{\frac{x}{gs}} - \frac{\pi}{4}\right)\cos\omega t \qquad \text{for} \quad \sigma \to \infty. \tag{32}$$

Keller (17) and Mei et al. (28) matched this result to linear deep-water theory, by which they found that

$$\frac{R}{a_\infty} = \left(\frac{2\pi}{s}\right)^{1/2} \qquad \text{and} \qquad \delta = -\frac{\pi}{4},$$

where a_∞ is the wave amplitude in deep water. However, tsunamis typically propagate as shallow-water waves in practically any depth, hence we prefer to match (32) to the linear shallow-water far-field solution (13), by which we obtain

$$\frac{R}{a_0} = 2\sqrt{\pi}\left(\frac{h_0\omega^2}{gs^2}\right)^{0.25} \qquad \text{and} \qquad \delta = -\frac{\pi}{4}. \tag{33}$$

This defines the run-up for given shallow-water far-field conditions h_0, a_0, and ω and for a given bed slope s.

2.6 Run-up in terms of the surf similarity parameter

Waves on beaches and their type of breaking depend on the beach slope, the wave period, and a representative wave height (either offshore or nearshore). Iribarren & Nogales (15), Galvin (7) and Battjes (1) characterized the beach processes in terms of the non-dimensional *surf similarity parameter* (also known as the *Iribarren number*) defined by

$$\xi_\infty \equiv \frac{s}{\sqrt{H_\infty/L_\infty}}, \tag{34}$$

where H_∞ is the deep-water wave height and L_∞ is the linear deep-water wave length. In terms of this parameter, Galvin (7) classified breaker types on plane impermeable beaches by

$$\begin{aligned}
&\xi_\infty < 0.5 && \Longrightarrow \text{spilling},\\
&0.5 < \xi_\infty < 3.3 && \Longrightarrow \text{plunging},\\
&3.3 < \xi_\infty && \Longrightarrow \text{surging}.
\end{aligned}$$

Because of their relatively short wave periods (and wave lengths) wind waves will typically have rather small ξ_∞ on most beaches (leading to spilling- or plunging-type breaking), while tsunamis will have rather large values (leading to surging and full reflection). Not surprisingly, the surf similarity parameter also plays an important role for the run-up of breaking and non-breaking wind waves on beaches and steep structures. Hughes (14) concluded that practically all present day empirical run-up expressions are given in terms of ξ defined by

$$\xi \equiv \frac{s}{\sqrt{H_0/L_\infty}}, \tag{35}$$

where H_0 is the local wave height at or near the toe of the slope, at depth h_0.

While the surf similarity parameter undoubtedly has been a successful tool in the classification of breaking and run-up of wind waves, we have so far not seen it used to characterize the impact of tsunamis on beaches. One reason could be that a tsunami is not really a periodic phenomena, but rather a transient phenomena of limited duration. Therefore, it is often approximated by the progression of either a Gaussian hump or a solitary wave, both of which do not really support the concept of a surf similarity parameter. On the other hand, if we accept the assumption of the tsunami being a quasi-periodic phenomena (a case which is certainly supported by some observations, see e.g. Figure 2, to be discussed later in more detail), we obtain the run-up solution (33). This solution can very easily be expressed in terms of the surf similarity parameter. To do so we first utilize that (35) leads to

$$\xi^2 = \frac{\pi}{a_0}\left(\frac{gs^2}{\omega^2}\right), \tag{36}$$

by which we can express (33) as

$$\frac{R}{a_0} = 2\pi^{3/4}\left(\frac{a_0}{h_0}\right)^{-1/4}\xi^{-1/2}. \tag{37}$$

The associated maximum flow velocity of the tsunami is determined by $U_m = \omega R/s$, which leads to

$$\frac{U_m}{\sqrt{ga_0}} = \frac{\sqrt{\pi}}{\xi}\left(\frac{R}{a_0}\right). \tag{38}$$

Finally, Carrier & Greenspan's breaking criterion from (35) can be expressed as

$$\frac{R}{a_0} \leq \frac{R_{\max}}{a_0} = \frac{1}{\pi}\xi^2. \tag{39}$$

Figure 1*a* shows the relative run-up determined by (37) up to the point of breaking and by (39) beyond the point of breaking. We notice that, the smaller the a_0/h_0, the larger the amplification R/a_0 can become before breaking takes over. As tsunamis typically initiate as very small waves on large water depths, the ratio a_0/h_0 can easily be as small as say 0.001 or less. The largest run-up will typically occur if ξ falls in the interval $3.5 \lesssim \xi \lesssim 6$, where the transition from non-breaking to breaking waves occurs. Figure 1*b* shows the associated maximum flow velocity determined by (38). Again we notice, that the smaller the a_0/h_0 ratio, the larger the maximum velocity $U_m/\sqrt{ga_0}$. In the following we shall utilize Figure 1 and equations (37), (38) and (39) to discuss the impact of tsunamis on various beach slopes.

2.7 Discussion of the impact of tsunamis

The impact of waves on shorelines turns out to be a critical balance between the steepness of the beach and the wave length (or wave period). Wind waves are generally quite short and they will experience substantial wave breaking on mild as well as on steep beaches, which dissipates most of their energy before they reach the shoreline. Tides, on the other hand, are extremely long, and from their point of view

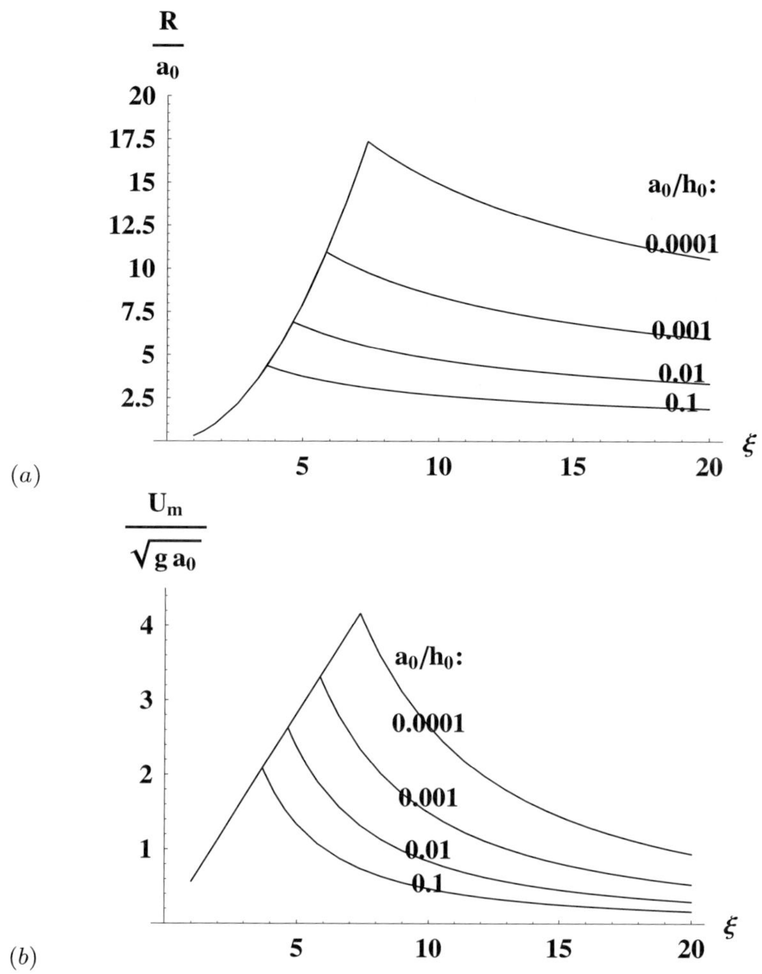

Fig. 1. Plot of the non-dimensional (*a*) run-up and (*b*) maximum horizontal velocity as a function of ξ for various a_0/h_0.

even the mildest beach acts like a vertical wall. Hence, although tides may result in significant run-up elevations, the associated beach velocities are almost nil. As an example, let us consider a tidal wave with period $T = 12$ hr and wave amplitude $a_0 = 5$ m in depth $h_0 = 100$ m, i.e. $a_0/h_0 = 0.05$. On a mild slope $s = 1/120$ this leads to $\xi = 142$, and (37), (38) and (39) yield $R = 4.2$ m and $U_m = 0.07$ m/s, i.e. an extremely small flow velocity. An exception is the formation of tidal bores, which occur if extreme tides are further amplified by special bathymetric conditions such as rapidly decreasing cross sections and water depths. As an example, the Huangzhou bore south of Shanghai in China, can move with a speed of up to 40 km/hr, with a steep front where the surface level jumps 3–5 m within 5 min, and with the maximum fluid velocities exceeding 5–6 m/s (see e.g. Madsen et al. (27)).

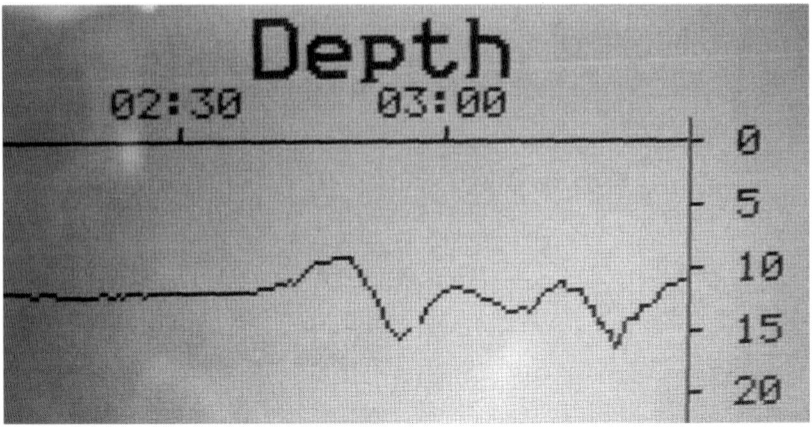

Fig. 2. Sound-meter registration of the Indian Ocean tsunami on December 26, 2004 at a location approximately one mile off Phuket, Thailand. Source: Thomas Siffer on the yacht Mercator (www.thomassiffer.be). See also the website of the Royal Dutch Meteorological Institute (www.knmi.nl).

However, such extreme tidal events cannot happen at random on beaches, but will occur only in funnel-shaped bays and attached rivers.

It is the combination of huge run-up elevations and extreme run-up velocities which makes the tsunami such a devastating event. The maximum impact will occur whenever the breaking criterion (39) is close to being satisfied, i.e. for ξ in the range of 3 to 7. For typical tsunami periods of 2 min $\leq T \leq$ 30 min, it turns out that the beach slope is a very critical parameter in the determination of the tsunami impact. As one example, let us consider a wave with period $T = 13$ min, and with amplitude $a_0 = 0.75$ m in $h_0 = 2000$ m, i.e. $a_0/h_0 = 0.000375$. On a steep slope of $s = 1/15$ this leads to $\xi = 53$, and (37), (38) and (39) yield $R = 3.5$ m and $U_m = 0.42$ m/s, i.e. quite a moderate impact on the shoreline. In contrast, on a slope of $s = 1/120$ the same tsunami leads to $\xi = 6.6$, and (37), (38) and (39) yield $R = 9.9$ m and $U_m = 9.5$ m/s. In this case the impact would be devastating! As another example let us consider again a period of $T = 13$ min, but this time with a wave amplitude of $a_0 = 3.3$ m in $h_0 = 12$ m, i.e. $a_0/h_0 = 0.275$. On a steep slope of $s = 1/15$ this leads to $\xi = 25$, and (37), (38) and (39) yield $R = 4.3$ m and $U_m = 0.5$ m/s. In contrast, on a slope of $s = 1/110$ the same tsunami leads to $\xi = 3.4$, while (37), (38) and (39) yield $R = 11.6$ m and $U_m = 10.3$ m/s. Again, a devastating impact!

In fact these two examples are both inspired by observations from the recent Indian Ocean tsunami from December 26, 2004. On this occasion Thomas Siffer from Belgium had anchored his yacht *Mercator* about a mile off the coast of Phuket, Thailand. During the passage of the tsunami, Mercator's fish-finder (sound-meter) made the recording shown in Figure 2, where the vertical axis shows the depth in meters, while the horizontal axis shows the time in hours. Note that the average depth at the location is 12 m, and that the initial disturbance is a wave trough of -2.7 m followed by a wave crest of +3.9 m, i.e. a wave height of 6.6 m. The second trough is only -0.5 m followed by a crest of +1.7 m, while the third trough is -0.6 m followed by a crest of +4.9 m. Although the crest and trough heights vary over

Fig. 3. A map of a selected region of Thailand affected by the Indian Ocean tsunami on December 26, 2004. The arrow shows the incoming tsunami direction.

time, the time span between successive crests or troughs is approximately 13-14 min during the registration. (This measurement also corresponds reasonably with findings of Liu et al. (20), who report an approximate 10 min duration between two successive peaks in the same event observed in Sri Lanka). Hence, although the event is basically a transient phenomena, based on Figure 2, we may approximate it by a regular wave event with a period of 13-14 min and with an average wave amplitude of $a_0 = 3.3$ m in the depth of $h_0 = 12$ m.

It is a fact that some beaches in Thailand were severely damaged while other nearby beaches were almost untouched. Obviously, local refraction and diffraction effects played an important role, but the most important factor was undoubtedly the local beach slope. The following observations from Thailand were made by Rex Towers from BBC Guernsey (see the map in Figure 3):

> Just two miles round the headland north of Kalim is Kamala Bay. There is awful devastation there. With hardly any rise in the seabed approaching the land there was nothing to cushion the wall of water coming onto the land. The coastline is flat too, going inland for nearly a mile, so the water went that far in places. Just a mile round the headland from Kamala is Surin beach, and what a difference here. It's like nothing ever happened, and very little did. The steep incline of the beach took the force out of the wall of water and only a few feet came up under the treeline, moving beach chairs and restaurant furniture about.

These observations agree remarkably well with the simple calculations made above on the basis of (37), (38) and (39).

3 Extension of a high-order Boussinesq-type model

In this section we will now present numerical simulations of wave run-up based on a recently extended highly accurate Boussinesq-type approach. The corresponding system of partial differential equations, and a brief discussion of their numerical solution, is presented in §3.1. A description of the newly incorporated moving boundary algorithm is provided in §3.2, and a comparison between numerical results involving tsunami run-up and the theory presented in §2 is made in §3.3.

3.1 The Boussinesq formulation

The Boussinesq-type formulation used herein was originally derived by Madsen et al. (24; 25) for slowly varying bathymetries, and has recently been extended to allow rapidly varying bathymetries by Madsen et al. (26). A brief summary of this extended method follows. This method utilizes the following exact expressions for the kinematic and dynamic free surface conditions

$$\frac{\partial \eta}{\partial t} = \tilde{w}\,(1 + \nabla\eta \cdot \nabla\eta) - \tilde{\mathbf{U}} \cdot \nabla\eta, \tag{40}$$

$$\frac{\partial \tilde{\mathbf{U}}}{\partial t} = -g\nabla\eta - \nabla\left(\frac{\tilde{\mathbf{U}} \cdot \tilde{\mathbf{U}}}{2} - \frac{\tilde{w}^2}{2}(1 + \nabla\eta \cdot \nabla\eta)\right), \tag{41}$$

where $\tilde{\mathbf{U}} = (\tilde{U}, \tilde{V}) \equiv \tilde{\mathbf{u}} + \tilde{w}\nabla\eta$. Here $\tilde{\mathbf{u}} = (\tilde{u}, \tilde{v})$ and $\tilde{w}$ are the horizontal and vertical velocities directly on the free surface $z = \eta$, $g = 9.81\ \mathrm{m/s^2}$ is gravitational acceleration, $\nabla = (\partial/\partial x, \partial/\partial y)$ is the horizontal gradient operator, and t is time.

By assuming that the expansion level $\hat{z}$ (but not the depth h) varies slowly in space, Madsen et al. (26) derived the following expressions for the vertical velocity distributions, which are given in terms of pseudo-velocity variables $\hat{\mathbf{u}}^*$,$\hat{w}^*$ located at $\hat{z}$

$$\mathbf{u}(x, y, z, t) = (J_{01} + \nabla\hat{z}J_{11})\hat{\mathbf{u}}^* + (J_{02} + \nabla\hat{z}J_{12})\hat{w}^*, \tag{42}$$

$$w(x, y, z, t) = (J_{01} + \nabla\hat{z}J_{11})\hat{w}^* - (J_{02} + \nabla\hat{z}J_{12})\hat{\mathbf{u}}^*, \tag{43}$$

where

$$J_{01} \equiv 1 + \left(-\frac{\psi}{2} + \frac{\hat{z}^2}{18}\right)\nabla^2 + \left(\frac{\psi^4}{24} - \frac{\hat{z}^2\psi^2}{36} + \frac{\hat{z}^4}{504}\right)\nabla^4, \tag{44}$$

$$J_{02} \equiv \psi\nabla + \left(-\frac{\psi^3}{6} + \frac{\hat{z}^2\psi}{18}\right)\nabla^3 + \left(\frac{\psi^5}{120} - \frac{\hat{z}^2\psi^3}{108} + \frac{\hat{z}^4\psi}{504}\right)\nabla^5, \tag{45}$$

$$J_{11} \equiv \psi\nabla + \left(-\frac{\psi^3}{2} - \frac{\hat{z}\psi^2}{9} + \frac{\hat{z}^2\psi}{18}\right)\nabla^3 + \left(\frac{\psi^5}{24} + \frac{\hat{z}\psi^4}{54} - \frac{\hat{z}^2\psi^3}{36} - \frac{\hat{z}^3\psi^2}{126} + \frac{\hat{z}^4\psi}{504}\right)\nabla^5, \tag{46}$$

$$J_{12} \equiv \left(\psi^2 + \frac{\hat{z}\psi}{9}\right)\nabla^2 + \left(-\frac{\psi^4}{6} - \frac{\hat{z}\psi^3}{18} + \frac{\hat{z}^2\psi^2}{18} + \frac{\hat{z}^3\psi}{126}\right)\nabla^4, \tag{47}$$

and $\psi = (z - \hat{z})$. These were combined with the following representation of the kinematic bottom condition

$$
\begin{aligned}
&\left(1+\left(\frac{\hat{z}^2}{18}-\frac{\psi_b^2}{2}\right)\nabla^2+\left(\frac{\hat{z}^4}{504}-\frac{\hat{z}^2\psi_b^2}{36}+\frac{\psi_b^4}{24}\right)\nabla^4\right)\hat{w}^*- \\
&\left(\psi_b\nabla+\left(\frac{\hat{z}^2\psi_b}{18}-\frac{\psi_b^3}{6}\right)\nabla^3+\left(\frac{\hat{z}^4\psi_b}{504}-\frac{\hat{z}^2\psi_b^3}{108}+\frac{\psi_b^5}{120}\right)\nabla^5\right)\hat{\mathbf{u}}^*+ \\
\nabla h\cdot\ &\left[\left(1+\left(\frac{\hat{z}^2}{18}-\frac{\psi_b^2}{2}\right)\nabla^2+\left(\frac{\hat{z}^4}{504}-\frac{\hat{z}^2\psi_b^2}{36}+\frac{\psi_b^4}{24}\right)\nabla^4\right)\hat{\mathbf{u}}^*+\right. \\
&\left.\left(\psi_b\nabla+\left(\frac{\hat{z}^2\psi_b}{18}-\frac{\psi_b^3}{6}\right)\nabla^3+\left(\frac{\hat{z}^4\psi_b}{504}-\frac{\hat{z}^2\psi_b^3}{108}+\frac{\psi_b^5}{120}\right)\nabla^5\right)\hat{w}^*\right]+ \\
\nabla\hat{z}\cdot\ &\left[\left(\psi_b\nabla+\beta_{13}\left(\frac{\hat{z}^2\psi_b}{18}-\frac{\hat{z}\psi_b^2}{9}-\frac{\psi_b^3}{2}\right)\nabla^3+\right.\right. \\
&\left.\beta_{15}\left(\frac{\hat{z}^4\psi_b}{504}-\frac{\hat{z}^3\psi_b^2}{126}-\frac{\hat{z}^2\psi_b^3}{36}+\frac{\hat{z}\psi_b^4}{54}+\frac{\psi_b^5}{24}\right)\nabla^5\right)\hat{w}^*- \\
&\left.\left(\beta_{12}\left(\frac{\hat{z}\psi_b}{9}+\psi_b^2\right)\nabla^2+\beta_{14}\left(\frac{\hat{z}^3\psi_b}{126}+\frac{\hat{z}^2\psi_b^2}{18}-\frac{\hat{z}\psi_b^3}{18}-\frac{\psi_b^4}{6}\right)\nabla^4\right)\hat{\mathbf{u}}^*\right] \\
&=0, \qquad (48)
\end{aligned}
$$

where $\psi_b = -(h+\hat{z})$, $\beta_{12} = 0.95583$, $\beta_{14} = 0.51637$, $\beta_{13} = 0.72885$, and $\beta_{15} = 0.28478$. Note that if the $O(\nabla\hat{z})$ terms are neglected and $\hat{z} = -h/2$ in the above, this system simplifies to the original formulation of Madsen et al. (24; 25). Madsen et al. (26) also included $O((\nabla\hat{z})^2)$ terms in their extended derivation, however, as they are not used in the present work they are left out for brevity.

Analysis and applications of the current method have shown that it retains highly accurate linear and nonlinear properties for (wave number times depth) $kh \leq 25$, provides accurate shoaling for $kh \leq 30$, and gives accurate velocity kinematics for $kh \leq 12$. Hence, over a large range of water depths this system may be considered as a highly accurate approximation of the exact Laplace problem governing nonlinear water waves. As such, depth limitations conventionally associated with Boussinesq-type approaches are effectively eliminated by the present method for most practical applications. For various deep-water applications see e.g. (4; 5; 6).

When a single horizontal dimension is considered, the system of partial differential equations outlined above is solved using a finite difference scheme, in a similar manner as done previously by Madsen et al. (24). In two horizontal dimensions numerical solutions are considerably more complex, and the finite difference solution strategies developed by Fuhrman & Bingham (3) are utilized. In both cases the classical fourth-order, four-stage, explicit Runge-Kutta method is adapted for time stepping. High-order Savitzky-Golay (29) smoothing is also applied to remove high wave number disturbances that can arise from the discretization of the nonlinear and variable depth terms.

3.2 The moving boundary algorithm

The moving boundary algorithm, necessary e.g. for computing wave run-up, employed within the current high-order Boussinesq model is the so-called "extrapolating boundary" method, which was first introduced by Sielecki & Wurtele (33), and later modified and utilized by Kowalik & Bang (18) within models based on nonlinear shallow-water equations. More recently, Lynett et al. (23) have also used this technique within a Boussinesq-type model, see also (21; 22).

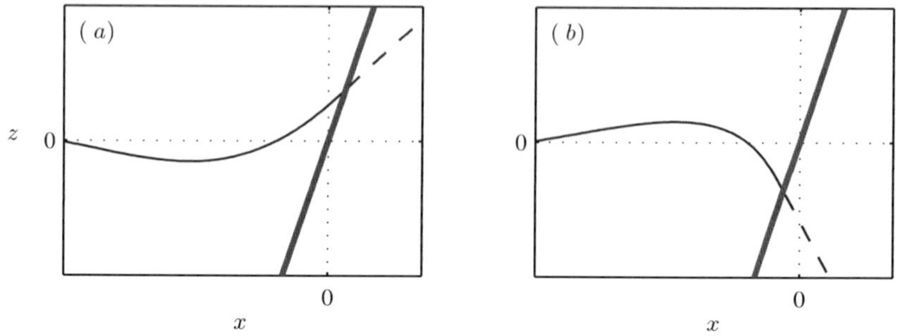

Fig. 4. Examples of the extrapolating boundary technique during (*a*) run-up and (*b*) draw-down. The sloping sea bottom is shown as the bold line, with wet free surface points indicated by the full lines. The linearly extrapolated (dry) free surface values are shown as dashed lines.

We will here only describe the algorithm in a single horizontal dimension, as the developed method in two horizontal dimensions is not utilized in the present work. The implemented algorithm employed herein is essentially the same as described previously by Lynett et al. (23). It begins by first determining the grid-point location of the current wet-dry boundary. This is done by finding the last grid point where the total water depth $d = h + \eta$ exceeds some threshold δ, which is assumed small. Computed values at the dry points (i.e. those where $d < \delta$) are then replaced with those from a linear extrapolation based on the last two wet points. Computed examples of a free surface after being modified by the moving boundary algorithm are shown in Figure 4 as demonstrations during both run-up and draw-down phases. The algorithm is applied within the present model after each time-stepping stage evaluation, for both time stepping variables η and $\tilde{U}$. Testing has indicated that the results are not very sensitive to the chosen value of δ, and throughout the present work we use $\delta = a_0/100$, where a_0 is the incident wave amplitude.

The adapted technique is attractive, as it allows the moving wet-dry boundary to exist at any location, i.e. it need not be located precisely at any single grid point, as previously noted by (23). Moreover, it does not require any special treatment of derivatives calculated in the vicinity of the moving boundary, and is therefore particularly attractive within the present model, which involves a number of high-derivative terms. We have found this technique simple to implement, as once the extrapolation described above is performed, the rest of the code may be run as usual without further modification.

3.3 Numerical simulation of tsunami run-up

We will now use the developed numerical model to simulate nonlinear wave run-up on a sloping beach. The model domain is constructed by connecting a flat region of depth h_0 to a sloping region with constant slope s, which is then smoothed to create a gentle transition. This is illustrated in the sample bathymetry provided in Figure 5. The origin $(x, z) = 0$ is defined as the crossing of the sloping region and the still

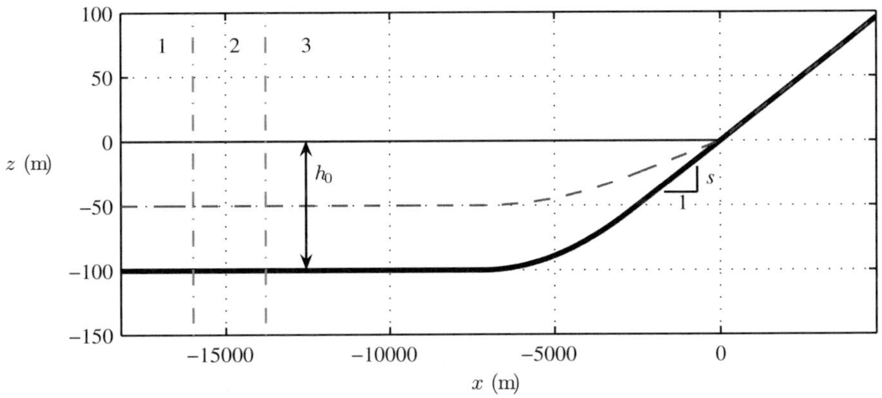

Fig. 5. Model bathymetry used for the run-up simulation with $T = 280.1$ s, $a_0 = 0.5$ m, $s = 1/50$, and $h_0 = 100$ m ($\xi = 7$, $a_0/h_0 = 0.005$), with incoming waves traveling from left to right. The bold line indicates the sea bottom $-h(x)$, the dashed line indicates the variation of $\hat{z}$, and the thin full line indicates the still water level $z = 0$. The vertical dash-dotted lines separate (1) the wave generating region, (2) the relaxation zone used to absorb the wave field reflected off the beach, and (3) the purely numerical region.

water level, with incoming waves at depth h_0 (and having wave length L_0, which should not be confused with the deep-water wave length $L_\infty = gT^2/2/\pi$) traveling in the $+x$-direction. The flat region (prior to smoothing) has total length $1.5L_0$. The left-most $L_0/4$ consists of a relaxation zone used for the analytical wave generation. This is followed by an additional relaxation zone of equivalent length, where the analytical wave signal is gradually relaxed to the computed solution, which absorbs the wave field reflected off the slope. This leaves roughly a full wave length for the generated wave to propagate before encountering the toe of the slope. The expansion level $\hat{z}$ (see again §3.1) is set according to

$$\hat{z}(x) = \max\left(-\frac{h(x)}{2}, -h(x)\right). \tag{49}$$

This makes it the optimal (in terms of accurate velocity kinematics, see (25)) $\hat{z} = -h/2$ when the sea bottom is below the still water level (i.e. for $x < 0$), while it follows the bottom for $x > 0$. Hence, the expansion point is never allowed to be below the sea bottom. These various features are also illustrated in Figure 5.

For validation of the numerical model we will consider incoming periodic waves (based on linear wave theory) with $a_0/h_0 = 0.005$, with constant $h_0 = 100$ m and slope $s = 1/50$. All cases considered are discretized such that $R/s/\Delta x = 7$, where R is determined from (33), which gives a constant discretization of the run-up region. The time step is varied based on the Courant number constraint $Cr = c_0\Delta t/\Delta x = 2$, where $c_0 = L_0/T$ is the incident wave celerity. Simulations are run for a sufficient length of time to essentially reach a repeating periodic state. We also note that in all cases dispersive terms incorporated in the Boussinesq-type formulation remain

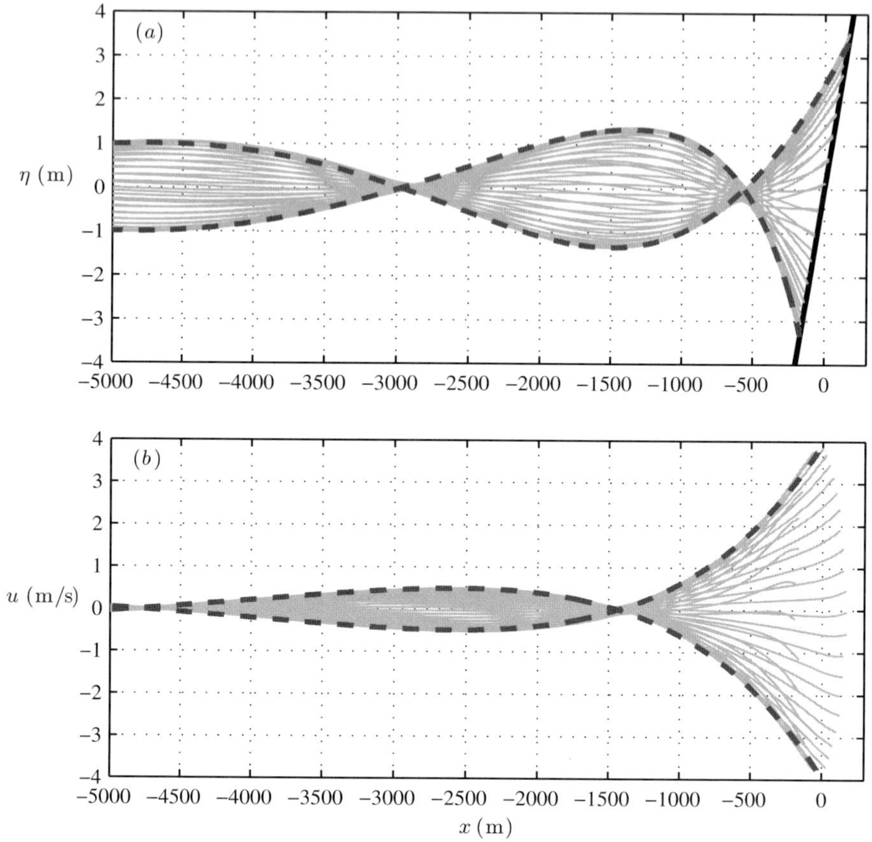

Fig. 6. Computed (thin grey lines) and analytical (bold dashed lines, from the theory of Carrier & Greenspan (2)) (*a*) free surface and (*b*) velocity envelopes for the run-up case with $T = 280.1$ s, $a_0 = 0.5$ m, $s = 1/50$, and $h_0 = 100$ m ($\xi = 7$, $a_0/h_0 = 0.005$).

small, hence it is reasonable to compare against the theory outlined in §2, which is again based on shallow-water equations.

Note, however, that there are differences between our numerical setup, which is based on a flat bottom connected to a sloping region, and the analytical solution described previously in §2, which assumes an infinitely sloping plane beach. We find that reasonable agreement can be obtained when the ratio of the slope length h_0/s and the incident wave length L_0 satisfy

$$\frac{h_0}{sL_0} > \frac{1}{4}, \tag{50}$$

i.e. the sloped region should cover at least one quarter of an incident wave length. From the definition of ξ in (36), after invoking the shallow-water expression $\omega = ck_0 = 2\pi\sqrt{gh_0}/L_0$, we obtain

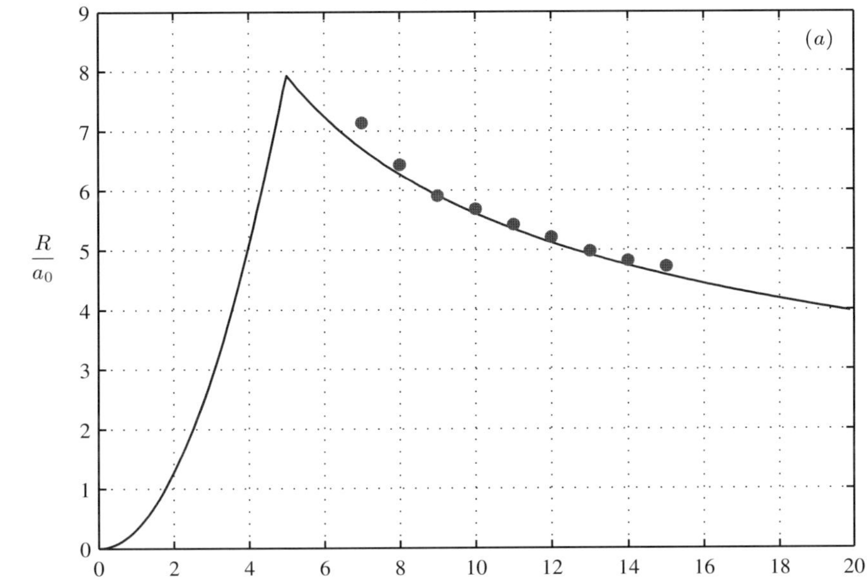

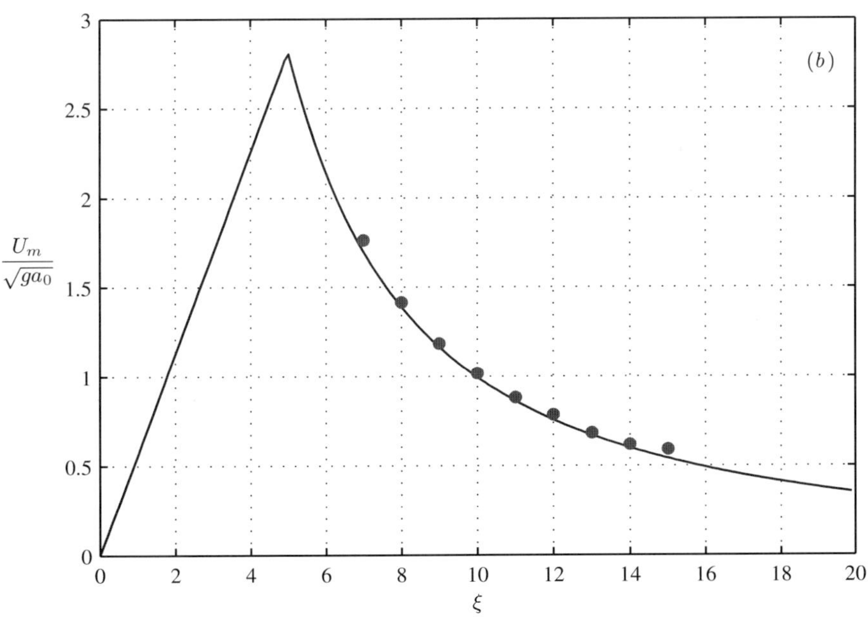

Fig. 7. Maximum dimensionless (*a*) run-up and (*b*) horizontal velocity from $a_0/h_0 = 0.005$. The circles indicate computed values, whereas the solid lines are from the theory presented in §2.

$$\frac{h_0}{sL_0} = \frac{1}{2\sqrt{\pi}} \frac{1}{\xi} \sqrt{\frac{h_0}{a_0}}. \tag{51}$$

Equating (50) and (51) then leads to the following practical limit for the surf similarity parameter ξ for a given ratio a_0/h_0

$$\xi < \frac{2}{\sqrt{\pi}} \sqrt{\frac{h_0}{a_0}}, \tag{52}$$

and we restrict our comparison to within this range. For the present case with $a_0/h_0 = 0.005$ this gives roughly $\xi < 16$. There is also a lower limit of ξ that can be considered within the present model, since reducing ξ can result in nearly-breaking events during the draw-down phase, which can in turn cause a computational breakdown. Due to these constraints, we have managed to compute cases within the range $7 \leq \xi \leq 15$.

The computed envelopes for the free surface and horizontal velocity (here simply taken as $\hat{u}^*$, since the vertical distribution of the velocity field is nearly uniform) near the moving wet-dry boundary are provided in Figure 6 for the numerical case nearest breaking (with $\xi = 7$), which is the most difficult of those simulated. Also shown for comparison are the theoretical envelopes based on the solution of Carrier & Greenspan (2). As can be seen, an excellent match is observed for the surface elevation, as well as for the corresponding velocity kinematics.

The maximum computed run-up and horizontal velocities for each of the cases considered are additionally shown in Figure 7, which are compared with curves based on the theory presented in §2. Again, a generally excellent match is observed over the full range of ξ. Actually, the case depicted in Figure 6 is the most inaccurate of those considered, as the run-up is overestimated. This is more exaggerated on Figure 7*a* than on Figure 6*a* due the vertical scaling. As can be seen from Figure 7, we have managed to compute values reasonably near, but not quite up to, the empirical breaking limit (39).

This comparison validates the developed numerical model for wave run-up in a single horizontal dimension for both surface elevation as well as velocity kinematics over a range of the surf similarity parameter. The model therefore seems a useful tool for engineering applications involving wave run-up, e.g. for the investigation of tsunami impact on various beaches.

4 Three-dimensional simulation of a hypothetical sound

In this section we will finally use the Boussinesq-type model described previously to investigate some three-dimensional effects relevant to tsunamis generated from nearshore landslides. In this connection we use will use an idealized bathymetry for a hypothetical sound given by

$$h(x, y) = h_s + \frac{h_d - h_s}{4} (\cos(k_s x) - 1)^2, \tag{53}$$

where $h_s = 50$ m is the depth of the shallow shelfs at both ends of the sound, $h_d = 500$ m is the largest depth occurring midway across the sound, $k_s = 2\pi/L_s$, with $L_s = 20$ km the total length across the sound. This bathymetry is plotted in

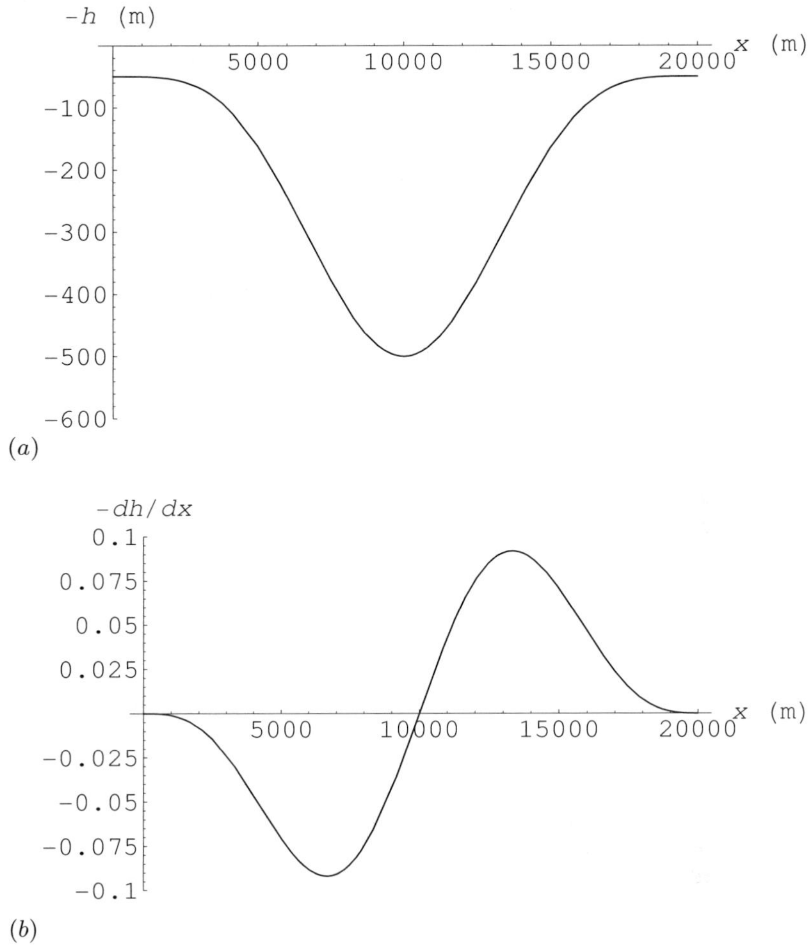

Fig. 8. Plot of (*a*) the bathymetry used for the hypothetical sound and (*b*) the bottom slope $-dh/dx$.

Figure 8. There the bottom slope is also shown, which is seen to vary within ±10%. Note that in this example the bathymetry has no variation in the y-direction.

As an idealized approximation to a tsunami generated near the shore, e.g. from a submerged or subarial landslide, we consider waves generated from a point source with period $T = 60$ s. Thus, via the linear dispersion relation the dimensionless depth in the present problem will be within the range $0.239 \leq kh \leq 0.825$, i.e. shallow to intermediate depths. Hence, a model that includes dispersive effects is necessary to accurately simulate this case.

For simplicity, and for later comparison against a linear refraction solution, the linearized form of the Boussinesq-type model is used. The results may therefore be scaled according to the wave height, though the linear assumption of a small amplitude obviously applies. Additionally, no run-up feature is utilized in the simulation, and the computational domain is instead surrounded by fully-reflecting numerical walls. Due to symmetry, only the lower half (in plan) is modeled, and results will simply be mirrored across the y-axis for presentation purposes. We use a 301×376 computational domain, with $\Delta x = \Delta y = 66.67$ m, and $\Delta t = T/40 = 1.5$ s.

For the point source generation, the free surface η at the four grid points nearest the origin (a 2×2 square) are specified to vary sinusoidally according to

$$\eta = a_{\text{source}} \cos(\omega t), \tag{54}$$

where $a_{\text{source}} = 5$ m. We have found that this results in periodic ring waves with amplitude $a = 1$ m at a radius of approximately one wave length from the origin. The model is simulated to a time $t = 16T$, which is sufficiently long for the generated waves to have traveled across the sound, and to have reflected off the far wall. We note that the numerical wave tank is sufficiently wide that reflections off the side-walls do not significantly affect the model results in the interior domain. The final

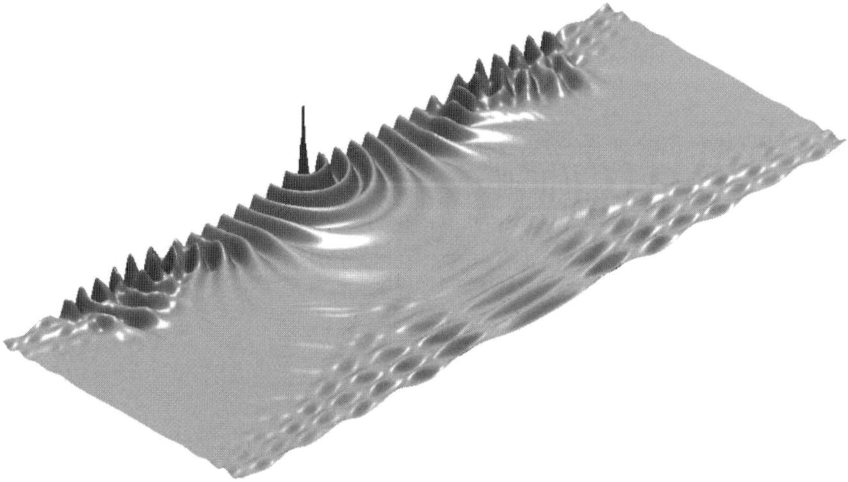

Fig. 9. Free surface computed by the high-order Boussinesq model for the case of the hypothetical sound, at $t = 16T$.

computed free surface is shown in Figure 9. We note that a reasonably similar pattern can be seen in Figure 2 of Lynett & Liu (22), there created from a simulation of a landslide generated tsunami.

For validation purposes, we will compare the Boussinesq results with those computed from a linear refraction model, which describes the evolution of individual wave orthogonals, as described in Appendix 5. In this setup the initial orthogonal positions (x_o, y_o) are specified along a circle of radius r according to

$$x_o(t=0) = r\cos\alpha, \qquad y_o(t=0) = r\sin\alpha, \tag{55}$$

with initial refraction angles α ranging from -50° to 50° in increments of 2°. We use $r = 1316.44$ m (matching the wave length at depth h_s), with each orthogonal initially having unit amplitude $a = 1$ m. Hence, this setup consists of 51 wave orthogonals propagating outwardly normal from the origin along a circle of radius r, approximating the periodic ring waves generated by the point source in the Boussinesq-type model. We use a temporal discretization $\Delta t = T/10 = 6$ s, which is sufficient to obtain converged results, and the wave orthogonal evolution is simulated to a time $t = 15T$. This is roughly equivalent to the $16T$ simulated in the Boussinesq-type model, since the initial wave orthogonals in the refraction model are again located one wave length from the point source.

The computed evolution of the wave orthogonals is shown in Figure 10. The physical region considered within the Boussinesq model is also enclosed by the bold rectangular region for comparison. As the simple refraction model does not take into account reflection s off the exterior walls, the results will inevitably differ somewhat where these are important e.g. near the back wall. These differences are kept minor for most of the domain, however, as will be seen.

A comparison of the computed surface elevation envelopes taken from both the Boussinesq solution and the refraction solution along the centerline $y = 0$ is provided in Figure 11, where an excellent match is observed. Similarly, computed envelopes along various lines of constant x are also shown in Figure 12, generally resulting in a similarly good agreement. Note that along the far wall (i.e. along $x = L_s$), Figure 12*d*, the refraction solution has simply been doubled to approximately account for the reflection, which indeed results in a reasonable match with the envelope computed by the Boussinesq model. Thus, despite their obvious differences in complexity, the two models validate each other in this idealized case, providing confidence in both.

We will now return our attention to the free surface shown in Figure 9, where a number of interesting features are observed. The most impressive feature is the clear development of trapped edge waves that do not escape to the other side of the sound. Rather, these initially outgoing waves simultaneously turn around and steepen, ultimately resulting in a region of enhanced impact on the nearshore wall, some distance (in this case approximately 13 km) away from the generating point source. This is more clearly illustrated in the envelope plot along the back wall (i.e. along $x = 0$) provided in Figure 13. This qualitatively indicates that there are locations on the same shore of a landslide-generated tsunami, but away from the source, that are potentially at risk. For a more in depth discussion of edge waves see e.g. Schäffer & Jonsson (30).

By comparing Figures 9 and 10 we can see that the steepened trapped waves correspond to areas where the wave orthogonals are close together, indicating a focusing of their energy. Indeed, part of the refraction solution near the location of

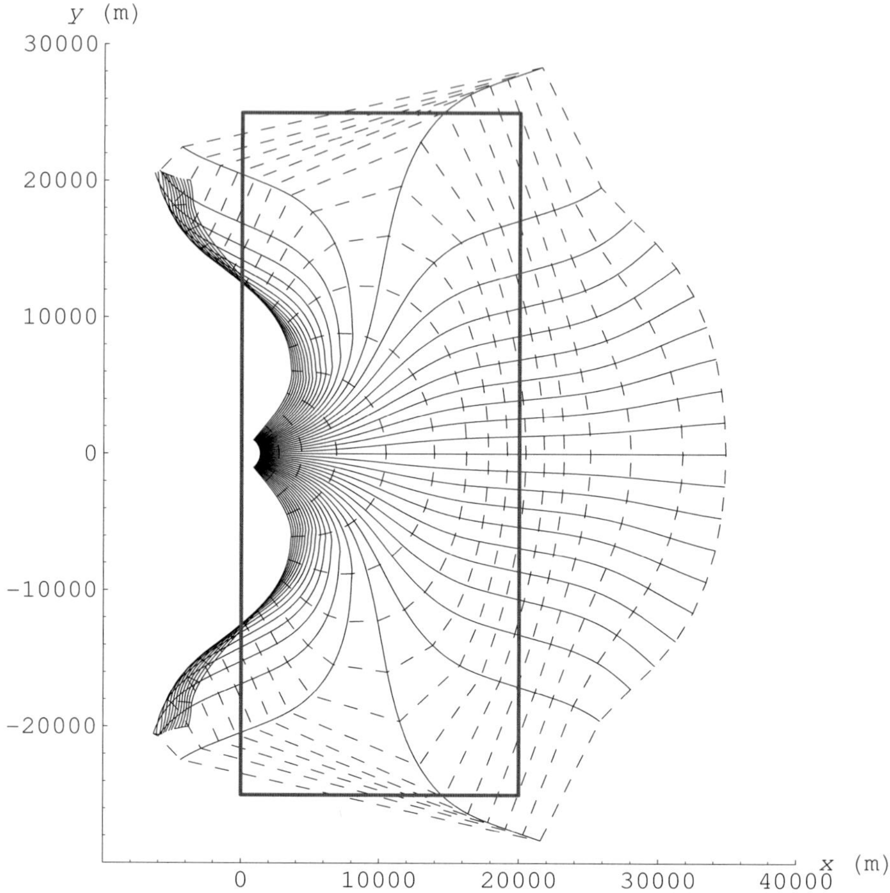

Fig. 10. Plot of the linear refraction solution. The solid lines indicate the path traveled by individual wave orthogonals, while the dashed lines indicate every 10th wave front. The physical region considered within the high-order Boussinesq model is outlined by the bold rectangular region.

the wall results in a family of crossing orthogonals, i.e. a *caustic*. At these locations the wave heights predicted by the linear refraction model are in fact infinite, indicating that the theory has locally broken down. Obviously, diffraction effects within the more-realistic Boussinesq-type model inhibit such an occurrence, resulting in steep (but finite-amplitude) waves. We finally note that nonlinear simulations

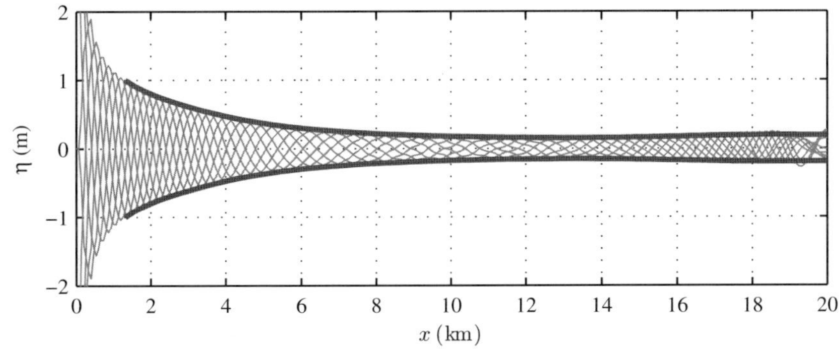

Fig. 11. Computed free surface envelopes from the Boussinesq results (thin grey lines) and the refraction solution (bold lines) along the centerline $y = 0$ at $t \approx 10T$.

have additionally been performed with the Boussinesq-type model, which result in qualitatively similar behavior.

5 Conclusions

In this work we present analytical and numerical models for the determination of the run-up of tsunamis and long waves on beaches. The analytical model is based on non-dispersive theory, while the numerical model is based on dispersive theory and a high-order Boussinesq-type formulation.

The first part of this work is based on the linear and nonlinear shallow-water equations, which are solved to obtain an analytical solution for the run-up of periodic long waves on an infinitely long slope. The theoretical formulation is based on classical earlier work by Green (8), Lamb (19) and Carrier & Greenspan (2), and leads to simple expressions for the maximum run-up and its associated flow velocity as a function of the surf similarity parameter and the amplitude to depth ratio at an offshore location. We use these expressions to analyze and discuss the impact of tsunamis on beaches, and conclude that it is the combination of huge run-up elevations and extreme run-up velocities, which makes the tsunami such a devastating event compared to other types of wave events such as tides and wind-waves. On the other hand, the impact of a given tsunami depends very much on the beach slope, and for a typical period of say 13 min (which was observed in connection with the Indian Ocean tsunami on December 26, 2004 off the coastline of Phuket, Thailand), the impact on flat beaches with slopes of the order 1/100 is huge compared to the impact on steep beaches with slopes of the order 1/15. Qualitatively, this analysis explains the observations from Thailand, where the flat Kamala beach was completely destroyed, while the neighboring steep Surin beach was left almost untouched.

In the second part of this work, we extend a recently developed high-order Boussinesq-type model to include moving shorelines. This allows for the assessment of a wide class of tsunami events and their associated run-up onto the shore.

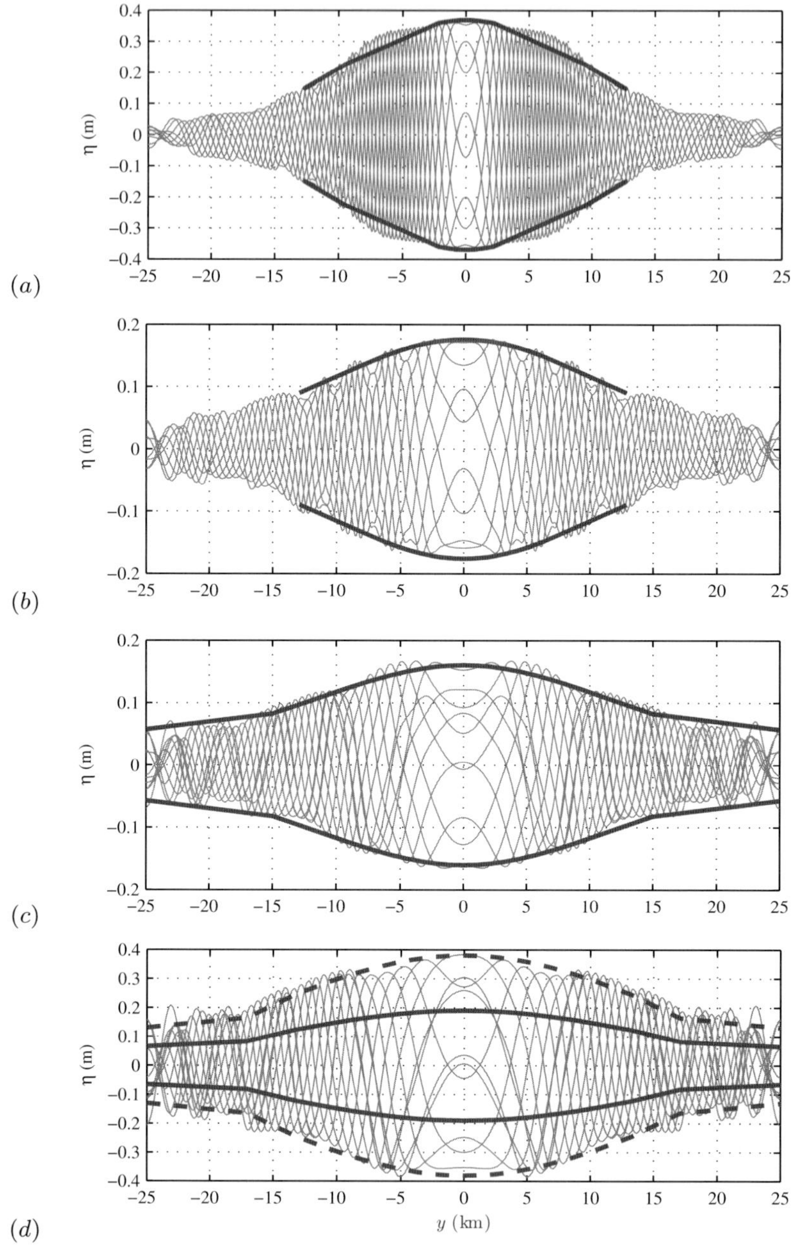

Fig. 12. Computed free surface envelopes along (a) $x = 5$ km at $t \approx 12T$, (b) $x = 10$ km at $t \approx 12T$, (c) $x = 15$ km at $t \approx 13T$, (d) $x = L_s = 20$ km at $t \approx 16T$. The lines are as in Figure 11, with the additional dashed bold lines in (d) representing the refraction solution multiplied by 2, to account for the reflection off the vertical wall at $x = L_s$.

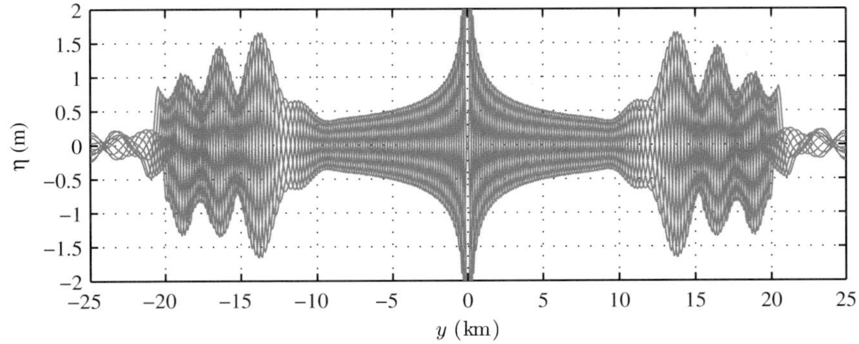

Fig. 13. Computed free surface envelope from the Boussinesq model along $x = 0$ at $t \approx 16T$.

The model is successfully validated against the previously derived analytical expressions for the surface elevation, as well as the run-up velocities in a single horizontal dimension, over a range of the surf similarity parameter.

Finally, in the third part of this work, the Boussinesq-type model is utilized to simulate an idealized periodic point-source representation of a nearshore landslide-induced tsunami on a hypothetical sound. Due to the combined larger depths and shorter period involved, dispersive effects are important in this case. Comparisons against a linear refraction solution demonstrate the accuracy of the simulation. Additionally, the emergence of shore-trapped edge waves is demonstrated, which lead to significantly enhanced wave impacts on the generating shoreline, some distance away from the generating source.

Acknowledgements

We thank the Danish Center for Scientific Computing for providing the supercomputing time used in the numerical simulations. This work was financially supported by the Danish Technical Research Council (STVF grant no. 9801635). This support is greatly appreciated.

References

[1] J. A. Battjes. Surf similarity. In *Proc. 14th Int. Conf. Coastal Engineering*, volume 1, pages 466–480. ASCE, 1987.

[2] G. F. Carrier and H. P. Greenspan. Water waves of finite amplitude on a sloping beach. *J. Fluid Mech.*, 4:97–109, 1958.

[3] D. R. Fuhrman and H. B. Bingham. Numerical solutions of fully non-linear and highly dispersive Boussinesq equations in two horizontal dimensions. *Int. J. Numer. Meth. Fluids*, 44:231–255, 2004.

[4] D. R. Fuhrman, H. B. Bingham, and P. A. Madsen. Nonlinear wave-structure interactions with a high-order Boussinesq model. *Coast. Eng.*, 52:655–672, 2005.

[5] D. R. Fuhrman and P. A. Madsen. Short-crested waves in deep water: a numerical investigation of recent laboratory experiments. *J. Fluid Mech.*, 559:391–411, 2006.

[6] D. R. Fuhrman, P. A. Madsen, and H. B. Bingham. A numerical study of crescent waves. *J. Fluid Mech.*, 513:309–341, 2004.

[7] C. J. Galvin. Breaker type classification on three laboratory beaches. *J. Geophys. Res.*, 73:3651–3659, 1968.

[8] G. Green. On the motion of waves in a variable canal of small depth and width. *Trans. Camp. Philos. Soc.*, 6:457–462, 1838.

[9] S. T. Grilli, S. Vogelmann, and P. Watts. Development of a 3D numerical wave tank for modeling tsunami generation by underwater landslides. *Eng. Anal. Bound. Elem.*, 26:301–313, 2002.

[10] S. T. Grilli and P. Watts. Modeling of waves generated by a moving submerged body. Applications to underwater landslides. *Eng. Anal. Bound. Elem.*, 23:645–656, 1999.

[11] S. T. Grilli and P. Watts. Tsunami generation by submarine mass failure. I: Modeling, experimental validation, and sensitivity analysis. *J. Waterw. Port C-ASCE*, 131:283–297, 2005.

[12] S. Hibberd and D. H. Peregrine. Surf and run-up on a beach: A uniform bore. *J. Fluid Mech.*, 95:323–345, 1979.

[13] D. V. Ho and R. E. Meyer. Climb of a bore on a beach, Part 1: Uniform beach slope. *J. Fluid Mech.*, 14:305–318, 1962.

[14] S. A. Hughes. Estimation of wave run-up on smooth, impermeable slopes using the wave momentum flux parameter. *Coast. Eng.*, 51:1085–1104, 2004.

[15] C. R. Iribarren and C. Nogales. Protection des ports. In *XVIIth International Navigation Congress*, pages 31–80, 1949. Section II, Communication.

[16] H. B. Keller, D. A. Levine, and G. B. Whitham. Motion of a bore over a sloping beach. *J. Fluid Mech.*, 7:302–316, 1960.

[17] J. B. Keller. Tsunamis–water waves produced by earthquakes. In *Tsunami Hydrodyn. Conf.*, pages 154–166, Honolulu, 1963. Int. Union Geod. Geophys. Monogr. 24.

[18] Z. Kowalik and I. Bang. Numerical computation of tsunami run-up by the upstream derivative method. *Sci. Tsunami Haz.*, 5:77–84, 1987.

[19] H. Lamb. *Hydrodynamics.* Cambridge University Press, Cambridge, 1932.

[20] P. L.-F. Liu, P. Lynett, H. Fernando, B. E. Jaffe, H. Fritzz, B. Higman, R. Morton, J. Goff, and C. Synolakis. Observations of the International Tsunami Survey Team in Sri Lanka. *Science*, 308:1595, 2005.

[21] P. Lynett and P. L.-F. Liu. A numerical study of submarine-landslide-generated waves and run-up. *Proc. R. Soc. Lond. A*, 458:2885–2910, 2002.

[22] P. Lynett and P. L.-F. Liu. A numerical study of the run-up generated by three-dimensional landslides. *J. Geophys. Res.*, 110:C03006/1–16, 2005.

[23] P. J. Lynett, T.-R. Wu, and P. L.-F. Liu. Modeling wave runup with depth-integrated equations. *Coast. Eng.*, 46:89–107, 2002.

[24] P. A. Madsen, H. B. Bingham, and H. Liu. A new Boussinesq method for fully nonlinear waves from shallow to deep water. *J. Fluid Mech.*, 462:1–30, 2002.

[25] P. A. Madsen, H. B. Bingham, and H. A. Schäffer. Boussinesq-type formulations for fully nonlinear and extremely dispersive water waves: derivation and analysis. *Proc. R. Soc. Lond. A*, 459:1075–1104, 2003.

[26] P. A. Madsen, D. R. Fuhrman, and B. Wang. A Boussinesq-type method for fully nonlinear waves interacting with a rapidly varying bathymetry. *Coast. Eng.*, 53:487–504, 2006.
[27] P. A. Madsen, H. J. Simonsen, and C.-H. Pan. Numerical simulation of tidal bores and hydraulic jumps. *Coast. Eng.*, 52:409–433, 2005.
[28] C. C. Mei, M. Stiassnie, and D. K.-P. Liu. *Theory and Applications of Ocean Surface Waves*. World Scientific, Singapore, 2005.
[29] A. Savitzky and M. J. E. Golay. Smoothing and differentiation of data by simplified least squares procedures. *Anal. Chem.*, 36:1627–1639, 1964.
[30] H. A. Schäffer and I. G. Jonsson. Edge waves revisited. *Coast. Eng.*, 16:349–368, 1992.
[31] M. C. Shen and R. E. Meyer. Climb of a bore on a beach, Part 2: Nonuniform beach slope. *J. Fluid Mech.*, 16:108–112, 1962.
[32] M. C. Shen and R. E. Meyer. Climb of a bore on a beach, Part 3: Run-up. *J. Fluid Mech.*, 16:113–125, 1962.
[33] A. Sielecki and M. G. Wurtele. The numerical integration of the nonlinear shallow-water equations with sloping boundaries. *J. Comput. Phys.*, 6:219–236, 1970.
[34] G. Watson, D. H. Peregrine, and E. F. Toro. Numerical solution of the shallow water equations on a beach using the weighted average flux method. In *Computational Fluid Dynamics*, volume 1, pages 495–502. Elsevier, 1992.
[35] P. Watts, S. T. Grilli, D. R. Tappin, and G. J. Fryer. Tsunami generation by submarine mass failure. II: Predictive equations and case studies. *J. Waterw. Port C-ASCE*, 131:298–310, 2005.
[36] H. H. Yeh, A. Ghazali, and I. Marton. Experimental study of bore run-up. *J. Fluid Mech.*, 206:563–578, 1989.

Appendix: The linear refraction model

This Appendix provides details of the developed linear refraction model for monochromatic waves utilized within §4. In this model the evolution of individual wave orthogonal positions (x_o, y_o) is described by the system of ordinary differential equations

$$\frac{dx_o}{dt} = c\cos\alpha, \qquad \frac{dy_o}{dt} = c\sin\alpha, \tag{56}$$

where α is the refraction angle measured positively from the x-axis to the orthogonal and $c = \omega/k$ is the local wave celerity, which is determined from the linear dispersion relation

$$\omega^2 = gk\tanh kh. \tag{57}$$

Equations (56) are integrated numerically using the (first-order accurate) forward Euler method, for simplicity. We assume that the bathymetry has no y-variation (i.e. that it has straight and parallel depth contours), hence the refraction angle along individual orthogonals is determined from Snel's law

$$\frac{\sin\alpha}{c} = \text{constant}. \tag{58}$$

Wave fronts are created by connecting orthogonal locations at equivalent times. Changes in the wave height $H(= 2a)$ are computed via conservation of energy flux E_f along individual orthogonals, defined by

$$E_f = bEc_g, \qquad E = \frac{1}{16}\rho g H^2, \qquad c_g = c\left(1 + \frac{2kh}{\sinh(2kh)}\right). \tag{59}$$

Here b is the local mean front distance between adjacent orthogonals (estimated as half the distance between left- and right-neighboring orthogonals), E is the energy density per unit front width, and c_g is the local group velocity. Wave height values along a particular x- or y-line are subsequently obtained by linear interpolation along individual orthogonals.

We finally note that in some instances, namely where an orthogonal becomes nearly parallel to a depth contour, the resulting refraction angle α from a solution of (58) can become complex, typically having $\mathrm{Re}\{\alpha\} = 90^\circ$, but with small imaginary part. When this occurs the absolute value of the imaginary part is simply added (or subtracted, depending on the sign) to the real part according to

$$\alpha \rightarrow \mathrm{Re}\{\alpha\} + \mathrm{sign}\{\mathrm{Re}\{\alpha\}\}|\mathrm{Im}\{\alpha\}|. \tag{60}$$

This forces α to again become real, while making its absolute value greater than 90°, thus initiating the turn of the orthogonal back towards shallower depths.

Large waves caused by oceanic impacts of meteorites

Robert Weiss[1] and Kai Wünnemann[2]

[1] Joint Institute for the Study of the Atmosphere and Ocean, University of Washington–NOAA Center for Tsunami Research, 7600 Sand Point Way NE, Seattle WA 98115, U.S.A. weiszr@u.washington.edu

[2] Institut für Mineralogie, Museum für Naturkunde, Humboldt-Universität zu Berlin, Invalidenstrae 43, 10115 Berlin, Germany kai.wuennemann@museum.hu-berlin.de

1 Introduction

Impact craters can be observed on all terrestrial planets and their larger satellites. Basically every body in the solar system with a solid crust, no matter how small it is, exhibits evidence of impacts in the past. For example, the Moon provides an excellent data base of impact craters. However, the major fraction of impact events occurred between 4.6 and 3.9 Billion years ago. The impact frequency at that time was $\sim$ 100 times larger than it has been ever since. Figure 1 shows the craters Ptolomaeus, Alphonsus, Arzahchel and Albetegnius. The image depicts that impact craters vary form large basins of several 100 kilometres in diameter (the largest impact basin is Valhalla with 4000 km in diameter on the Jovian satellite Callisto) to structures that are only several 10's of meters in diameter.

On Earth impact craters are much less common. A total of about 170 impact craters have been confirmed (e.g. (1; 3)). In principle the impact record of the Earth should not differ significantly from that of the Moon since the flux rate of colliding bodies is the same for both. Due to the dynamic nature of the Earths crust (plate tectonics, volcanism, long-term erosional processes) the traces of impact events are much harder to identify and often completely erased from the surface. Among the known crater structures on Earth there exist only 27 which show clear evidence of an impact into a marine environment (marine impacts; (2)). Due to the fact that two thirds of the Earth surface is covered by water the number of marine impact structures should be much larger. Glikson (4) estimated a total of 8,104 marine impact craters larger than 20 km in diameter that should have formed during the Earth history. The imbalance between the continental and marine impact craters can be ascribed to: (i) the coarse resolution of the investigated marine basins, (ii) plate–tectonic destruction of oceanic crust (oceanic crust is generally much younger than continental one and is therefore much shorter exposed to the meteorite flux), and (iii) small meteorites do not necessarily produce impact craters in the oceanic crust.

Fig. 1. Picture of the craters (1) Ptolemaeus, (2) Alphonsus, (3) Arzachel and (4) Albategnius. (Courtesy NASA/JPL-Caltech)

2 Meteorite impact processes

2.1 Shock waves

The kinetic energy of the projectile is transmitted to the target as an abrupt impulse causing large stress waves at the contact between projectile and target. In this section, we concentrate on the characteristics of the generated stress waves, how they propagate through the target, and how they may change their properties while they travel. But before these aspects are considered it is worthwhile to provide an outline of the fundamentals of stress waves. Firstly, a basic knowledge of the mechanics and thermodynamics of large stress waves is crucial for the understanding of impact cratering and it offers a convenient way to introduce the main equations and theoretical concepts of crater mechanics as well. A more extensive presentation of the stress waves are given in Zel'dovich and Raizer in (26) and Melosh in (7). Both books are the basis of the following considerations.

In general, there are three different types of stress waves: *elastic, plastic*, and *shock waves*. A distinction between these waves can be drawn by the amplitude (stress or for simplicity the hydrostatic pressure) and the propagation velocity. A relatively weak pressure impulse results in the more familiar elastic wave type that travels through a solid. The elastic stress in a solid is limited by the yield strength Y. Figure 2 shows the course of Y (expressed as the shear stress τ) as a function of hydrostatic pressure P, often called the yield envelope (*Coulomb Failure Law*). The

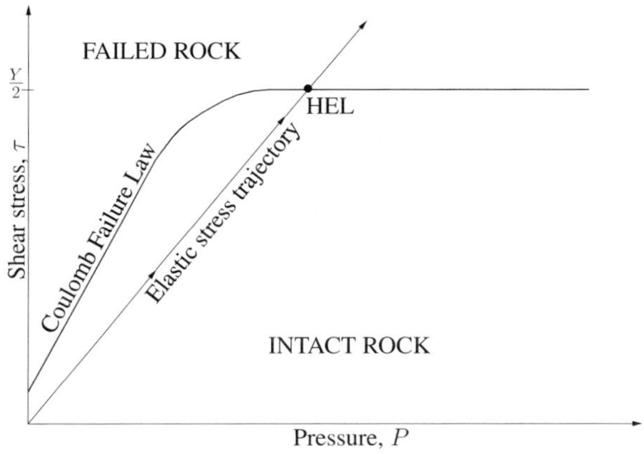

Fig. 2. Shear stress as a function of the mean pressure to show the failure of rocks when the pressure exceeds a certain shear stress. The trajectory is linearly increasing and intersects the failure envelope at the Hugoniot elastic limit (HEL). Adopted from (7).

specified trajectory of an elastic stress increases linearly with mean pressure, which means that the ratio between shear stress and the mean pressure is constant. It rises slightly less steeply than the illustrated failure envelope for a *typical* rock. Because of the reversible character of elastic waves, the solid, through which the wave travels, remains intact and the rock properties after the shock has passed correspond to the initial conditions before the waves arrival.

Once the elastic trajectory crosses the yield envelope (the corresponding longitudinal stress is called the *Hugoniot Elastic limit*, HEL) where the shear stress is $Y/2$), shear stress can no longer be accumulated. The pressure impulse of the stress wave that is strong enough to exceed the yield envelope results in irreversible changes in the rocks (plastic deformation). All further strain from this point on is purely compressive since the maximum shear stress is limited by the yield strength. The mean pressure P can be expressed by the longitudinal σ_L and perpendicular σ_P stresses (if we assume that two components of the stress tensor σ_{ij} are equivalent),

$$P = -\frac{\sigma_L + 2\sigma_P}{3}$$

Respectively, the shear stress τ is given by

$$\tau = \frac{\sigma_L - \sigma_P}{2}$$

With the help of these expressions, the Hugoniot Elastic Limit (HEL) can be expressed by

$$\sigma_L = -\sigma_{HEL} = -\left(\frac{1-\nu}{1-2\nu}\right) Y \qquad (1)$$

with ν as the Poison ratio. The distinction between elastic and plastic waves is also illustrated in Fig. 3a, here the particle velocity is plotted versus propagation speed U of stress waves. For elastic waves the propagation speed is constant and both the shear modulus and the bulk modulus contribute to the longitudinal velocity

$$U \equiv c_L = \sqrt{\frac{(K_0 + \frac{4}{3}\mu)}{\rho_0}}$$

If the stress becomes larger than the Hugoniot Elastic Limit σ_{HEL}, which correspond to certain particle velocity u_p, the propagation speed goes down to nearly the bulk propagation speed

$$U \equiv c_B = \sqrt{\frac{K_0}{\rho_0}}$$

So far the bulk modulus K_0 was assumed to be a constant material property.

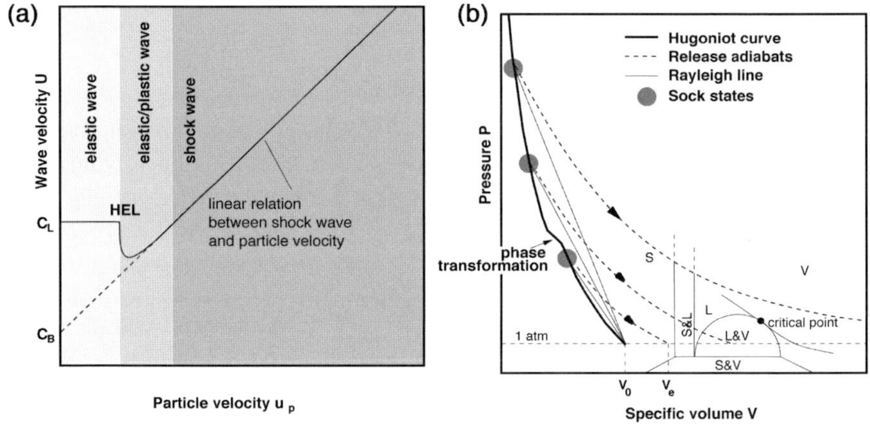

Fig. 3. (a) The propagation speed of stress waves as a function of the particle velocity. (b) Hugoniot curve in the $P - V$ and $U_p - u_p$ plane. S, L, V mark the phases solid, liquid and vapor. Modified after (26; 7).

Although K is a function of the pressure and increases under compression. K_0 needs to be replaced by $K = f(P)$ in the equation for the bulk propagation speed c_B. At the onset of elastic-plastic transition, close to the HEL, the plastic wave travels significantly slower than the elastic wave (Fig. 3a) . However, in the high–pressure regime (large longitudinal stress $-\sigma_L$ or high particle velocities u_p) $K > (K_0 + 4/3\mu)$ and thus $U > C_L > C_B$ (with $U = \sqrt{K(P)/\rho_0}$) and the plastic wave begins to overtake the elastic wave front. Such a wave is considered as a shock wave because it travels faster than the speed of sound (Fig. 3a).

Shock waves propagate not only faster than elastic waves they also show a different characteristics than elastic and plastic waves. The wave front (Fig. 4a) corresponds to an abrupt pressure rise. The best mathematical approximation for such a sudden rise is a discontinuous jump.

The fundamental equations describing the conditions before and after the shock front were derived in 1887 by P.H. Hugoniot and are based on the conservation of mass energy and momentum (see e.g. (26)). Figure 4c shows the setup with ρ_0, P_0, T_0, E_0, and u_0 as the density, pressure, specific internal energy, and particle velocity in the uncompressed area, while ρ_s, P_s, T_s, E_s, and u_p denotes the same parameters behind the shock front in the compressed regime. The parameters ρ, P, E, T, and u describe the state of the material after the passage of the shock wave. Assuming

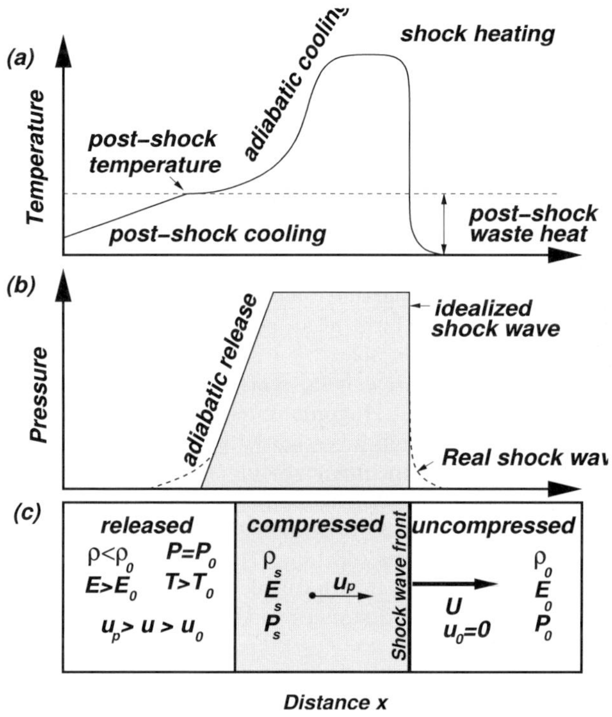

Fig. 4. (a)Showing a cross-sectional intersect of the temperature, (b) of the pressure. (c) Parameters describing the pre–shock, shock and release conditions.

that V_s is the compressed and V_0 the uncompressed specific volume ($\rho = 1/V_s$, $\rho_0 = 1/V_0$) and that the shock front travels at a velocity U, the Hugoniot equations are:

$$\begin{aligned} \rho\,(U - u_p) &= \rho_0 U \\ P_s - P_0 &= \rho_0 u_p U \\ E_s - E_0 &= \frac{1}{2}\,(P_s + P_0)\,(V_0 - V_s) \end{aligned} \tag{2}$$

The shock wave compression is an isentropic process where the state of the material changes almost instantaneously. The reverse process, the release from high pressure, happens under thermodynamic equilibrium conditions. The release (Fig. 4b) is much smoother and corresponds to an adiabatic decompression path. Shock compression is a thermodynamic irreversible process during which more internal energy is transferred to the material than it is released by decompression. The remaining residual energy is turned into heat (post-shock temperature) as illustrated in Fig. 4a and a residual particle velocity. The induced velocity field behind the shock wave plays an important role in crater excavation and will be addressed in the section 2.2 on impact cratering.

Post-shock conditions of a solid material are basically a function of the shock intensity (pressure amplitude). In other words, the stronger the shock wave the more the material changes its thermodynamic (or mechanical) state after release. Different shock states are illustrated by grey dots in Fig. 3b where specific volume is plotted versus peak shock pressure. The curve connecting these states is called the *Hugoniot curve* and is not to be mistaken with the thermodynamic path the material goes through to reach a certain shock state (grey dot). For instance if a material has an initial specific volume V_0 then it is compressed along the so called *Rayleigh line.* After release from shock state along adiabatic decompression paths it may cross phase boundaries on the thermodynamic phase diagram and end up in a molten or vaporized state (see Fig. 3b). By the usage of the Hugoniot equations (Eqs. 2) the $P - V$ diagram can be turned into a $u_p - U$ diagram (Fig. 3b). Obviously there exists a linear relation between particle velocity and propagation speed for shock waves which represents a simple form of an equation of state already. This will be discussed in more detail in section 3.2.

Finally, shock waves can cause a solid state phase transition, known as shock wave modifications (see kink in the Hugoniot curve in Fig. 3a). In some minerals, for example Quartz, this process is well investigated and the presences of high pressure phases like Coesite or Stishovite unequivocally identify shock wave events, which can be only induced by meteorite impact events. Specific shock wave features (e.g. planar deformation features, shatter cones, Stishovite, Coesite, diaplectic glasses) can be assigned to certain pressure ranges (43; 46; 47; 48) and therefore it is possible to reconstruct the shock wave decay in a natural impact crater by the mineralogical analysis of samples from different distances to the point of impact.

2.2 Impact cratering

The physical processes during the formation of a crater are summarized in the term *Impact cratering.* According to Gault et al. in (5), impact cratering can be subdivided into three different stages, which are characterized by suite of spatio–temporal processes. However, the transition between subsequent stages is not sharp; it is more a continuous change from one stage to another, which may not occur everywhere at the same time. In other words it is possible that across the forming crater structure at different locations, a later stage has already begun while at another location the earlier stage has not yet been completed (e.g. (6; 7)). The three stages of impact cratering are: (i) Contact– and Compression stage, (ii) Excavation stage, and (iii) Modification stage.

The processes described within the concept of impact cratering (according to Gault et al. (5)) focus on the interaction between the solid target (continental, or oceanic crust) and the impacting body. However, technically the impact on Earth begins when the impacting body enters the atmosphere. While traveling through the atmosphere, the body continuously looses energy. Similar is the process in the case of an oceanic impact, where the projectile is not only decelerated by a gas layer but also by the water column. In both cases the kinetic energy loss can be estimated by computing the drag a body experiences at supersonic speed while moving through a comparatively low density medium (atmosphere or water; see e.g. (32; 49)). Depending on the structural and chemical composition, and the mechanical properties (strength) of the impactor, it might explode, burn out, or be decelerated to the gravity controlled "fall velocity", in which case no hypervelocity impact crater would be formed. The distinction between hypervelocity impact craters and ordinary low velocity ($\ll$speed of sound) impact craters is very important and will be clarified in the following section

Contact– and Compression Stage

The contact– and compression stage begins when the impactor touches the surface of the target. The velocity of the projectile at the time of impact may vary between the escape velocity and 70 km/s (for comets). The average impact velocity on Earth is 18 km/s (e.g. (29)). Due to the high velocities, shock waves are generated at the interface between the impacting body and target; although there occur two shock fronts: One shock travels upwards into the projectile and another one travels downwards into the target. In the region between the shock fronts (the shock plateau) the material is compressed to high pressures at constant amplitude. Other properties, such as densities in the target and projectile or the specific internal energy and the propagation velocity of the shock front depend on the material properties of the projectile and the target. The conditions during shock wave compression can be determined by using the Hugoniot equations (e.g. (26; 7)) for each material, separately.

The snapshot series in Fig. 5a-f illustrates the contact and compression stage by means of a 1 km seized projectile that has an impact velocity of 10 km/s. Fig. 5b shows the two shocks, one is traveling upwards in the projectile and a second is traveling downward in the target. Depending on the composition of projectile and target, and the impact velocity, the pressure amplitude, the induced particle velocities and the propagation speed can be estimated by the usage of the Hugoniot equations (e.g. (26; 7)). Assuming that the pressure is equal in both shocks (for projectile and target) and that the projectile is moving downward with the impact velocity v_i, there can be given a linear relation between the particle velocity in the target and the projectile

$$u_t = v_i - u_p \tag{3}$$

where u_t and u_p are the particle velocities in the target and projectile. If the projectile and target consist of the same material. u_t and u_p are equal and correspond to $1/2v_i$. With the help of this linear relation of the particle velocities, the propagation speed of shocks (see 3a) U can be calculated

$$U = C_B + Su \tag{4}$$

where C_B is approximately the bulk speed of sound and S a material parameter (e.g. $S = 1.6$ for water or $S = 1.24$ for granite; for more parameters see (7)). Finally with the Hugoniot equation (e.g. (26; 7)) the peak shock pressure can be determined from the initial density ρ_0, the particle velocity u and the propagation speed of shock U. In a similar manner the same computations can be done if projectile and target material are not the same (see Ahrens et al. in (25)). Note that these calculations are only valid for a planar wave, which work in the case of a spherical impactor only for the very first moment of impact. However, these equations give good results in

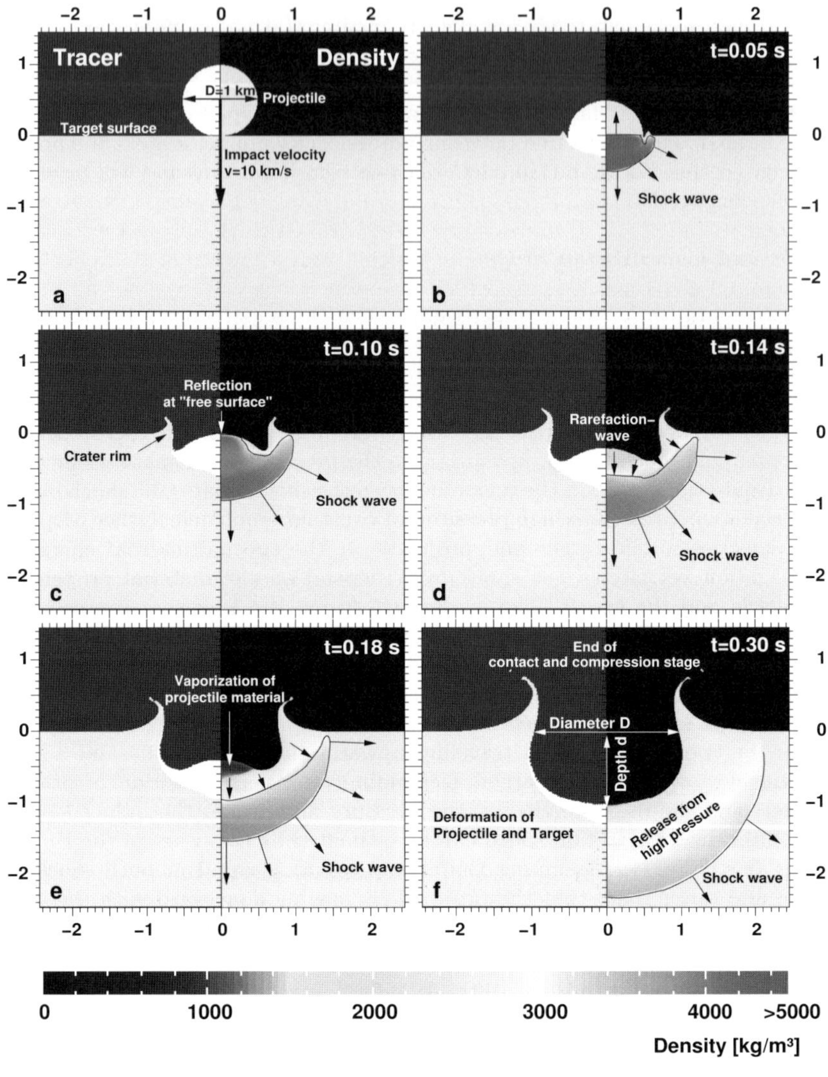

Fig. 5. Snap shot series of the contact and compression stage. The left side of each panel gives the tracer distribution at a certain time and the right side shows the densities. The scale on the axes is in km

(52).

order to estimate the initial peak shock pressure in a moderate pressure range (the linear relation in Eq. 4 holds only in a certain pressure range, see e.g. (44).

The pressures that occur during a meteorite impact reach several hundreds of GPa, which easily exceeds the Yield strength Y of any solid. As shown in section 2.1, such pressure impulses result in plastic deformations that include the target and the projectile. Owing to the plastic deformation, the crater initially begins to grow, and the base of the projectile fits more and more the shape of the growing crater cavity (Fig. 5b-f). The shock wave that travels in the projectile is reflected at the free surface at the back of the projectile (Fig. 5c). During the reflection the pressure impulse changes its sign, which means that the wave is now characterized by decompression rather than by compression. The reflected wave propagates as a rarefaction wave toward the target. The adiabatic decompression can propagate faster in compressed material than the shock wave itself. The material may be vaporized, molten, or shattered, after the passage of the shock during rarefaction (Fig. 5d-e). Figure 5f depicts the final phase of the contact and compression stage. The projectile is now completely unloaded from pressure, and in most cases it is also completely vaporized (e.g. (7)). This snapshot (Fig. 5f) can also be seen as an early stage of subsequent excavation stage. This fact underlines that there is no sharp boundary between two subsequent stages in space and time in the concept of impact cratering according to Gault et al. The duration of this stage is between $\sim 2.5 \cdot 10^{-3}$ and $\sim 2.5 \cdot 10^{-1}$ (7). This interval corresponds to a diameter of the projectile of between 10 m for the lower boundary and 1 km for the upper boundary value (7).

Excavation stage

The excavation stage is characterized by two processes: (i) the propagation and weakening of the shock wave, and (ii) the excavation of the crater cavity. During the contact and compression stage the shock wave is traveling downwards into the target and decays only due to the spherical spreading of the wave (the same energy has to be distributed over an increasing area), which yields to an amplitude attenuation proportional to $1/r$. The important difference of the propagating shock wave during the excavation stage compared to the contact and compression stage is that it now interferes with rarefaction wave coming down from the back side of the projectile and the free surface of the target. The rarefaction wave attaches to the shock wave, which leads to a much more rapid decay of the shock wave ($\geq 1/r^2$ depending on the impact velocity (25)). The propagation process does no longer directly interact with the crater formation. The combination of the shock wave with an anchored rarefaction wave is called a detached shock wave (7).

An important aspect, which distinguishes hypervelocity impacts ($v \gg C_B$) from the low velocity strike of a projectile, is that the passage of a shock wave leaves a residual velocity component in the target material (see section 2.1 about shock-waves). The value of the velocity component typically corresponds to one-fifth of the peak particle velocity during shock wave compression (44), and it therefore is large enough to induce a material flow (excavation flow) directed outwards, away from the point of impact. Due to interferences with free-surface, an upwards component is added to the velocity field, which causes the ejection of material out of the crater (excavation flow). The geometry of the excavation flow is illustrated in Fig. 6a. The lateral and vertical growth of the initially spherical, later approximately parabolic cavity can be described by scaling laws in case of a vertical impact (38):

$$\frac{x}{r} \sim \left(\frac{vt}{r}\right)^{\alpha} \tag{5}$$

The parameter x can be replaced by either the crater depth d or the diameter of the crater D. The exponent α is ~ 0.36 for d or ~ 2.0 for D. Parameter r is the radius or the projectile. The term in brackets can be interpreted as a scaled time t which allows to compare similar stages of crater formation independently to the size and velocity of the projectile.

Eventually, the growth of the crater is stopped by the strength properties of the surrounding target material (dry friction or any cohesive resistance) and primarily (for most natural craters on planetary surfaces) by gravity. The diameter D_t of final crater is called the *transient crater* (42) and can by estimated by the following equation (38):

$$D_t = 1.161 \left(\frac{\rho_p}{\rho(t)}\right)^{1/3} (2r)^{0.78} v^{0.44} g^{-0.22} \tag{6}$$

Equation 6 is based on experimental observations and numerical modeling (28).

Because of the fact that the excavation flow does not cease in all directions at the same time and does not grow laterally and vertically at the same rate the maximum extent of the crater cavity is best described by a paraboloid. However this may hold only as an idealized description since this situation never really exist during crater formation in reality. In fact crater modification launches already in vertical direction while the crater diameter is still growing. However small bowl-shaped craters (simple craters; see next paragraph) come quite close to the theoretical construct of a transient crater.

Modification Stage

The transient cavity is not the final form of the crater. As already mentioned, the target strength and gravity work against the excavation flow, until the growth of the crater eventually ceases. Both parameters have an essential influence on the modification of the transient cavity and the balance of the two forces is crucial for the subsequent modification processes. As already stated in section 1, the crater morphology is a function of the crater size. Simple bowled-shaped craters occur up to a certain transition diameter D_{trans} (45). Above the threshold value D_{trans}, morphology becomes more complex and can be subdivided in to the different categories of central peak, peak-ring craters and multi-ring basins. Assuming that the controlling properties (strength) on surfaces of different planetary bodies do not significantly differ, it can be concluded that the transition diameter D_{trans} between simple and complex craters is only a function of gravity (45), which means that D_{trans} varies for different planetary bodies. For example, on the Moon, D_{trans} is $15km$ and on Earth $2-4km$. The variation on Earth can be explained by varying strength properties of the Earth's crust (sedimentary targets bear less resistance against deformation than crystalline targets do).

In order to quantify the dependence of crater morphology on gravity and crater diameter it is convenient to introduce non-dimensional parameters. Like in Hydrodynamics, the ratio between internal force and gravitational influence can be expressed by the *Froude number*, which represents a parameterization of gravity (29):

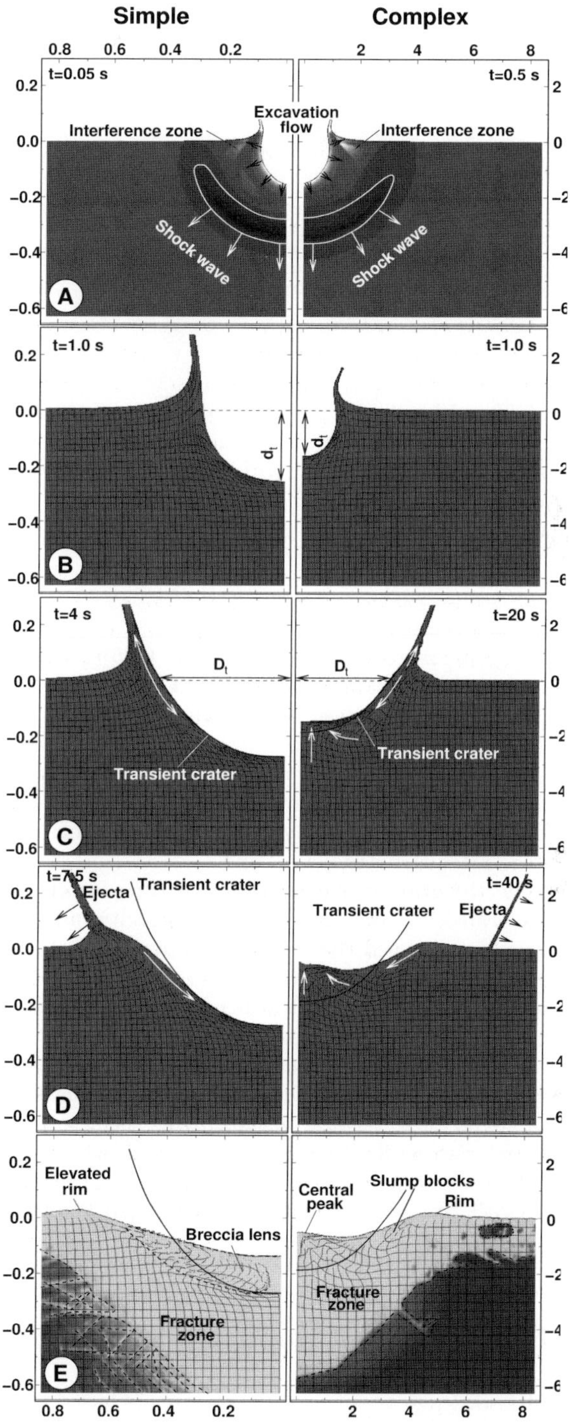

Fig. 6. (A) Snapshot of the excavation after $t = 0.05s$ for the simple crater and $t = 0.5s$ for the complex crater. (B-D) Depict differences in the crater modification for simple and complex craters. (E) The internal structure of the a simple and complex crater.

$$F = \frac{v_i^2}{gd_t} \tag{7}$$

Analogous to Eq. 7, the strength of the target can be expressed by the *Cauchy* number which is given by the ratio of stresses to strength (29):

$$C = \frac{\rho v_i^2}{Y} \tag{8}$$

The ratio of these two important numbers poses a good parameter to quantify the balance between gravity and strength (50):

$$S = \frac{Y}{\rho g d_t} \tag{9}$$

where d_t can be calculated by $d_t/r = 1.2\left(gr/v_i^2\right)^{-0.22}$ (e.g. (29)). To determine the value S_{trans} for the transition between simple and complex craters, Melosh in (51) assumed the growth of a parabolic cavity, which results in an upward–directed force $F_A = (1/8)\pi\rho g D_t^2 d_t$ (buoyancy). The force F_A acts on the area $A = (1/2)\pi D_t^2$. The strength is $Y = F_A/A$, which eventually gives:

$$S_{trans} = \frac{Y}{\rho g d_t} = 0.25 \tag{10}$$

If the ratio S $(= f(Y, g, d_t)$ is greater than S_{trans} $(= 0.25)$ then the strength of the target is predominant; a simple crater is generated. The material lining the transient cavity slides downwards to the deepest point of the cavity (Fig. 6C-E) and gathers together to a so called breccia lens consisting of heavily fractured rocks and little melt. An elevated crater rim with an overturned stratigraphy marks the rim of the crater. Vertically and laterally the evidence of the passage of a shock wave extends beyond the boundaries of the crater cavity. A large zone of fractures and faults surrounds the visible crater structure (Fig. 6E).

If $S < S_{trans}$ then the gravity dominates and a complex crater will be formed. This means that the transient cavity is big and the gravity induces lithostatic pressure that is high enough to exceed the strength properties of the surrounding rocks and collapse occurs. First the crater floor starts to rise (Fig. 6C). Simultaneously the rim slumps into the cavity and an upwards rising central peak emerges. For very large impact craters the central peak overshoots a certain limit, becomes unstable and collapses again, resulting in the formation of a peak-ring or multi-ring basins in the case of an oscillating behavior. The entire process appears to be quite similar to the collapse of a cavity in water and in fact, rocks seem to behave on the given scale much more like a viscous fluid than a brittle solid, which appears to support the simplifications made in section 2.1 on physics of shock waves.. A more detailed description of crater collapse is given in Melosh in (7) and Wünnemann and Ivanov in (50).

3 Numerical modelling of impact processes

Most of our present knowledge about impact cratering is based on (i) field studies of natural impact structures and remote sensing of planetary surfaces, (ii) large-scale explosion tests and laboratory impact experiments, and (iii) numerical computer simulations. Due to the large increase in the capability of modern computers the latter has gained importance in the last decades in particular with respect to shock wave propagation (e.g. (19)), ejection of material (15; 16)) and crater formation, and late stage modification (e.g. (17; 18)).

3.1 Equation of motion

The numerical simulation of impact processes is a classical fluid dynamic problem, which is usually solved by assuming that the matter (fluid, vapor, or solid state body) is a continuous media which can be plastically deformed. The dynamics can then be described by a set of differential equations (equations of motion, EM) based on the principles of mass, energy and momentum conservation. Most fluid dynamic problems deal with subsonic, incompressible flows (flow speed $v \ll$ speed of sound c), like classical convection problems where the mass transport is driven for instance by a temperature or density contrast. This limitation allows some simplification of the equations of motion, e.g. the Boussinesq approximation for an incompressible, constant density problem (20). The extremely high velocities of meteorite impacts induce shock waves which propagate by nature much faster than the speed of sound ($v > c$, see previous section on shock waves) and thus have to be described as compressible supersonic flow. The derivation of the full set of conservation equations of a super- or hypervelocity ($v \gg c$) flow is beyond the scope of this chapter and we refer to (20) or (21), where the so called Navier-Stokes equations for a compressible media are described in detail. This set of equations relates bulk density ρ, velocity v_i, specific internal energy E, the stress tensor σ_{ij}, which is composed of the hydrostatic part, the pressure P, and the deviatoric part Π_{ij} ($\sigma_{ij} = -P\delta_{ij}\Pi_{ij}$), and gravity g_i. Generally one has to distinguish between fundamentally different descriptions of the kinematic deformation/transport of a continuous media: the *Eulerian* (spatial) and *Lagrangean* (material) notation. The Cartesian form of the Navier-Stokes equations for both kinematic approaches is given by:

	Lagrangean	Eulerian
Mass:	$\frac{D\rho}{Dt} + \rho\frac{\partial v_i}{\partial x_i} = 0$	$\frac{\partial \rho}{\partial t} + \frac{\partial}{\partial x_i}(\rho v_i) = 0$
Momentum:	$\frac{Dv_i}{Dt} = g_i + \frac{1}{\rho}\frac{\partial \sigma_{ij}}{\partial x_j}$	$\frac{\partial v_i}{\partial t} + v_j\frac{\partial v_i}{\partial x_j} = g_i + \frac{1}{\rho}\frac{\partial \sigma_{ij}}{\partial x_j}$
Energy:	$\frac{DE}{Dt} = g_i v_i + \frac{p}{\rho}\frac{\partial v_i}{\partial x_j} + \frac{1}{\rho}\Pi_{ij}\dot{\epsilon}_{ij}$	$\frac{\partial E}{\partial t} + v_i\frac{\partial E}{\partial x_i} = g_i v_i + \frac{p}{\rho}\frac{\partial v_i}{\partial x_j} + \frac{1}{\rho}\Pi_{ij}\dot{\epsilon}_{ij}$

(11)

We adopt the Einstein convention which implies the summation of repeated indices. The differences between the two sets of equations are inherent in the definition of the total time (or substantial) derivative D/Dt:

$$\frac{D}{Dt} = \frac{\partial}{\partial t} + v_i\frac{\partial}{\partial x_i} \tag{12}$$

A less mathematical and more descriptive distinction is given by the assumption that in the Eulerian kinematic approach the independent space variables are associated with a coordinate system fixed in time and space, through which the fluid or matter is transported. In the Lagrangean formulation the coordinate system is connected with the matter and transport as warranted by the movement/deformation of the grid itself. This will become more apparent when describing the numerical implementation of these equations into a hydrodynamic computer model later in the text. For the full description of compressible plastic flow two more expressions are required.

3.2 Equation of state

The thermodynamic equation of state (EOS) relates pressure P to material density ρ and internal energy E ($E = e - 1/2v_i v_i$) or temperature T. Since most numerical models typically yield density ρ and internal energy E as a result of one computation cycle it is more convenient to express the pressure as function of these two quantities. The temperature T is then only indirectly related to the specific internal energy E.

$$P = f(\rho, E) \tag{13}$$

The equation of state poses a central role in the description of the thermodynamic behaviour of matter and is of particular importance for the treatment of shock wave compression. For simple materials (e.g. metals) equation of states can be derived from solid-state physics with a reasonable degree of accuracy. However, most materials of geological interests show a more complicated behaviour which can be described only by semi-analytic EOS which relate empirical data with meaningful physical assumptions (e.g. Thomas-Fermi asymptotic limit). For a large range of possible shock (pressure and temperature) states, phase transitions (solid state, melting, vaporization, dissociation, ionisation) must be defined by the EOS providing reasonable assumptions of material properties (speed of sound, heat capacity, heat conductivity, etc.) under the given conditions. Therefore more advanced EOS often require a long list of material parameters and do not allow a simple analytic solution for a particular state.

It is not the objective of this chapter to discuss this topic in detail and we refer to more appropriate technical literature (e.g. (26)). We want to mention here only a few commonly used EOS that the reader may come across in the context of shock wave compression and meteorite impact processes. The *perfect gas equation* is probably (do you have to say probably, how about it sounds too unsure) the simplest approach to describe the thermodynamic behaviour of an ideal gas: $P = (\gamma - 1)\rho E$, where γ is the ratio of specific heat. Despite its simplicity it provides reasonable good results for shock wave compression for instance in the atmosphere during the passage of a meteorite before it impacts the Earth surface. More complex equation of state, suitable for describing the thermodynamics of solid-state materials, are the *Murnaghan* and the *Mie-Grüneisen* EOS (e.g. (7)) which have been successfully used, e.g. for the description of the thermal state of the Earth interior and plastic waves at moderate pressure levels. For the treatment of shock waves, where a wide range of pressures and density is met, the *Tillotson* EOS (e.g. (23)) was especially designed. It duplicates the linear shock-particle velocity relation at low pressures and approaches the Thomas-Fermi limit for infinite pressure. Since this EOS is an important analytic expression for shock thermodynamics we provide here the full set of equations. For the compressed regime ($\rho > \rho_0$, the density at zero pressure) and cold expanded state ($E < E_{iv}$, the energy of incipient vaporization) P is given by:

$$P_c = \left(a + \frac{b}{(E/(E_0\eta^2)+1)}\right)\rho E + A\mu + B\mu^2 \tag{14}$$

The form used in the expanded state ($\rho_0 < \rho$), when the energy exceeds the energy of complete vaporization ($E > E_{cv}$) is:

$$P_e = a\rho E + \left(\frac{b\rho E}{(E/(E_0\eta^2)+1)} + A\mu e^{-\beta(\rho_0\rho - 1)^2}\right) e^{-\alpha(\rho_0/\rho - 1)^2} \tag{15}$$

For the partially vaporized region ($\rho_0 < \rho$ and $E_{iv} > E > E_{cv}$), a hybrid formulation is a common approximation (7; 25):

$$P = \frac{(E - E_{iv})P_e + (E_{cv} - E)P_c}{(E_{cv} - E_{iv})}. \tag{16}$$

In these equations $\eta = \frac{\rho}{\rho_0}$, $\mu = \eta - 1$ and $A, B, a, b, \alpha, \beta, E_0$ are material parameters (see Tab.1). The Tillotson EOS generally provides good results for metals and for some relatively simple rock types where solid state phase modifications are negligible. A much more sophisticated description of the thermodynamic state is given by the *Analytic* EOS (*ANEOS*). This EOS is basically a package of analytical models describing the thermodynamic state over restricted phase space areas. In order to achieve a continuous transition at phase boundaries complex mathematical techniques are required which are described in more detail in (24). In summary, EOS are essential for the treatment of shock waves and for the models presented in this chapter we refer to the Tillotson EOS and ANEOS.

Table 1. *Tillotson equation of state parameters (7)*

Material	ρ_0 kg/m^3	a	b	A (GPa)	B (GPa)	E_0 (MJ/kg)	α	β	E_{iv} (MJ/kg)	E_{cv} (MJ/kg)
Iron	7800	0,5	1,5	128	105	9,5	5	5	2,4	8,67
Aluminum	2700	0,5	1,63	75,2	65	5	5	5	3,0	13,9
Granite	2680	0,5	1,3	18	18	16	5	5	3,5	18
Gabbroic Anorthosite										
(lpp)	2940	0,5	1,5	71	75	478	5	5	4,72	18,2
(hpp)	3970	0,5	1,3	240	130	1800	5	5	3,19	16,8
Andesite	2700	0,5	1,3	18	18	16	5	5	3,5	18
Wet Tuff	1970	0,5	1,3	10	6	11	5	5	3,2	16
Dry Tuff	1700	0,5	1,3	4,5	3	6	5	5	3,2	18
Limestone	2700	0,5	0,6	40	67	10	5	5	2,5	14
Halite	2160	0,5	0,6	25	30	5	5	5	2	15
Alluvium	2600	0,5	0,8	30	10	6	5	5	3,5	18
Water (0^oC)	998	0,7	0,15	2,18	13,25	7	10	5	0,419	2,68
Ice (hpp, -10^oC)	1293	0,3	0,1	10,7	65	10	10	5	0,773	3,04

3.3 Constitutive equations

During shock wave compression material can be treated as fluid-like, since the hydrodynamic pressures exceed by far the material strength. After decompression the material returns to a brittle or ductile properties. Whilst the contact and compression stage and the excavation stage are mainly controlled by hydrodynamic material flow, the late stage modification like crater collapse is dominated by the balance between gravity driven stresses and the shear strength of the material. Similar to the concept of an EOS constitutive equations (CE) are used to relate shear stresses σ, to strain ϵ, temperature T, strain rate $\dot{\epsilon}$, and damage D:

$$\sigma_{ij} = f(\epsilon_{ij}, \dot{\epsilon}_{ij}, T, D) \tag{17}$$

For the purely elastic regime this general expression can be replaced by Hook's law, wherein strain is a linear function of stress: $\sigma = E\epsilon$, where E is the Young's modulus (Fig. 7a). If the stresses are in excess of a certain yield point (strength), the material starts to behave plastically and offers no further resistance to stress. This idealized material behaviour represents reasonably well the response of many metals to large stresses. Rocks exhibit a much more complicated rheology when entering the plastic regime of deformation. It is therefore necessary to distinguish brittle from ductile failure. Strength may be different when subject to compression, tension, or shear (e.g. (9)). In the following we focus on shear strength Y of rocks as a function of pressure P, temperature T, strain rate $\dot{\epsilon}$, and the previous deformation history D (damage), since it has turned out to be of most importance to crater collapse (modification stage). The influence of confining pressure is illustrated in Fig. 7b. The solid line describes the yield strength Y as a function of confining pressure P from purely brittle behaviour to ductile flow regime. The course of the yield strength may be described by (10):

$$Y(p) = Y_C + \frac{\mu P}{1 + \frac{\mu P}{Y_m - Y_C}} \tag{18}$$

where Y_C is the cohesion at zero pressure, μ the internal friction coefficient of intact material, and Y_m the yield strength at infinite pressure (*Von Mises plastic limit*). The black dots mark experimental values for intact Berea sandstone (8). If the same material has experienced already some sort of deformation history it might have been fractured (damaged) due to shearing or tensile failure in the brittle regime. In this case the material loses its cohesive strength at zero pressure and the increase of shear strength with confining pressure becomes linear (*Mohr-Coulomb* strength model; (11)). This material model represents a simplified description of the response of rocks to elasto-plastic deformation. The influence of temperature, the dynamic computation of shear- and tensile fracturing, and strain-rate are described in more detail for example at (12; 14; 13). Generally the strength models have become quite sophisticated; however many processes like *shear localization* (faulting), *shear bulking* (dilatency) are not reproduced by these models and further development are required to match the observed crater morphology.

3.4 Implementation of a hydrocode

A number of models have been proposed to compute the high velocity strike of an asteroid on both solid (e.g. (27; 28; 30; 17)) and oceanic targets (e.g. (32; 31)). All hydrocodes that have been developed in the past (e.g. CTH, (33), and SOVA, (34)) are based on the solution of the EM, EOS, and CE using different numerical solution

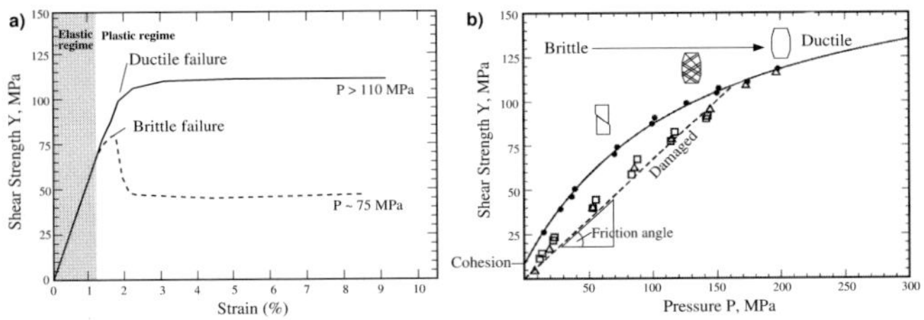

Fig. 7. a) Sketch of a stress-strain curve as measured in triaxial tests at different confining pressures. b) Shear strength as a function of confining pressure for Berea sandstone (intact, black dots, solid line), (damaged, open symbols, dashed line). The original measured values are from (8).

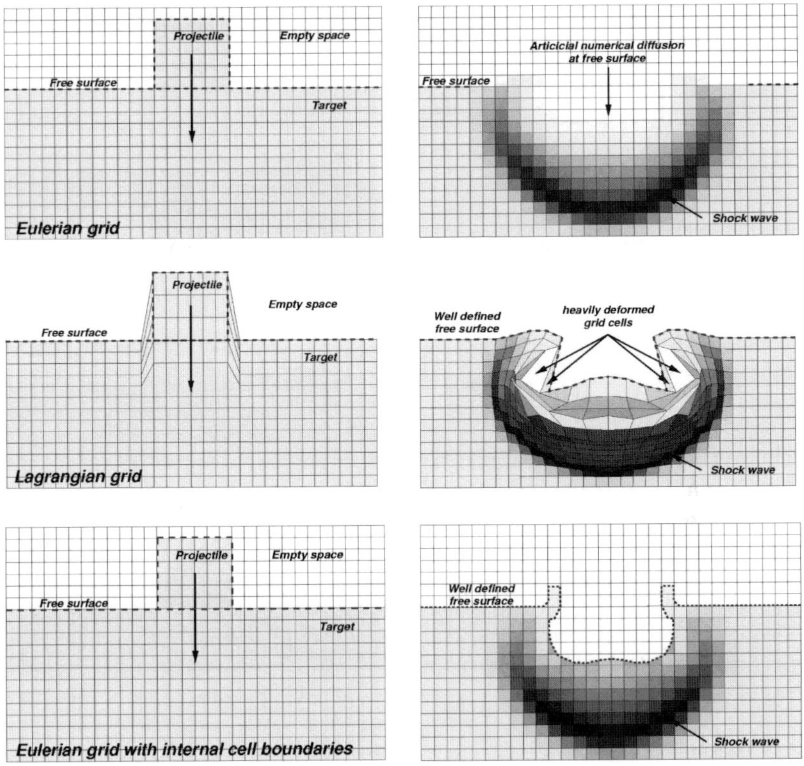

Fig. 8. Comparison between the different kinematic models (Eulerean, a; Lagrangean, b; and Eulerian with interface tracking, c) by means of a cylinder impacting on the surface of a half-space. Projectile and target consist of the same material

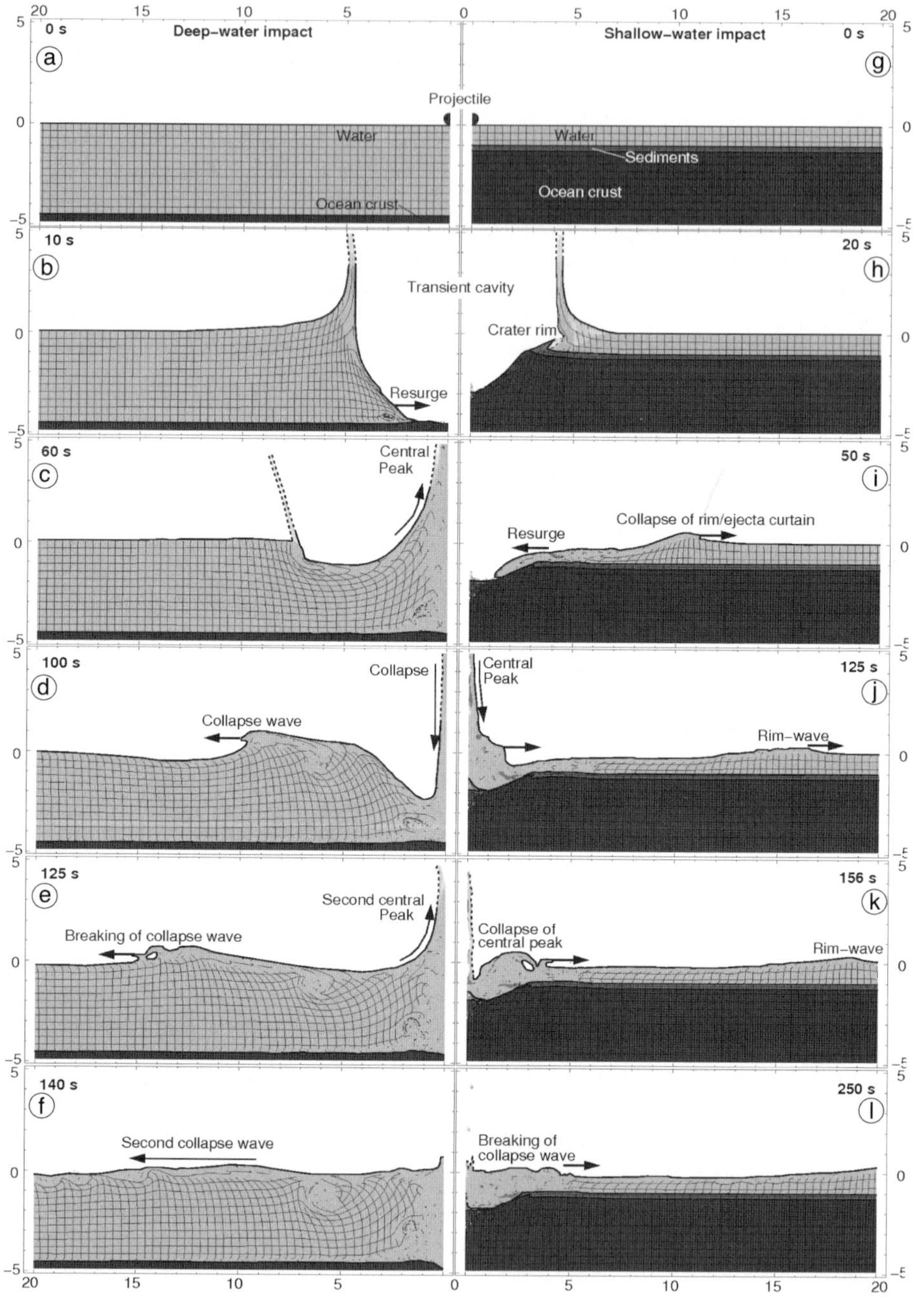

Fig. 9. Two snapshot series showing the evolution of the water column for an impact into deep water (a–f) and into shallow water (g–l). Scales are in km.

techniques (finite difference, FD; finite volume, FV; smooth particles, SPH; etc) and kinematic approaches (Eulerian/Lagrangean). A general overview of hydrocode modeling is given in (35) or (36), respectively. All codes have in common that the computational domain is discretized into a grid of cells with a specified size (dx, dy) to replace the spatial derivatives in the differential equations by simple (finite) difference equations. Analogously, time is discretized by the timestep dt which defines the length of each integration cycle. During each cycle the flow and the position of the grid due to external and internal forces is computed for each cell of the grid. The two different formulations of the EM, the Eulerian and the Lagrangean description, result in differences in the numerical approaches. As mentioned already above, the difference lies in the way in which mass, energy and momentum are transported through the grid. In the Lagrangean kinematic model (see Fig. 8b), matter is associated with a computational cell. Material motions are calculated by the deformation of the cell, yielding a distortion of the entire grid. Mass within a cell is invariant; changes in density are exclusively due to changes in a cell's volume. If the deformations become too large the finite difference approximation may become inaccurate and in extreme cases it may even happen that a cell folds over itself resulting in the computation of a negative volume. In contrast, the dynamic motions of mass, momentum and energy are realized in the Eulerian approach by computing the transport between adjacent cells in a grid that remains fixed in space (Fig. 8a). Impact processes cause large deformations and even fluid and gas flow due to phase transformation may have to be taken into account. Therefore, the Eulerian description appears to be the most approach for impact calculations. However, the transport of mass across cell boundaries gives rise to artificial mixing of different kinds of matter along material boundaries, referred to as artificial numerical diffusion. For example, even the simplest assumption of a model set-up implies a boundary between a half-space consisting of water or solid rock and void (empty space or vacuum) above. The material transport from a full cell to an adjacent empty cell yields mixing of vacuum and matter. The sharp boundary at the free surface becomes more diffuse, the more time steps have been integrated. This artificial numerical diffusion process is illustrated in Fig. 8a. To avoid material mixing and to preserve sharp material boundaries one has to introduce a procedure to track the boundaries between different kinds of matter in cells containing a mixture. This is a rather complicated procedure and a description in-depth is beyond the scope of this paragraph. We refer to (21) for further details. Figure 8 shows a simple numerical simulation of a block impacting on a target using the three different solution schemes: Eulerian, Lagrangean, and the Eulerian schema including a surface tracking routine. For simplicity there is only one boundary at the free surface involved; in a more realistic example different kinds of matter (water, rock or different rock types) have to be taken into account and for each of those the boundary tracking routine has to be applied.

3.5 Limitations and numerical stability

Generally numerical models can be used to simulate shock waves and deformation processes on any scale (micro-scale: intergranular deformations and thermodynamic phase transformations; meso-scale: shearing and compaction of grains and rock fragments; macro-scale: crater formation and deformation of entire rock units). The important flow variations, which may be thermodynamically or mechanically driven, must be resolved by a sufficient number of computational cells in the underlying grid. For this reason it is impossible to run simulations that can resolve both micro-scale processes and large scale deformations like crater formation, in one model at the same time. As this chapter is about ocean impacts, associated with the generation of large water wave and crater formation the objective of the computational grid is to resolve the projectile, the water layer, and the ocean bottom and crust underneath properly. This means, if the proportion between two of these units becomes extreme (e.g. if the water depth H is much smaller than the projectile diameter d) the smaller object cannot be resolved by a sufficient number of cells. In this case the simplified approach namely that the smallest units do not significantly affect the overall processes, and therefore can be neglected, is the usual procedure of solution. The model requirements have to be balanced with the available hardware configuration, which also depends on the dimensionality of the problem under investigation. Whether two, or three-dimensional modeling is essential depends on whether obliquity of the impact is important, and whether the target consists of planer layers (water, ocean-bottom sediments, crust). If a cylindrical axial symmetric condition is applied, then two-dimensional modeling of vertical impacts is often sufficient. This approach is widely used due to the very high costs in terms of computer memory and run-time of 3D computations. The total computer storage requirements are proportional to the number of grid cells. If the resolution in each dimension is N, the total storage is N^k, where k represents the dimensionality of the problem. The computation time depends on N^k times the number of integration cycles n, which is given by the total model time τ divided by the time-step dt. According to the *Courant-Friedrich-Lewy*-condition numerical stability is warranted if $dt < dx/c$ (35), where dx is the size of the spatial increment and c is the speed of sound. In the next section we want to discuss by means of examples of impact simulation in marine environment the capabilities of numerical modeling. The hydrocode we have used for this computations is the well know iSALE (impact Simplified Arbitrary Lagragean Eulerian) code (e.g. (54; 53)) which has been used in several other studies of crater formation before.

4 Simulation of deep– and shallow–water impacts

The simulation of oceanic impacts with varying impact velocities, projectile diameters, and water depths provides an insight into how many waves are generated and their characteristics. In this section we only focus on the different categories of waves generated. Principally there are two different categories of waves caused by oceanic impacts: (i) the rim wave, and (ii) the collapse wave. The rim wave has its origin in the excavation stage when water and crust is ejected. The ejected material forms a rim, which collapses under gravity and generates the rim wave. During the collapse of the crater rim, the water begins the fill the excavated cavity. The water masses meet at the impact center and generated a central high. This central high collapses and eventually results in the generation of the collapse wave.

In general, impacts into oceanic environments can be categorized into three different types: impacts in deep, intermediate–deep, and shallow water. Figure 9 contains snapshots of an impact in deep water (Fig. 9a-f) and in shallow water (Fig. 9g-l). Both impacts are based on the same projectile diameter ($500m$) and impact velocity ($10km/s$). The water depth for the impact in deep water is $4500m$ (Fig.

9a). Figure 9b shows that at the same time the ejecta is still moving, the water starts to fill the cavity. The generation of the central high is depicted by Fig. 9c. The collapse of it results in the generation of the collapse wave (Fig. 9d). The water forms a second central high that results in the generation of a second collapse wave (Fig. 9e, f). The first collapse wave experiences severe wave breaking close to the impact center.

The water depth for the impact in shallow water is $1km$. The oceanic crust is covered by a thin layer ($100m$ thick) of sediment (Fig. 9g). Figure 9h illustrates that the water column and the crust generated two different crater rims. The crater rim of the water collapses and generates the rim wave (Fig. 9i,j). The water surges into the cavity and generates a central high as in the case of an impact into deep water. However in case of the impact into deep water, the oceanic crust did not generate a significant crater, whereas the impact cratering processes resulted in the generation of a crater structure in case of the impact in shallow water. Comparing Fig. 9c with Fig. 9g reveals that the volume of water in the central is much less in case of the impact into shallow water. The diameter of the central high at the bottom is less than or very close to the diameter of the generated crater structure. This means that the collapse of this central high, especially the generation and the early propagation of the collapse wave, is affected by the crater structure. Figures 9j-l demonstrate that the collapse of the central high results in the generation of a huge bore and early wave breaking results in severe weakening of the wave in its early stage.

The characteristics of the wave generated during oceanic impact are tremendously different from those wave generated by earthquakes, these waves are commonly very long waves of more than 100 km length. For instance the first collapse of the impact in deep water as presented here has a length of about $18km$ in a water depth of $4500m$. This means that this wave must not be considered as a tsunami wave, since the water depth to wave length ratio violates the definition of a long wave. Furthermore, it is assumed that because of the long–wave characteristic, the horizontal component of the particle velocity in the water due to the wave passage does not vary with depth. The grid showed to be regular in Fig. (9a,g) indicates by its deformation that this horizontal velocity component does vary with depth for collapse wave generated during the impact in deep water. On the opposite, this grid shows only minor horizontal deformations for the rim wave in case of the impact in shallow water. (Fig. 9a) shows that the length of the rim wave is about $15km$. The water depth to wave length ratio would indicate a long wave. Another interesting point is the symmetry between wave crest and trough. While earthquake-generated wave are in general symmetrical, Fig.9k shows that the wave crest is short compared to the wave trough.

5 Synopsis

Impacts have been important for the evolution of our solar system, and therefore for the Earth, as well. They even had influence on the evolution of the life as the rise of the mammals and the extinction of the dinosaurs is linked to the Chicxulub impact event, 65 Ma ago. The way this impact had influence on the extinction of the dinosaurs and other species is vividly discussed. Considering the Earth, impacts in water are much more likely than impacts on continents, owing to the distribution of water and land masses.

In the previous sections, the physical concepts of impact cratering, such as the generation and propagation of the shock waves induced during the contact between the projectile and target were inroduced and discussed. Furthermore, the three stages of impact cratering according to Gault et al. in (5) showed what processes take place in the continuum from the contact between the target and projectile to the final crater. These concepts of impact cratering were then discussed in relation to the numerical modeling of the crater formation; however, it is clear that numerical models need more specific requirements than simplified theoretical considerations.

By the modeling of impacts into an oceanic environment, it could be shown what waves are generated, and that considerable difference exist between impacts into deep water and into shallow water. The generated wave differs from those generated, for instance, by earthquakes. The propagation over the open ocean and their run-up with regard to the hydrodynamic stability and the simulation of impacts in extremely shallow water are challenges for future research on meteorite impacts in oceanic environments.

Acknowledgement

This publication is partially funded by the Joint Institute for the Study of the Atmosphere and Ocean (JISAO) under NOAA Cooperative Agreement No. NA17RJ1232, Contribution No. 1393 and DFG-Grant WU 355/5-1.

References

[1] Grieve RAF, Robertson PB and Dence MR (1981) Constraints on the formation of ring impact structures, based on terrestrial data. In: Constraints on the formation of ring impact structures, based on terrestrial data.
[2] Gersonde R, Deutsch A, Ivanov BA, Kyte F T (2002) Oceanic impacts – a growing field of fundamental science. Deep Sea Research 49:951–957
[3] Abels A, Plado J, Pesonen LJ, Lehtinen M (2002) The impact cratering record in Fennoscandia. In: Impacts in Precambrian Shields. Springer, Berlin Heidelberg New York
[4] Glikson AY (1999) Oceanic mega impacts and crustal evolution. Geology 27:387–390
[5] Gault DE, Quaide WL Oberbeck VR (1968) Impact cratering mechanics and structures In: French BM, Short NM (eds) Shock Metamorphism of Natural Materials. Mono Book Co.
[6] Turtle EP, Pierazzo E, Collins GS, Osinski GR, Melosh HJ, Morgan JV, Reimold WU, Spray JG (2005) Impact structures: What does crater diameter mean? Geological Society of Americal special paper 384: *in press.*
[7] Melosh HJ (1989) Impact Cratering: A Geological Process. Oxford University Press, New York
[8] Scott TE, Nielsen KC (1991) The effects of porosity on the brittle-ductile transition in sandstone. J. of Geophys Res 96:405–414
[9] Jaeger JC, Cook NGW (1969) Fundamentals of reock mechanics. Chapman and Hall.

[10] Lundborg N (1968) Strength of rock-like materials. Int. J. Rock Mech. Min Sci. 5:427–454
[11] Stesky RM, Brace WF, Riley DK, Robin PYF (1974) Driction in faulted rock at high temperature and presuure. Tectonphysics 23:177–203
[12] Collins GS, Melosh HJ, Ivanov BA (2004) Modeling damage and deformation in impact simulations. Meteoritics Planet. Sci. 39:217–231
[13] Melosh HJ, Ryan EV, Asphaug E (1992) Dynamic fragmentation in impacts: Hydrocode simulations of laboratory impact. J Geophys Res. 97:14735–14759
[14] Ivanov BA, Deniem D, Neukum G (1997) Implementation of dynamic strength models into 2D hydrocodes: Applications for atmosphereic breakup and impact cratering. Int. J. Impact Engin. 20:411–430
[15] Artemieva NA, Ivanov BA (2004) Launch of martian meteorites in oblique impacts. Icarus 171:84–101
[16] Stöffler D, Artemieva NA, Pierazzo E (2004) Modeling the Ries-Steinheim impact event and the formation of the moldavite strewn field. Meteoritics Planet. Sci. 37:1893–1908
[17] Wünnemann K, Morgan JV, Jödicke H (2005) Is Ries crater typical for its size? An analysis based upon old and new geophysical data and numerical modeling. In Kenkmann T, Hörz F, Deutsch A (eds) Large meteorite impacts III. Geol. Soc. America Special Paper 384:67–83
[18] Melosh HJ, Ivanonv BA (1999) Impact crater collapse. Annual Review of Earth and Planetary Sci. 27:385–415
[19] Pierazzo E, Melosh HJ (2000) Melt production in oblique impacts. Icarus 145:252–261
[20] Chandrasekhar S (1981) Hydrodynamic and hydromagnetic stability. Dover Publications Inc., (New York) Chap. 2, p9-16
[21] Ferziger JH, Peric M (1997) Computational methods of fluid dynamics. Springer (Berlin, Heildeberg) Chap. 1, p1-20
[22] Melosh HJ (1989) Impact cratering: Geological process. Oxford University Press (New York) 245pp
[23] Tillotson JH (1962) Metallic equation of state for hypervelocity impacts. Technical Report General Atomic Report GA-3216
[24] Thompson SL, Lauson HS (1972) Improvements in the chart-D radiation-hydrodynamic code III: Revised analytical equation of state. Technical Report SC-RR-61 0714. Sandia National Laboratories (Albuquerque, NM)
[25] Ahrens TJ, O'Keefe JD (1977) Equation of state and impact-induced shock-wave attenuation on the moon. In Roddy DJ, Pepin DJ, Merrill RB (eds) Impact and Explosion Cratering. Pergamon (New-York) 639–656
[26] Zel'dovich YB, Raizer YP (2002) Physics of shock waves and high-temperature hydrodynamic phenomena. In Hayes WD, Probstein RF (eds) Dover Publications (Mineola, New York), 916pp
[27] Roddy DJ, Schuster SH, Rosenblatt M, Grant LB, Hassig PJ, Kreyenhagen KN (1987) Computer simulation of large asteroid impacts into oceanic and continental sites - preliminary results on atmospheric cratering and ejecta dynamics. Int. J. Impact. Engin 5:525–541
[28] O'Keefe JD, Ahrens TJ (1999) Complex craters: Relationship of stratigraphy and rings to impact conditions. J. Geophys. Res. 104:27091–27104
[29] O'Keefe JD, Ahrens TJ (1994) Impact-induced melting of planetary surfaces. In Dressler BO, Grieve RAF, Sharpton VL (eds) Large Meteorite Impacts

and Planetary Evolution, Boulder, Colorado, Geological Society of America, Special Paper 293: 103-109

[30] Ivanov BA, Artemieva NA (2002) Numerical modeling of fromation of large impact craters. In Koeberl C, MacLeod KG (eds) Catastrophic events and mass extinctions: Impacts and beyond.Geol. Soc. America Special Paper 356:619–630

[31] Shuvalov VV, Dypvik H, Tsikalas F (2002) Numerical simulations of the Mjolnir marine impact crater. J Geophys. Res. 107:DOI10.1029/2001JE001698

[32] Wünnemann K, Lange MA (2002) Numerical modeling of impact-induced modifications of the deep-sea floor. Deep-Sea Res. II 49:669–981

[33] McGlaun JM, Thompson SL (1990) CTH: A three-dimensional shock wave physics code. Int. J. Impact Engin. 10:351–360

[34] Shuvalov VV (1999) Multi-dimensional hydrodynamic code SOVA for interfacial flow: Applications to the thermal layer effect. Shock Wave 9: 382–390

[35] Anderson Jr CE (1987) An overview of the theory of hydrocodes. Int J Impact Engin 5:33–59

[36] Knowles CP, Brode HL (1977) The theory of cratering phenomena, an overview. In Roddy DJ, Pepin DJ, Merrill RB (eds) Impact and Explosion Cratering. Pergamon (New-York) 369–395

[37] Holsapple KA, Schmidt RM (1982) On the scaling of crater dimensions II - Impact. J Geophys Res 87:1849-1870

[38] Holsapple KA, Schmidt RM (1987) Point-source solution and coupling parameters in cratering mechanics. J. Geopys. Res. 92:6350-6376

[39] Holsapple KA (1993) The scaling of impact processes in planetary science. Annual Review of Earth and Planetary Sciences 21:333-373

[40] Schmidt RM, Holsapple KA (1987) Some recent advances in the scaling of impact and explosion cratering. Int. J. Imp. Engin 5:543-560

[41] Gault DE (1974) Impact cratering. In: Greeley R, Schulz PH (eds) A primer in lunar geology. Moffett Field: NASA Ames Research Center. 137-175

[42] Dence MR (1968) Shock zoning at Canadian craters: Petrography and structural implications, in French BM, Short NM (eds) Shock metamorphism of natural materials. Baltimore, Maryland, Mono Book Cooperation 169-184

[43] Stöffler D, Langenhorst, F (1994) Shock metamorphism of quartz in nature and experiment: I. Basic observations and theory. Meteoritics 29:155-181

[44] Turtle EP, Pierazzo E, Collins GS, Osinski GR, Melosh HJ, Morgan JV, Reimold WU (2005) Impact structures: What does crater diamter mean? In Kenkmann T, Hörz F, Deutsch A (eds) Large meteorite impacts III. Geol. Soc. America Special Paper 384:1–24

[45] Pike RJ (1988) Geomorphology of Impact Craters on Mercury. In: Mercury, University of Arizona Press. 165-273

[46] Stöffler D (1972) Deformation and transformation of rock-forming minerals by natural and experimental shock processes. Fortschritte der Mineralogie 49: 50-113

[47] Jeanloz R (1980) Shock effects in olivine and implications for Hugoniot data. J Geophys Res 85:3163-3176

[48] Roddy DJ, Davis LK (1977) Shatter cones formed in large-scale experimental explosion craters. In: Roddy DJ, Pepin, RO, Merrill RB (eds) Impact and explosion cratering, Pergamon Press, New York

[49] O'Keef JD, Ahrens TJ (1982) The interaction of the Cretaceous/Tertiary extinction bolide with the atmosphere, ocean and solid Earth. Geol Soc Amer, 190:103-120
[50] Wünnemann, K, Ivanov B. (2003) Numerical Modelling of impact crater depth–diameter dependence in an acoustically fluidized target. Planetary and Space Science 51:831-845
[51] Melosh HJ (1977) Crater modification by gravity: A mechanical analysis of slumping. In: Roddy DJ, Pepin, RO, Merrill RB (eds) Impact and explosion cratering, Pergamon Press, New York
[52] Wünnemann K (2001) Die Numerisch Behandlung von Impaktprozessen – Kraterbildung, stosswelleninduzierte Krustenmodifikationen und ozeanische Enschlagsereignisse. PhD Thesis at Institut of Geopyhsics, Westfälische Wilhelms Universtität.
[53] Amsden AA, Rupel HM, Hirt CW (1080) SALE: a simplified ALE computer program for fluids at all speeds. Los Alamos National Laboratories,LA–8095
[54] Wünnemann K, Collins GS, Melosh HJ (2006) A strain-based porosity model for use in hydrocode simulations of impacts and implications for the transient-crater growth in porous targets. Icarus 180:514-527
[55] Buttkus B (2000) Spectral analysis and filter theory in applied geophysics. Springer, Berlin Heidelberg New York
[56] Snieder R, Trampert J (2000) Linear and nonlinear inverse problems. In: Dermanis A, Grün A, Sanso F (eds) Geomatic methods for the analysis of data in the earth sciences. Lecture notes in earth sciences, vol 95. Springer, Berlin Heidelberg New York
[57] Montagner, JP, Nataf HC (1986) On the inversion of the azimuthal anisotropy of surface waves. J Geophys Res 91:511–520
[58] Ross DW (1977) Lysosomes and storage diseases. MA Thesis, Columbia University, New York

Retracing the tsunami rays

R. Shankar

The Institute of Mathematical Sciences,
C.I.T Campus, Chennai 600113, INDIA.shankar@imsc.res.in

1 Introduction

The propagation of the tsunami waves can be modelled with varying degrees of accuracy. The more accurate methods, which involve the numerical solution of the shallow water equations, are costly in terms of simulation times. With present day technology, the simulations are too slow for many of the requirements of a tsunami early warning system. The ray diagram technique is one where accuracy is sacrificed for computational speed. Also referred to as refraction diagrams, they were first used to analyse tsunami propagation by Kenji Satake (1; 2).

The ray diagram technique models the propagation of the "leading edge" of the tsunami waves. It can therefore be used to compute the travel times of the waves from the source to any desired location. It does not compute the amplitudes of the waves. Thus not only accuracy but also all amplitude information is sacrificed in this approach. However, this sacrifice results not only in increased computational speed but also, as described in section 4, in an easy solution of the inverse problem. Again, while the source region can be mapped out using this technique, there is no information about the initial amplitudes.

In section 2, we will review the derivation of the ray equations from the linearised shallow water equations using the eikonal approximation.Section 3 discusses the smoothening of the bathymetry which is required to apply the ray equations to tsunami waves. The method of backward ray tracing, which provides the solution to the inverse problem to determine source region is described in section 4. Section 5 reports on some work (3) which uses the above techniques to delineate the northern extent of the 2004 Indian Ocean tsunami source region.

2 The ray equations

2.1 The approximations

The tsunami waves are described by the non-linear shallow water equations on a spinning sphere. Thus, in general, the non-linearities, the curvature of the earth and the Coriolis force have to be incorporated in the equations. However, we will argue below, if we are using the model in a limited region and for short periods of time, neglecting these features is a reasonably good approximation.

We first address the nonlinear and dispersive terms in the equation. The dispersive terms are suppressed by a factor of $(kh)^2$ and the nonlinear terms by a/λ, where k is the wave vector, λ the wavelength, a the amplitude and h the depth. For a tsunami with a period of 20 min, amplitude 1 m and propagating at a depth of 3500 m, we have $(kh)^2 \approx 10^{-2}$ and $a/\lambda \approx 5 \times 10^{-5}$. Thus these are indeed small. The nonlinear and dispersive effects could build up over time, but given the magnitude of the suppression factors, it is reasonable to assume that for the first few hours of propagation, the linear theory is a good approximation. The linear approximation of course will not be valid very near the shore.

To get a rough idea of the error involved in neglecting the curvature of the earth, we note that the approximation involved in measuring distances is,

$$d\theta^2 + \cos^2\theta \; d\phi^2 \to d\theta^2 + d\phi^2 \tag{1}$$

Where θ and ϕ are the latitudes and longitudes respectively. If, as in the work (3) reported in section 5, we are restricting our modelling to the Bay of Bengal, then the range of latitudes we are interested in is $0 - 20°$. Since $\cos(20°) = 0.94$, we can see that the error involved will not exceed a few percent.

The Coriolis force will modify the wave equation (6) derived in section 2.2 and introduce a dispersion. For a flat ocean bed, the modified dispersion relation is,

$$\omega = \sqrt{c^2k^2 + \Omega^2} \tag{2}$$

where ω is the frequency of the wave, k its wave vector, $c = \sqrt{gh}$, g the accelaration due to gravity, h the depth and $\Omega = \sin\theta \; 2\pi/24 \; hr^{-1}$ is the component of the earths angular velocity normal to the surface. The wave speed $\tilde{c} \equiv d\omega/dk$ is now frequency dependent and is given by

$$\tilde{c} = c\sqrt{1 - \frac{\Omega^2}{\omega^2}} \tag{3}$$

Even for tsunami waves with time periods of 90 min, the factor $\Omega^2/\omega^2 = 4.5 \times 10^{-4}$. So the Coriolis force can indeed be neglected to a very good approximation.

2.2 The wave equation and rays

The linearised shallow water equations that model the tsunami propagation are

$$\frac{\partial}{\partial t}\mathbf{v} = -g\boldsymbol{\nabla}\zeta \tag{4}$$

$$\frac{\partial}{\partial t}\zeta = -\boldsymbol{\nabla}.(h\mathbf{v}) \tag{5}$$

where, as mentioned earlier, we have neglected the curvature of the earth and the Coriolis force. $\mathbf{v}$ is the depth averaged velocity field, ζ the height of the water above its static equilibrium level, h the depth of the water and g the acceleration due to gravity. The velocity field can be eliminated from the above equations to obtain a wave equation for the amplitude ζ,

$$\left(\frac{\partial^2}{\partial t^2} - \boldsymbol{\nabla}.c^2\boldsymbol{\nabla}\right)\zeta(\mathbf{x}, t) = 0 \tag{6}$$

where $c(\mathbf{x}) \equiv \sqrt{gh(\mathbf{x})}$. Since $c(\mathbf{x})$ is time independent, the solutions are linear combinations of waves with specific frequencies of the general form

$$\zeta(\mathbf{x},t) = A_1(\mathbf{x},\omega)cos(\omega t) + A_2(\mathbf{x},\omega)sin(\omega t) \tag{7}$$

where $A_1(\mathbf{x},\omega)$ and $A_2(\mathbf{x},\omega)$ satisfy the so called Helmholtz equation,

$$(\omega^2 + \mathbf{\nabla}.c^2(\mathbf{x})\mathbf{\nabla})A_{1(2)}(\mathbf{x},\omega) = 0 \tag{8}$$

Equation (7) can be written as,

$$\zeta(\mathbf{x},t) = A(\mathbf{x})cos(L(\mathbf{x}) - \omega t) \tag{9}$$

We will refer to $A(\mathbf{x})$ as the amplitude and $S(\mathbf{x},t) \equiv (L(\mathbf{x}) - \omega t)$ as the phase of the wave. The set of points with the same phase are called wavefronts. A wavefront is thus mathematically defined by $S(\mathbf{x},t)$=constant. Rays are normals to the wavefronts. They define how the wavefront propagates in time. They can be mathematically defined in terms of the phase function as follows. Consider the wavefront of zero phase,

$$S(\mathbf{x},t) = 0 \quad \Rightarrow \quad L(\mathbf{x}) = \omega t \tag{10}$$

Consider two wavefronts, one at time t and the other at time $t + \Delta t$. Let $\mathbf{x}_R(t)$ be a point on the wavefront at time t, i.e.

$$L(\mathbf{x}_R(t)) = \omega t \tag{11}$$

The normal to the wavefront at $\mathbf{x}_R(t)$, $\mathbf{k}(\mathbf{x}_R(t))$ is given by,

$$\mathbf{k}(\mathbf{x}_R(t)) = \mathbf{\nabla} L(\mathbf{x})|_{\mathbf{x}=\mathbf{x}_R(t)} \tag{12}$$

We define $\mathbf{x}_R(t + \Delta t)$ to be the point where the line starting from $\mathbf{x}_R(t)$ in the direction of this normal intersects the wavefront at $t + \Delta t$. We then have,

$$\begin{aligned} \omega(t + \Delta t) &= L(\mathbf{x}_R(t + \Delta t)) \\ &= L(\mathbf{x}_R(t) + \Delta l\ \hat{k}(\mathbf{x}_R(t))) \end{aligned} \tag{13}$$

Equations (11), (12) and (13) imply,

$$\frac{\Delta l}{\Delta t} = \frac{\omega}{|\mathbf{k}(\mathbf{x}_R(t))|} \equiv c_R(\mathbf{x}_R(t)) \tag{14}$$

Thus we have,

$$\frac{d}{dt}\mathbf{x}_R(t) = c_R(\mathbf{x}_R(t))\hat{k}(\mathbf{x}_R(t)) \tag{15}$$

Given the phase function $S(\mathbf{x},t)$, we can determine $\mathbf{k}(\mathbf{x})$ and hence the RHS of equation (15). It can then be integrated to get the ray trajectory.

$\mathbf{x}_R(t)$ can be be visualised as the trajectory of a corpuscule travelling with speed $c_R(\mathbf{x}_R(t))$ along the ray. We will next show that, in the geometrical optics limit (GOL), $c_R(\mathbf{x}_R(t)) = c(\mathbf{x}_R(t))$.

2.3 The geometric optics limit

We substitute the form of $\zeta(\mathbf{x}, t)$ in equation (9) into the wave equation (6),

$$\left(\partial_t^2 - \boldsymbol{\nabla}.c^2\boldsymbol{\nabla}\right) A(\mathbf{x})cos(L(\mathbf{x}) - \omega t) = 0 \tag{16}$$

It is convenient to use the complex notation. Defining $\alpha = lnA$, we have

$$\left(\partial_t^2 - \boldsymbol{\nabla}.c^2\boldsymbol{\nabla}\right) e^{(\alpha + iL(\mathbf{x}) - i\omega t)} = 0 \tag{17}$$

The real and imaginary parts of equation (17) can then be written as,

$$\omega^2 - c^2\boldsymbol{\nabla}L.\boldsymbol{\nabla}L + c^2\boldsymbol{\nabla}\alpha.\boldsymbol{\nabla}\alpha + \boldsymbol{\nabla}.c^2\boldsymbol{\nabla}\alpha = 0 \tag{18}$$

$$2c^2\boldsymbol{\nabla}\alpha.\boldsymbol{\nabla}L + \boldsymbol{\nabla}.c^2\boldsymbol{\nabla}L = 0 \tag{19}$$

Upto now we have made no approximations. The GOL is the limit where the derivatives of c and α can be neglected. Physically it is the limit when c varies very slowly in comparison to the wavelength. In this limit we have,

$$\boldsymbol{\nabla}L.\boldsymbol{\nabla}L = \frac{\omega^2}{c^2} \tag{20}$$

This is the basic equation of geometrical optics. Comparing with equations (12) and (14) we have,

$$c_R(\mathbf{x}_R(t)) = c(\mathbf{x}_R(t)) \tag{21}$$

Note that the GOL approximation is exact when c is constant.

2.4 The ray equation on the plane

In the GOL, in principle we could solve equation (20), obtain $\mathbf{k}(\mathbf{x})$ and then integrate equation (15) to get the rays. However, it is possible to derive a second order ODE for the rays which does not involve $\mathbf{k}(\mathbf{x})$. This equation is easier to solve than (20). To obtain this ray equation, we use equation (20) to write equation (15) as,

$$\frac{\omega}{c^2}\frac{d\mathbf{x}_R}{dt} = \boldsymbol{\nabla}L(\mathbf{x}_R) \tag{22}$$

Differentiating the above equation with respect to t and doing some algebra we can obtain the ray equation,

$$\frac{d}{dt}\frac{1}{c^2}\frac{d\mathbf{x}_R}{dt} = -\boldsymbol{\nabla}ln\ c \tag{23}$$

This can be cast in the first order form suitable for numerical solution in the standard way,

$$\frac{d\mathbf{x}_R}{dt} = c^2\mathbf{p}_R \tag{24}$$

$$\frac{d\mathbf{p}_R}{dt} = -\boldsymbol{\nabla}ln\ c \tag{25}$$

From these equations we can get the relation,

$$\frac{d}{dt}\left(p_R^2 - \frac{1}{c^2}\right) = 0 \tag{26}$$

Assuming the physical initial condition $p_R(0) = 1/c$, we have

$$p_R(t) = \frac{1}{c} \tag{27}$$

Equation (27) can be used to reduce the system of 4 equations (24,25) to three. We put

$$\mathbf{p}_R(t) \equiv \frac{1}{c}\hat{v}_R(t) \tag{28}$$

where

$$\hat{v}_R(t) \equiv (cos\chi, sin\chi) \tag{29}$$

The ray equations (24,25) can then be written as,

$$\frac{dx_R}{dt} = c\ cos\chi \tag{30}$$

$$\frac{dy_R}{dt} = c\ sin\chi \tag{31}$$

$$\frac{d\chi}{dt} = -sin\chi\frac{\partial c}{\partial x} + cos\chi\frac{\partial c}{\partial y} \tag{32}$$

These are to be solved with the initial conditions that $\mathbf{x}_R(0)$ is a point on the initial wavefront and $\hat{v}_R(0)$ is the normal to the wavefront at that point.

The important feature of the ray equations derived above is that they are *independent of the frequency* (ω) *of the wave* and only depend on the bathymetry. Thus when applied to tsunamis, which are typically wave packets containing a range of frequencies, we do not have to compute a different ray diagram for each frequency. This is basically a consequence of the fact that for a flat bathymetry, the wave solutions of the shallow water equations are dispersionless.

3 Smoothening the bathymetry

The ray equations (30-32) were derived in the limit of the wavelength being much smaller than the length scale at which the bathymetry varies. This is not true for tsunamis, where we are interested in wavelengths $\sim 100\ km$. Nevertheless, as we will argue in this section, the ray equations can be applied to describe the propagation of the tsunami waves if the bathymetry is smoothened.

Consider an undersea feature of linear extent L in an otherwise flat ocean bed. If a wave of wavelength λ passes above this feature, then the wavefront will distort significantly if $\lambda << L$ but will hardly be affected if $\lambda >> L$. The ray equations will describe the behaviour of the $\lambda << L$ waves. An *ad. hoc.* way to describe the behaviour of the $\lambda >> L$ waves using ray equations is to use an "effective" bathymetry which is obtained by smoothening out the real bathymetry to the extent of the wavelength of interest.

There are many ways of smoothening the bathymetry. In our work, presented in section 5, we choose to smoothen the wave speeds, $c(\mathbf{x}) = \sqrt{gh(\mathbf{x})}$. We do this by replacing the speed at any point by the average of the speeds of all the points in a box of length $\sim \lambda$ around it. i.e. If the bathymetry is given on a grid whose points are labelled by (n, m), we compute the corresponding speeds, $c_{nm} = \sqrt{gh_{nm}}$ and define the smoothened bathymetry to be

$$c_{nm}^{eff} = \frac{1}{(2N+1)^2} \sum_{i=n-N}^{n+N} \sum_{j=m-N}^{m+N} c_{ij} \tag{33}$$

Where N is chosen such that the physical length of the box of side $2N + 1$ grid points is $\sim \lambda$. Care has to taken for the near shore points to ensure that all the land points in the box are removed.

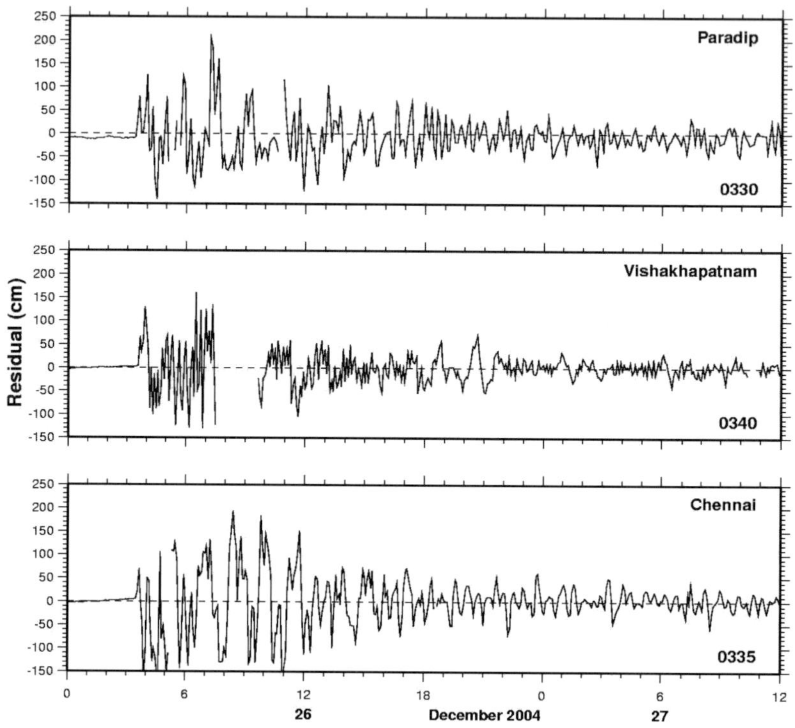

Fig. 1. The residual gauge record. The tsunami arrival time (UTC) is in the bottom right corner.

4 Backward ray tracing

The solutions to the ray equations tell us the time a disturbance will take to travel from one location to another. The equations are reversible which implies that the time taken for a disturbance to travel from point A to B will be the same as the time taken for it to travel from B to A. This fact is exploited by the so called backward ray tracing method to put constraints on the location of the initial tsunami wavefronts.

Typically, we will know the travel times of the tsunami from the source to some specific locations (tidal gauge stations presently and bottom pressure recorder locations also in the future). The ray equations can be integrated starting from these locations, in all possible directions, for the known travel time. We will call the set of the end points of the rays thus obtained as the backward wave front . It is clear that the initial wavefront has to include at least one point from each backward wavefront. Also, it cannot include any of the points on any of the rays which are not on the backward wavefront. Thus the envelope of all the backward wavefronts defines a curve which bounds the source region. Given the travel time data from enough stations one can more or less completely specify the source region.

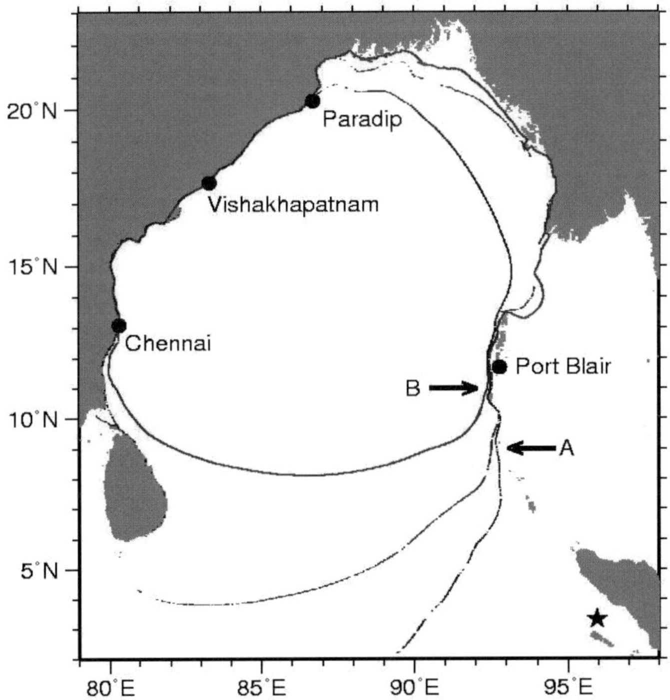

Fig. 2. The backward wavefronts from Chennai, Vishakapatnam and Paradip . The star is the epicenter of the earthquake. The arrow marked A indicates the northern extent as estimated in Lay et. al. The arrow marked B indicates the extent from our analysis.

5 The source region of the 2004 Indian Ocean Tsunami

The earthquake that caused the 2004 Indian Ocean tsunami was analysed by Lay et. al. (4). They concluded that the rupture propagated norhtward along the fault in three stages. The first segment, which they call the Sumatra segment extends from the epicenter to about 420 km north of it. The second Nicober segment extends about 375 km further and the third Andaman segment for another 570 km upto about 14° N. They also mapped out region of the initial tsunami wavefront using the backward ray tracing method. They estimated that the source region extended to about 600 km north of the epicenter, upto 9° N. This is the point marked B in figure 2. Thus their analysis indicated that the tsunami waves were only excited in the Sumatra segment and a part of the Nicobar segment.

The northern extent of the source region in this analysis was determined by the travel times to the tidal gauge stations at Port Blair , Chennai and Vishakapatnam . However, the arrival time of 45 min they used for Port Blair turned out to be incorrect. A careful analysis by Singh et. al. (5), including the comparison of the tides recorded at the tide gauge there with tidal models showed that at the time of the tsunami the tide gauge clock was ahead by 46 min [1]. The ambiguity in the Port Blair arrival time was further compounded by a data gap of 24 min, $35 - 59$ min after the earthquake. We therefore did the backward ray tracing analysis without the Port Blair data but included instead the tidal gauge data from Paradip , which may not have been available to Lay et. al. (4). Our finding was that the source actually extended for about 200 km further north than estimated them, upto about 11° N.

5.1 Retracing the 2004 Indian Ocean tsunami rays

The tide gauge data along the east coast of India that are maintained by the Survey of India showed that the taunami travel times were 156 min to Chennai (80.30° E, 13.08° N), 161 min to Vishakapatnam (83.28° E, 17.68° N) and 151 min to Paradip (86.70° E, 20.26° N). The residual tide gauge data (with tides subtracted) from these three stations are shown in figure 1. This data shows that the tsunami arrived almost simulataneously at these three stations. In fact it reached Paradip a few minutes before the other two stations south of it indicating that the initial wavefront must have extended further north than what was estimated by Lay et. al. (4). To confirm this, we did a forward ray tracing from the point marked B in figure 2 and indeed found that the wave front was well short of Paradip after 151 min of propagation. This confirms that the travel time of 45 min to Port Blair is inconsistent with the travel times to the other three stations.

The backward wave fronts from these three locations were computed and the result is shown in figure 2 It can be seen that the backward wave front of Paradip lies in the disallowed regions of the backward wavefronts of Chennai and Vishakapatnam except at the point marked B which is about 11° N. Thus the initial tsunami wavefront must have extended to at least about 200 km north of the point marked A at about 9° N, which was the estimate of Lay et. al. (4). The 30% increase in the

[1] This analysis was independently done by the group at the National Institute for Oceanography, Goa, with concurring results.

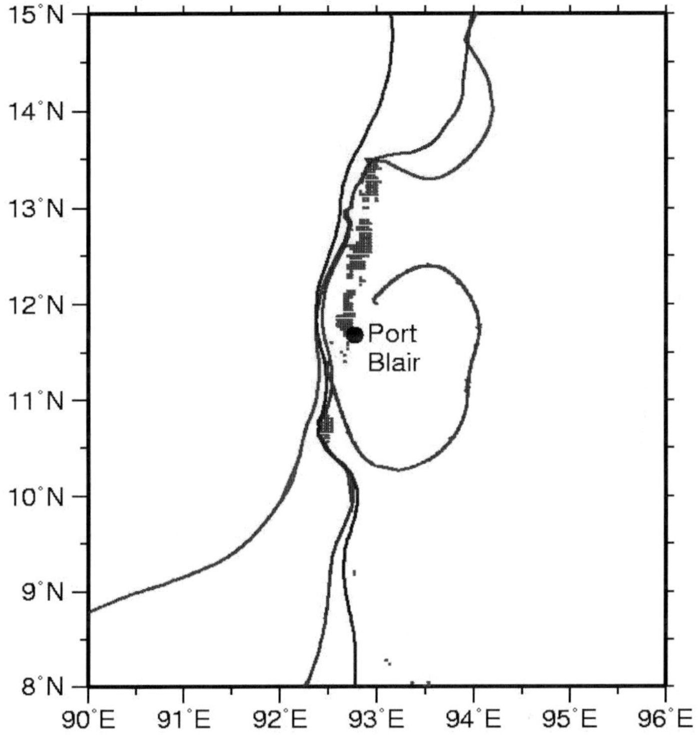

Fig. 3. The backward wavefronts from Port Blair , assuming a travel time of 30 *min*.

length of the initial tsunami wavefront implies that the full Nicobar segment as well as a part of the Andaman segment of the earthquake did excite tsumani waves.

5.2 Port Blair arrival time

To determine the tsunami arrival time at Port Blair, we did backward ray tracing from there. We varied the travel time until the backward wavefront became consistent with those of Paradip, Chennai and Vishakapatnam. This gave us a travel time of 30 *min*. The backward wavefront from Port Blair with a travel time of 30 *min* is shown in figure 3. This estimate is in agreement with the estimate of 31 *min* obtained in the analysis of Singh et. al. (5) who used different methods.

5.3 Conclusion

Backward ray tracing using the arrival times recorded by the Survey of India tidal gauges in Chennai, Vishakapatnam and Paradip shows that the northern extent of the initial wavefront of the 2004 Indian Ocean tsunami extended to about 11° N. This is about 30% longer than the estimate of Lay et. al. and implies that the entire Nicobar segment and a part of the Andaman segment of the earthquake excited tsunami waves. The requirement that the backward wavefront from Port Blair be consistent with the backward wavefront from the other three abovementioned stations yields a travel time of 30 min to Port Blair, consistent with the estimate of Singh et. al. (5).

References

[1] Kenji Satake (1988). PAGEOPH **126** 1.
[2] Kenji Satake (2000). In: D. J. Parker (ed) Floods, pp. 223-239. Routledge, London.
[3] S. Neetu, I. Suresh, R. Shankar, D. Shankar, S.S.C. Shenoi, S.R. Shetye, D. Sundar and B. Nagarajan (2005) Comment on "The Great Sumatra-Andaman Earthquake of 26 December 2004". Science **310** 1431a.
[4] T. Lay et. al. (2005) The Great Sumatra-Andaman Earthquake of 26 December 2004. Science **308** 1127.
[5] S.K. Singh, M. Ortiz, H.K. Gupta and D.G.A Ramadass (2006) Slow slip below Port Blair, Andaman during the great Sumatra-Andaman earthquake of 26 December 2004. Geophys. Res. Lett. **33** L03313.

Modeling and visualization of tsunamis: Mediterranean examples

Ahmat C. Yalciner[1], Effim Pelinovsky[2], A. Zaitsev[3], and A. Kurkin[4], C. Ozer[5], H. Karakus[5], G. Ozyurt[5]

[1] Department of Civil Engineering, Middle East Technical University, Ocean Engineering Research Center, 06531 Ankara Turkey, yalciner@metu.edu.tr

[2] Department of Nonlinear Geophysical Processes, Institute of Applied Physics, 46 Uljanov Street, 603950 Nizhny Novgorod, Russia, pelinovsky@hydro.appl.sci-nnov.ru

[3] Department of Applied Mathematics, Nizhny Novgorod State Technical University, 24 Minin Street, 603950 Nizhny Novgorod, Russia, aizaytsev@mail.ru

[4] Department of Applied Mathematics, Nizhny Novgorod State Technical University, 24 Minin Street, 603950 Nizhny Novgorod, Russia, kurkin@kis.ru

[5] Department of Civil Engineering, Middle East Technical University, Ocean Engineering Research Center, 06531 Ankara Turkey, cozer@metu.edu.tr, khulya@metu.edu.tr, gulizar@metu.edu.tr

Summary. There are numerous earthquakes and tsunamis that occurred in Eastern Mediterranean and are documented in historical records. Hellenic Arc is one of the important tsunami source regions in Eastern Mediterranean. In order to assess and visualize the tsunami propagation in the Eastern Mediterranean from the sources along Hellenic Arc, a new tsunami simulation/visualization software NAMI DANCE is used. In this study, five different tsunami cases related to selected rupture characteristics are used in modeling application. The results and comparisons are presented with discussions.

1 Introduction

The Mediterranean sea is one of the biggest marginal seas on the planet and located in between latitudes 30°N and 47°N, and longitudes -5°E and 43°W. It is bordered on the North and West by Europe, on the South by Africa, on the East by Asia. Sicily divides the sea into eastern and western basins. Numerous earthquakes and associated tsunamis in history in the Mediterranean sea seem as the precursor of the future similar events.

The fault zones around eastern Mediterranean basin are Hellenic Arc, North Anatolian Fault Zone (NAF), East Anatolian Fault Zone (EAF), Cyprus Arc, and Dead Sea Fault. At the centre of the Aegean Sea there is a series of volcanic systems almost parallel to the trench and forming the internal arc (Milos, Antimilos, Antiparos, Santorini, Christiana, Colombus, Kos, Yali, Nisiros and others).

Hellenic arc is one of the important tsunami prone areas for the far field propagation of tsunamis. In this study the estimated zones of tsunami sources in relation to the recent seismic data in the Eastern Mediterranean together with the simulations of different tsunami scenarios are presented.The numerical modeling is applied to five different tsunami cases and their propagation, arrival time and possible effected areas are indicated and discussed.

2 General characteristics and tsunamis in the Eastern Mediterranean

The Mediterranean Sea is 3900 km long and its maximum width is 1600 km and greatest depth is 4400 m (Fig. 1). The Mediterranean Sea extends to the Aegean Sea and connects with the Black Sea through the Dardanelles; the Bosphorus, and the Sea of Marmara; with the Atlantic Ocean through the Strait of Gibraltar; and with the Red Sea through the Suez Canal. Adriatic, Ionian, Tyrrhenian seas are its other main divisions.

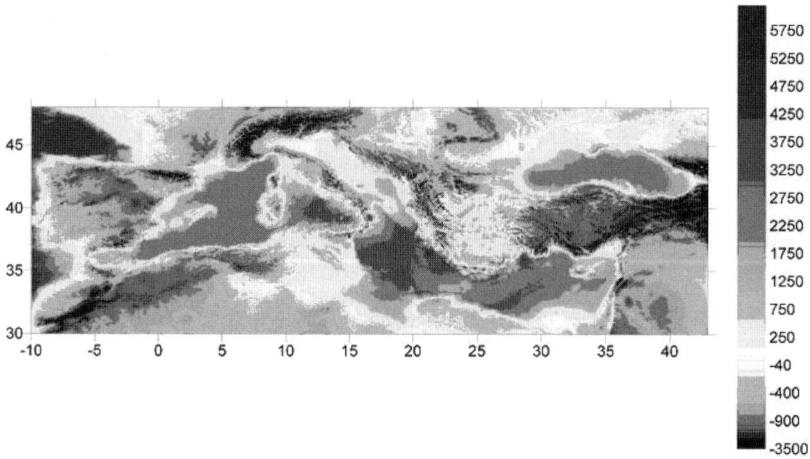

Fig. 1. General View of Topography and Bathymetry of the Mediterranean Sea

Mediterranean Basin is consisted of the Mediterranean Sea, Aegean Sea, Black Sea and the Sea of Marmara. With the opening of the Suez Canal (1869) the Mediterranean resumed its importance as a link on the route to the East.

The coastal areas at the Eastern Mediterranean Basin have experienced tsunamis many times in history. The generation mechanisms and their characteristics have not been well described. According to the historical information, or distribution of fault zones, volcanos, and other probable tsunamigenic sea bottom deformations, there are numerous source areas which may be considered responsible for those tsunamis.

One of the important source areas of tsunamis in the eastern Mediterranean is the fault zone named Hellenic Arc (Fig. 2) which is a subduction zone of about 1000 km in length starts from south west of Greek mainland and follows a curve at south of Crete and south east of Rhodes island and directs towards Anatolia along North East direction near Dalaman town. Two of the deepest regions of Mediterranean with a depth about 4300 m or more are i) in between Rhodes and Dalaman region at eastern end of Hellenic Arc and ii) from west of Crete Island at western end of Hellenic Arc. These regions can be called Rhodes - Dalaman trench and Kithira trench respectively. These trenches might be two of the most important tsunami prone areas in the Mediterranean Basin.

The epicenters of earthquakes (with magnitudes $M > 4$ and $M > 7$) occurred since 1900 in the Eastern Mediterranean are shown in Fig. 2 together with probable tsunami source areas.

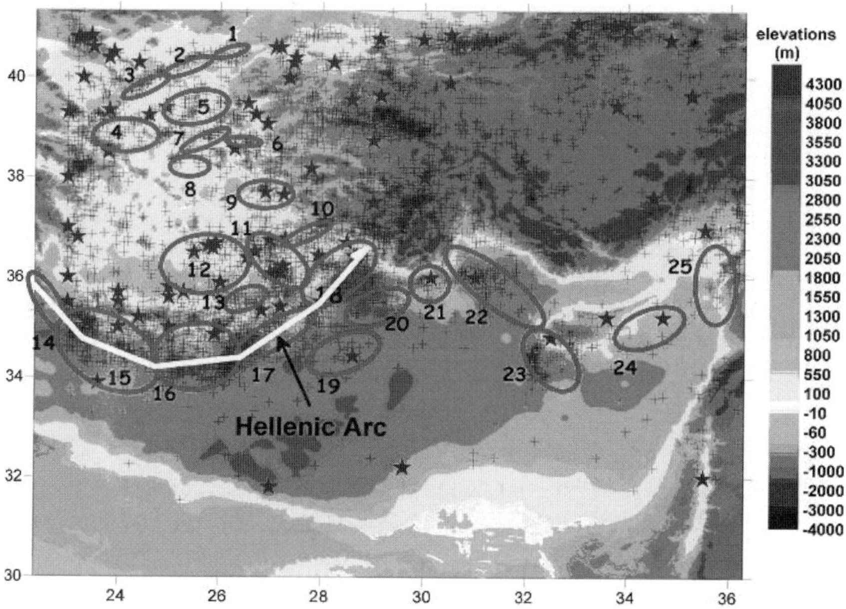

Fig. 2. Probable Tsunami Source Areas in the Eastern Mediterranean by using the distribution of Earthquake Epicenters (+ for magnitude $M > 4$ and * for magnitudes $M > 7$) occurred in the last century (8)

3 Historical tsunamis in Eastern Mediterranean

A comprehensive table of tsunamis in the vicinity of Anatolia is given in (14) by using the data from historical documents compiled from (5), (7), (1), (2). In regard to the present study for understanding tsunamis generated along Hellenic Arc, the historical tsunamis effected the coastlines near Hellenic Arc are selected from the respective document. Table 1 shows the historical tsunamis in the vicinity of Helenic Arc.

Table 1. The list of tsunamis occurred in the vicinity of Hellenic Arc since 22 A.D. Further info about the coastal regions is in (14)

No	Date	Coastal Region
1	22	Rhodes, Cyprus, Corinth
2	26	Paphos, Cyprus
3	46 A.D.	North east of Crete, Santorini Isl.
4	53/62/66	Cnossos-Crete, Leben
5	68	Demre, Patara-Lycia
6	142	Southeast Aegean Coasts
7	261-262	South coasts of Anatolia
8	365.07.21	Eastern Mediterranean
9	524/525	South coast of Anatolia, Anazarba-Adana
10	554.08.15	South-west coast of Anatolia, Kos Isl., Mandalya Gulf
11	1157.07.15	Hama-Homs, Chaizar Region
12	1202.05.22	Cyprus, Syrian coasts, Egypt
13	1222.05,11	Paphos, Limasol-Cyprus, Egypt
14	1304.08.08	Eastern Mediterranean
15	1403.11.16	South coasts of Anatolia, Syrian Coasts
16	1481.05.03	Rhodes, South west coast of Anatolia, Crete
17	1489	South coasts of Anatolia,Antalya
18	1494.07.01	Harekleion-Crete
19	1609.04	Rhodes , Eastern Mediterranean
20	1612.12.08	North of Crete
21	1659.09.29	Southeast Aegean
22	1667.11.30	Izmir Gulf
23	1672.02.14	Bozcaada, Kos Isl.
24	1741.01.31	Rhodes
25	1851.02.28	Fethiye Rhodes
26	1851.04.03	Fethiye Gulf
27	1851.05.23	Rhodes, Dodecanese
28	1855.02.13	Fethiye Gulf
29	1926.06.26	South and Southeast Aegean
30	1948.02.09	Karpathos-Dodocanese
31	1953.09.10	South coasts of Turkey

During the last 36 centuries at least 96 tsunamis were documented in the Eastern Mediterranean (14). Most probably 32 of them have occurred near or around Hellenic Arc. Among those the tsunami in 365 was one of the most important tsunamis in Eastern Mediterranean and has widely affected the region, from Sicily and Libya at east and South west of Turkey, Syria and Egypt. The coastal areas were not so densely populated and utilized in the past when these tsunamis have occurred. Therefore the effects of these tsunamis are limited only with the populations near the coasts. It must be clearly expected that any future similar tsunami will be much more destructive on the coasts. Besides examining the archive information, it is also helpful to use simulation technique to understand the generation, propagation and coastal amplification of tsunamis.

In the following sections some possible tsunamis to be generated along Hellenic Arc are simulated and discussed.

4 Modeling of Mediterranean tsunamis

Tsunami numerical modeling is based on solution of long wave equations with respect to related initial and boundary conditions. There were several numerical solutions of long wave equations for tsunamis. In general the explicit numerical solution of Nonlinear Shallow Water (NSW) Equations is preferable for the use since it uses reasonable computer time and memory, and also provide the results in acceptable error limit. They are very often used for case studies (9), (11), (13), (12), (3), (4), (17), (16), (19), (14), (18).

New tsunami simulation software called NAMI DANCE is developed for tsunami numerical modeling. It is developed by C++ programming language by following the staggered leap-frog scheme (6) numerical solution procedure of Nonlinear Shallow Water Equations. It simulates and computes i) tsunami source from either rupture characteristics or predetermined wave form, ii) propagation, iii) arrival time, iv) coastal amplification v) inundation (according to the accuracy of grid size), vi) distribution of current velocities and their directions at selected time intervals, vii) distribution of water surface elevation (sea state) at selected time intervals, viii) relative damage levels according to drag force and impact force, ix) time histories of water surface fluctuations, x) 3D plot of sea state at selected time intervals from different camera and light positions, and xi) animation of tsunami propagation (15). NAMI DANCE is very useful user friendly software to simulate, visualize and animate tsunami generation, propagation and coastal behavior processes (It can easily be applied to investigate several tsunami scenarios in the selected computational domain in order to obtain the useful database for assessment of tsunamis and also for the use of database in an effective tsunami warning system).

Five different tsunami scenarios for the Eastern Mediterranean Sea have been simulated by using the numerical model NAMI DANCE.

From Fig. 2 the ruptures 14 – 18 are selected for simulation of possible tsunamis generated along Hellenic Arc. The main input data set of the modeling is the rupture characteristics (fault location, length and width; strike, dip, slip angles; displacement and focal depth). There is no significant information available about the rupture characteristics of the historical tsunamis along Hellenic Arc. The only available data is earthquake epicenters and magnitudes measured in the last century. The similarities between subduction zone properties of Hellenic Arc (Mediterranean sea) and Sunda Arc (Indian Ocean) can also be used to estimate the rupture characteristics of probable tsunamis along Hellenic Arc.

The rupture characteristics of the December 26, 2004 Indian Ocean Tsunami has been determined as focal depth: 10 km; dip angle: 8°; slip angle: 110°; fault displacement: 22 m, with the earthquake magnitude as 9.3. The strongest earthquakes in Hellenic Arc region were about Magnitude 8. According to this comparison the rupture characteristics of possible tsunamis are estimated using this information. The five different tsunami cases are determined representing possible future tsunamis along Hellenic Arc. Their selected rupture characteristics for modeling are presented in the following. Additionally, the maximum positive and negative tsunami amplitudes of the initial tsunami wave for each rupture (14 – 18) are also calculated using (10).

1. Rupture 14. Coordinates of starting point of fault: 21°N 36.8^{o}E; coordinates of end point of fault: 23.31^{o}N 35°E; length of fault: 294 km; strike angle: 125°; width of fault: 30 km; focal depth: 30 km; dip angle: 45°; slip angle: 80°; fault displacement: 10 m; maximum positive tsunami amplitude at the source 2.35 m; maximum negative tsunami amplitude at the source: 0.16 m.
2. Rupture 15. Coordinates of starting point of fault: 23.5^{o}N 35°E; coordinates of end point of fault: 24.9^{o}N 34.4^{o}E; length of fault: 152 km; strike angle: 125°; width of fault: 30 km; focal depth: 30 km; dip angle: 45°; slip angle: 80°; fault displacement: 10 m; maximum positive tsunami amplitude at the source 2.25 m; maximum negative tsunami amplitude at the source: 0.12 m.
3. Rupture 16. Coordinates of starting point of fault: 24.9^{o}N 34.65^{o}E; coordinates of end point of fault: 26°N 35.85^{o}E; length of fault: 162 km; strike angle: 125°; width of fault: 30 km; focal depth: 30 km; dip angle: 45°; slip angle: 80°; fault displacement: 10 m; maximum positive tsunami amplitude at the source 2.27 m; maximum negative tsunami amplitude at the source: 0.13 m.
4. Rupture 17. Coordinates of starting point of fault: 21°N 34.65^{o}E; coordinates of end point of fault: 27.77^{o}N 35.41^{o}E; length of fault: 143 km; strike angle: 125°; width of fault: 30 km; focal depth: 30 km; dip angle: 45°; slip angle: 80°; fault displacement: 10 m; maximum positive tsunami amplitude at the source 2.22 m; maximum negative tsunami amplitude at the source: 0.11 m.
5. Rupture 18. Coordinates of starting point of fault: 28°N 35.4^{o}E; coordinates of end point of fault: 28.77^{o}N 36.33^{o}E; length of fault: 118 km; strike angle: 45°; width of fault: 30 km; focal depth: 30 km; dip angle: 45°; slip angle: 80°; fault displacement: 10 m; maximum positive tsunami amplitude at the source 2.30 m; maximum negative tsunami amplitude at the source: 0.13 m.

For modeling of each case, the model is used to compute the sea state at different times, and time histories of water surface fluctuations and the maximum positive amplitudes at every grid point. Snapshots (at $t = 0$, 15, 30, 40 min for rupture 14, and at $t = 0$, 15, 30, 60 min for ruptures 15 – 18) of the tsunami wave propagation are shown in Figures 3 – 7.

As seen from Figure 3 that when a tsunami is generated at Eastern Hellenic Arc near the location of rupture 14, its directivity will be toward southwest and northeast directions. The arrival time of this tsunami to Peloponnesus and Crete will be less than 10 minutes, and to north Africa will be less than 30 minutes. The main effect of tsunami in this case may be expected near west of Crete south of Peloponnesus.

As seen from Figure 4 that when a tsunami is generated at south of Crete near the location of rupture 15, its directivity will be towards south and north directions. The arrival time of this tsunami to north Africa will be about 30 minutes. The main effect of tsunami in this case may be expected near south of Crete, north of Libya and northwest of Egypt.

As seen from Figure 5 that when a tsunami is generated at southeast of Crete near the location of rupture 16, its directivity will be towards south and north directions. The arrival time of this tsunami to north Africa will be about 30 minutes. The main effect of tsunami in this case may be expected near south of Crete, northeast of Libya and north of Egypt.

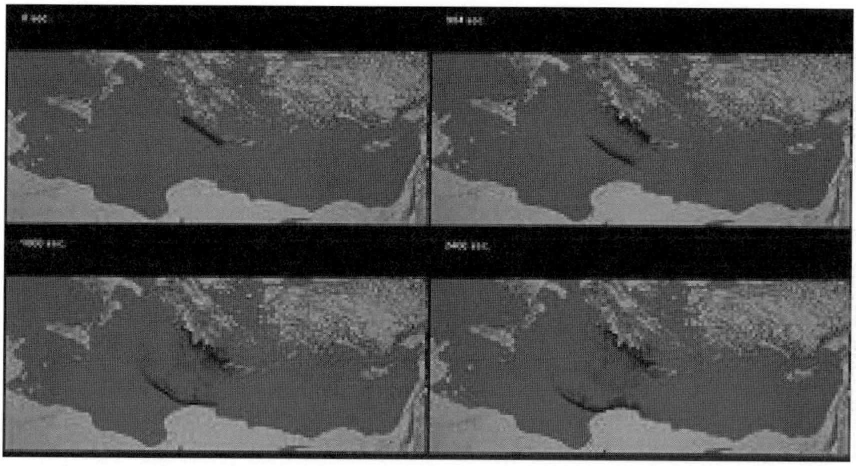

Fig. 3. The Sea State 0, 15, 30, 40 Minutes for Tsunami Rupture 14

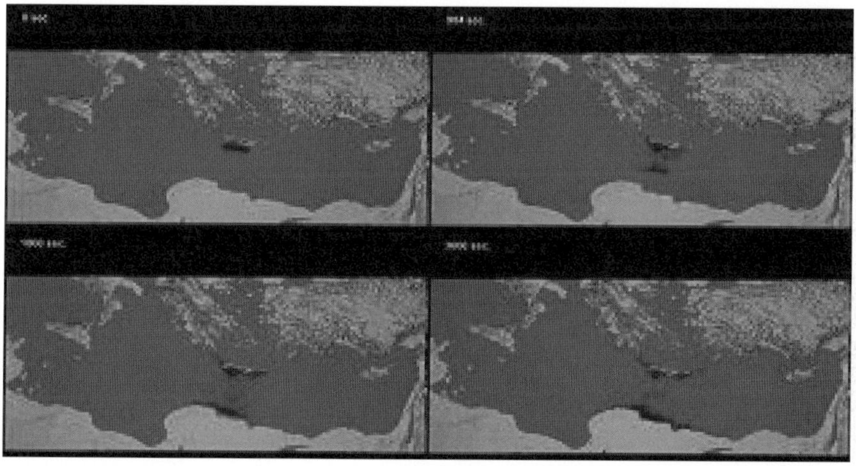

Fig. 4. The Sea State 0, 15, 30, 60 Minutes for Tsunami Rupture 15

As seen from Figure 6 that when a tsunami is generated at southeast of Crete near the location of rupture 17, its directivity will be towards southeast and northwest directions. The arrival time of this tsunami to northeast of Africa will be less than 50 minutes. The main effect of tsunami in this case may be expected in the region between west of Crete and south of Rhodes, and also north of Egypt.

As seen from Figure 7 that when a tsunami is generated at Rhodes-Dalaman trench near the location of rupture 18, its directivity will be towards southeast and northwest directions. The arrival time of this tsunami to north Africa will be about 45 minutes. The main effect of tsunami in this case may be expected near Karpathos, Rhodes and southwest coast of Turkey, and north of Egypt.

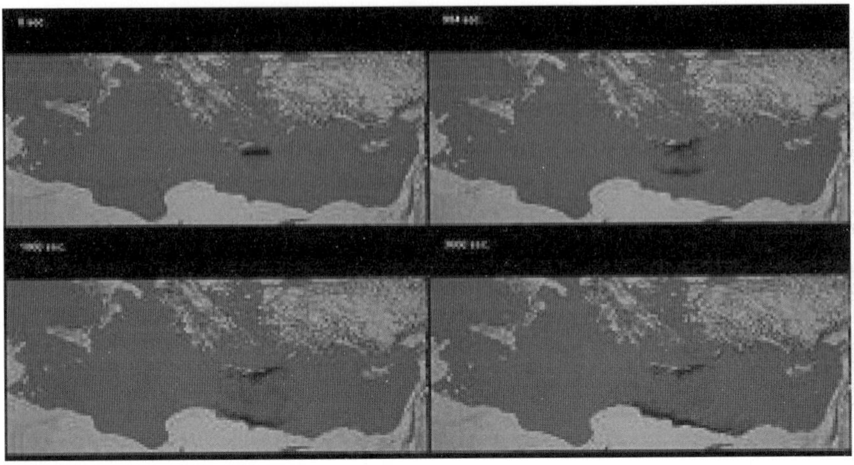

Fig. 5. The Sea State 0, 15, 30, 60 Minutes for Tsunami Rupture 16

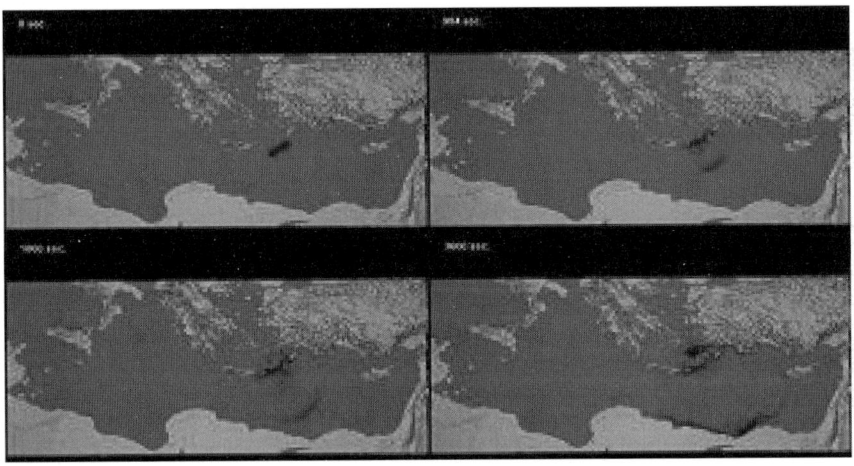

Fig. 6. The Sea State 0, 15, 30, 60 Minutes for Tsunami Rupture 17

5 Discussion and conclusions

The numerical modeling is applied by using the code NAMI DANCE to five different tsunami cases possible to generate in Eastern Mediterranean sea by using the only available data of earthquake epicenters and magnitudes measured in the last century and subduction zone properties of Hellenic Arc. Five different ruptures are assigned for each case with specially estimated rupture characteristics. According to the results of the simulations for five different tsunami cases, it is obtained that when a tsunami is generated at Eastern Hellenic Arc near the location of rupture 14 (case 1), the tsunami waves will arrive to Peloponnesus and Crete in less than 10 minutes, and to north Africa in less than 30 minutes. In this case, near west of Crete south of Peloponnesus is expected to be effected significantly. In cases 2 and

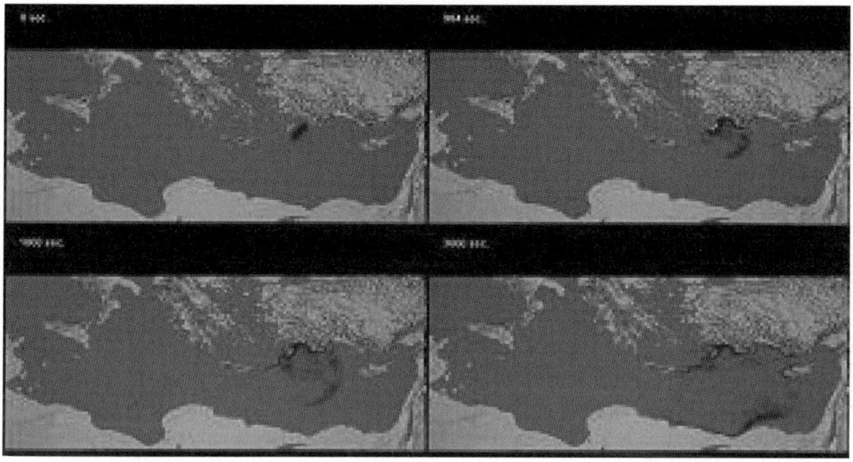

Fig. 7. The Sea State 0, 15, 30, 60 Minutes for Tsunami Rupture 18

3 in which a tsunami is generated at south of Crete, the arrival time of the waves to north Africa will be about 30 minutes, and south of Crete, north of Libya and northwest and north of Egypt will be effected. Near the location of ruptures 17 and 18 in cases 4 and 5, tsunami waves will arrive to northeast of Africa in less than 50 minutes. In these cases the main effect of tsunami waves may be expected from west of Crete to south of Rhodes, Karpathos, north of Egypt and southwest coast of Turkey.

Acknowledgements

Prof. Fumihiko Imamura, Prof. Costas Synolakis and Prof. Emile Okal are acknowledged because of their long term, valuable cooperation, discussion and collaboration during the development of the model used in this study. The authors also acknowledge partial support of UNESCO-IOC and INTAS Ref. Nr 05-109-5100 for development of the code. EP thanks the RFBR grant (05-05-64265).

References

[1] Altinok, Y, Ersoy, S (2000) Tsunamis observed on and near Turkish Coasts. Natural Hazards 21: 185–199.
[2] Altinok, Y, Ersoy, S, Yalciner, AC, Alpar, B, Kuran, U (2001) Historical tsunamis in the Sea of Marmara, Proc. Int. Tsunami Symposium (Seattle, August 7-9, 2001), Session 4, Paper 4-2, 527–535.
[3] Choi, BH, Pelinovsky, E, Kim, KO, Lee, JS (2003) Simulation of the trans-oceanic tsunami propagation due to the 1883 Krakatau volcanic eruption. Natural Hazards and Earth System Sciences 3: 321 – 332.

[4] Choi, BH, Pelinovsky, E, Hong, SJ, Woo, SB (2003) Computation of tsunamis in the East (Japan) Sea using dynamically interfaced nested model. Pure and Applied Geophysics 160: 1383–1414.
[5] Doumas, C (1980) Thera and the Aegean World. G. Tsiverotis Ltd., Athens Greece. Vol. 1–2.
[6] Goto, C, Ogawa, Y, Shuto, N, Imamura, N (1997) Numerical method of tsunami simulation with the leap-frog scheme (IUGG/IOC Time Project), IOC Manual, UNESCO, No. 35.
[7] Hardy, DA, Renfrew, AC (1990) Thera and the Aegean World Archaeology, Proc. Third Int. Congress (Santorini, Greece, September 3-9, 1989). The Thera Foundation London. Vol. 1–3.
[8] KOERI (2004) Earthquake database of Bosphorus University, Kandilli Observatory and Earthquake Research Institute (KOERI), The link of Earthquake Database http://www.koeri.boun.edu.tr/jeofizik/defaulteng.htm
[9] Kuran, U, Yalciner, AC (1993) Crack Propagations Earthquakes and Tsunamis in the Vicinity of Anatolia, In: Tinti S. (ed) Tsunamis in the World, Series: Advances in Natural and Technological Hazards Research. Kluwer Academic Publisher. 159–175.
[10] Okada, Y (1985) Surface deformation due to shear and tensile faults in a half-space, Bull. Seism. Soc. America 75: 1135–1154.
[11] Yalciner, AC, Kuran, U, Akyarli, A, Imamura, F (1995) An Investigation on the Generation and Propagation of Tsunamis in the Aegean sea by Mathematical Modeling, In: Tsuhiya Y, Shuto N (eds) Tsunami: Progress in Prediction, Disaster Prevention and Warning, Series: Advances in Natural and Technological Hazards Research. Kluwer Academic Publishers. 55–71.
[12] Yalciner, AC, Alpar, B, Altinok, Y, Ozbay, I, Imamura, F (2002) Tsunamis in the Sea of Marmara: Historical Documents for the Past, Models for Future, Marine Geology 190: 445–463.
[13] Yalciner, AC, Imamura, F, Synolakis, CE (2002) Simulation of tsunami related to caldera collapse and a case study of thera volcano in Aegean sea. Abstract Published and paper presented in EGS XXVII General Assembly, Nice, France, April 2002 Session NH8.
[14] Yalciner, AC, Karakus, H, Kuran, U (2006) Modeling of Tsunamis in the Eastern Mediterranean and Comparison with Caribbean. In: Mercado A, Liu PLF (eds) Caribbean Tsunami Hazard. World Scientific. 326–340.
[15] Yalciner, AC, Pelinovsky, E, Zaytsev, A, Kurkin, A, Ozer, C, Karakus, H (2006) NAMI DANCE Manual, METU, Civil Engineering Department, Ocean Engineering Research Center, Ankara, Turkey.
[16] Yalciner, A, Pelinovsky, E, Talipova, T, Kurkin, A, Kozelkov, A, Zaitsev, A (2004) Tsunamis in the Black Sea: comparison of the historical, instrumental and numerical data. J. Geophys. Research 109, No. C12, C12023 10.1029 2003JC002113.
[17] Zahibo, N, Pelinovsky, E, Yalciner, A, Kurkin, A, Koselkov, A, Zaitsev, A (2003) The 1867 Virgin Island Tsunami: observations and modelling. Oceanologica Acta 26: 609 – 621.
[18] Zahibo, N, Pelinovsky, E, Talipova, T, Kozelkov, A, Kurkin, A (2006) Analytical and numerical study of nonlinear effects at tsunami modelling. Applied Mathematics and Computation 174: 795–809.

[19] Zaitsev, AI, Kurkin, AA, Levin, BV, Pelinovsky, EN, Yalciner, A, Troitskaya, YuI, Ermakov, SA (2005) Numerical simulation of catastrophic tsunami propagation in the Indian Ocean. Doklady Earth Sciences 402: 614–618.

Characterization of Potential Tsunamigenic Earthquake Source Zones in the Indian Ocean

N. Purnachandra Rao

National Geophysical Research Institute, Hyderabad 500 007, India
raonpc@ngri.res.in

1 Introduction

The uniqueness of the Indian tectonic plate is manifested in the diverse nature of deformation along its plate margins. The magnificent Himalayan mountain belt to the north represents a continent-continent collision. To the east, a complex, oblique subduction takes place along the Burma and Andaman arc regions that continues further down south, to the Sumatra and Java trenches. A nascent diffuse deformation zone in the northeastern Indian Ocean further south, that formed during the last 7 Ma (Gordon et al., 1990), marks the southern plate margin separating the Indian plate from the Australian plate. A spreading ridge with respect to the African plate to the south-west and a strike-slip environment with respect to the Arabian plate to the west are the other plate boundaries (Fig. 1). The Indian plate has made a remarkably long journey in the geological time scale, with the initial break up from the Gondwanaland about 125 million years ago. Initially the Indian plate moved very rapidly north, at about 10 cm/y until about 53 Ma, when it made the initial contact with Asia. With the closure of the Neo-Tethys Sea during the Late Cretaceous to Early Tertiary period, the scenario changed to that of a continent-continent collision with a reduced rate of about 5 cm/y. The gigantic Himalayan mountain range formed from the scraped off Indian crust and is the site of some of the world's largest and most devastating earthquakes.

2 Seismotectonics of the Indian Ocean region

2.1 Burma-Andaman arc region

The Burma-Andaman arc lies on the eastern margin of the Indian plate, along which an oblique convergence between the India and Burma plates (also referred as the Burma platelet) has been suggested (Fitch, 1972; Curray et al., 1979). The transition between the two arcs is marked by an aseismic zone (Chandra, 1984). The major tectonic features along the arc (Fig. 2) are the N-S trending Indo-Burman ranges in the north and the Andaman Nicobar ridge in the south. The Sumatran fault system in the southeast, the Western Andaman fault and the Sagaing fault, further

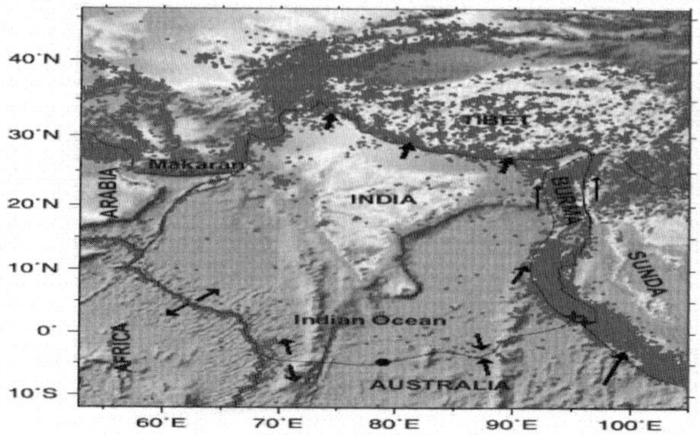

Fig. 1. Seismicity and topographic map of the Indian plate region depicting the varied plate kinematics along its boundaries, Arrows indicate the plate motion vectors along the boundaries.

east, are the features supporting major right lateral movements in this region. The Sagaing fault separates the central low lands from the eastern high lands of Burma and seems to continue into the Andaman sea rift system which is characterized by sea floor spreading and transform faulting (Curray et al., 1979). The presence of the major right lateral strike-slip faults in the east, running parallel to both the arcs, has been related to the oblique nature of convergence (Fitch, 1972; Maung, 1987). The development of the two arcs as distinct entities has been attributed by Maung (1987), to the northward drag of the Indo-Burman ranges and the Andaman-Nicobar ridge.

Studies of satellite imageries and focal mechanisms (Le Dain et al.,1984, Ni et al., 1989) suggest the possibility that the Indian plate no longer underthrusts the Indo-Burman ranges. In the Burmese arc region, almost all the focal mechanisms on the subducted lithospheric slab show P-axis azimuths oriented NNE. This implies that the tectonics of the slab is governed mostly by a NNE-directed stress field owing to the Indian plate motion. The focal mechanisms along the slab are distinctly segregated into predominantly strike-slip type, occurring in the upper half of the slab down to a depth of about 90 km, and thrust type, occurring exclusively in the lower half, below that depth level. This could be due to shearing of the Indian plate past the Burmese plate in the zone of contact, and compression of the lower portion of the slab against the asthenosphere in the NNE direction (Kumar and Rao, 1995; Kumar et al., 1996). A comparative study with world-wide subduction zones provides further evidence for cessation of subduction in the Burmese arc (Rao and Kumar, 1999). Other studies, however, support the idea that the Indian plate still continues to underthrust along the Burmese arc, based on seismicity trends (Verma et al., 1976), gravity studies (Mukhopadhyay and Das Gupta, 1988) and T axis orientations predominantly along the dipping slab (Satyabala, 1998).

Using stress inversion of focal mechanism data in the Burmese arc region, Rao and Kalpna (2005) demonstrated an unusual overturn and resistance to penetration at the 410 km discontinuity. They proposed a model of attempted slab detachment due to down-dip tensile forces that explains the reverse faulting on steeply dip-

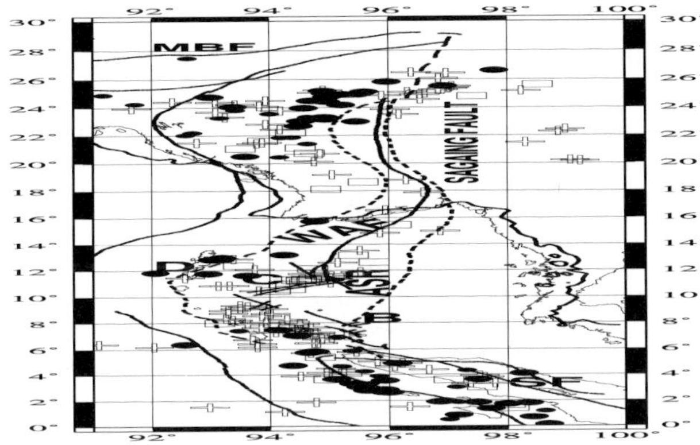

Fig. 2. Seismotectonic map of Burma-Andaman arc region showing the distribution of earthquakes of strike-slip (+), reverse (O), normal ([]) and odd (o) types (after Rao, 1999). SF- Sumatran Fault System, ANR- Andaman Nicobar Ridge, ASR-Andaman Spreading Ridge, WAF- West Andaman Fault and MBF- Main Boundary Fault

ping fault planes at the base of the lithospheric contact zone. The results are also confirmed by seismic tomography studies (Bijwaard, 1998).

In the northern Andaman arc region, a majority of the earthquake focal mechanisms correspond to sea floor spreading and transform faulting, indicated by normal or right lateral strike-slip earthquakes (Curray et al., 1979). The southern part of the Andaman arc region exhibits distinct tectonic characteristics of an active subduction zone, as evidenced by disposition of focal mechanisms along the slab and the P and T axis orientations (Kumar et al., 1996).

2.2 Northeastern Indian Ocean region

The northeastern Indian Ocean region (Fig. 3), to the south of the Indian peninsula has displayed anomalous seismicity, too high to be intra-plate. Sykes (1970) attributed this to the development of a nascent island arc between Sri Lanka and Australia. Stein and Okal (1978) suggested that most of the seismicity can be explained by a major left lateral strike-slip motion along the Ninetyeast ridge. This model was attributed to the greater resistance of the western part of the Indo-Australian plate against the Himalayan continent-collision zone, as compared to the eastern part, which subducts smoothly beneath the Sunda arc.

Plate motion modelling studies provided evidence for relative plate motion between the Indian and Australian plates (Minster et al., 1974; Minster and Jordan, 1978; Stein and Gordon, 1984). Wiens (1986) suggested that the diffuse plate boundary separating India and Australia plates extends equatorially from the Central Indian ridge up to the northern Ninetyeast ridge and further north, up to the Sumatran trench. The Euler vector model of plate rotation for Australia-India (Fig. 3) determined by Gordon et al. (1990) and DeMets et al. (1990, 1994) predicts a N-S divergence at the Chagos bank in the west, a N-S convergence in the east, west of the Ninetyeast ridge, and a left lateral strike-slip along the Ninetyeast ridge. Other studies like gravity, bathymetry, seismic profiling, heat flow, etc. have parallely indicated a diffused deformation zone with intense faulting and folding of sediments in approximately N-S direction (Eittreim and Ewing, 1972; Weissel et al., 1980; Geller et al., 1983; Neprechnov et al., 1988; Levchenko, 1989; Levchenko et al., 1992; Chamot-Rooke et al., 1993).

A model of 'wrench fault tectonics' was earlier proposed for the northeastern Indian ocean region based on seismic reflection and refraction studies in a portion of the deformation zone by Neprechnov et al. (1988). In this model relative motion between the Australia and India plates is accomplished through left lateral shearing along NE-SW trending blocks in the deformation zone, along transform faults from the central Indian ridge towards the northern part of the Ninetyeast ridge. This model, however, appears inadequate since it does not explain the observed large scale convergence and divergence in the eastern and western parts respectively of the deformation zone, which are well explained by the plate rotation models. In view of the above, Rao and Kumar (1996) proposed a combination of the rigid plate Euler pole rotation model in conjunction with strike slip faulting of book-shelf or wrench-fault type in the northern portion extending from the central Indian ridge in the west to the northern ninetyeast ridge in the east, to explain the India-Australia plate kinematics.

2.3 The Central Indian Ridge

The Central Indian ridge comprises the south-western boundary of the Indian plate. It is a spreading mid-oceanic ridge that separates the India and Africa plates. The earthquakes are of moderate to large size with focal mechanisms, generally of strike-slip and normal types. A large earthquake of magnitude 7.6 and focal depth 10 km occurred in this region near Carlsberg ridge on 15 July 2003.

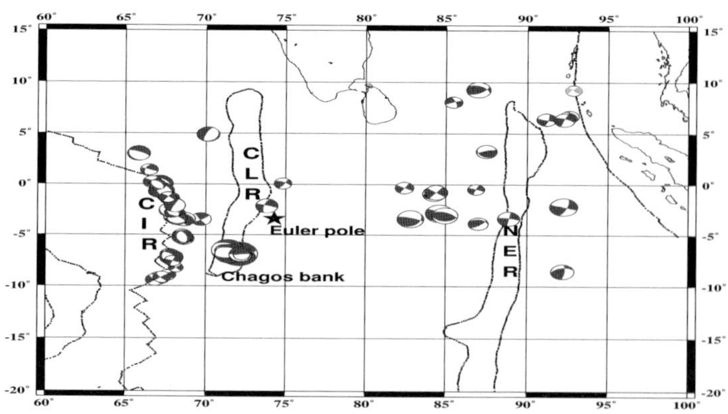

Fig. 3. Seismotectonic map of the Indian Ocean region along with the focal mechanism solutions, indicating the complex plate kinematics of the duffuse deformation zone separating India and Australia plates. Note the coexistence of convergent and divergent environments along the same plate boundary on the east and west respectively. Further, the relative plate motion is aided by strike-slip mechanism along the ninetyeast ridge and further north.

3 The 26 December 2004 Banda Aceh earthquake

An unusually large and devastating earthquake of magnitude greater than 9.0 occurred on the 26th December 2004, off the coast of northern Sumatra, close to Banda Aceh (Fig. 4). The mega-thrust earthquake that released about 20 x 10**17 Joules of energy triggered a killer tsunami, apparently quite uncommon to the Indian Ocean region, causing massive destruction with a death toll of about 200,000 in Indonesia, Thailand, India, Sri Lanka and other countries. The hypocentral parameters estimated by the US Geological Survey (latitude 3.298 N, longitude 95.779 E, depth 30 km) indicate that the earthquake occurred along the zone of subduction between the India and the Burma plates, along the Andaman arc. The moment magnitude estimated based on complete waveform analysis of data from hundreds of seismometers worldwide, is 9.3 (Stein and Okal, 2005; Tsai et al., 2005) making it the third largest earthquake during the last century. However, this has been regarded as an overestimate by some in view of the very long period energy associated with this earthquake, which was never considered for any of the previously recorded earthquakes, due to limitations of instrumentation. The United States Geological Survey placed its estimate of moment magnitude at 9.1. The focal mechanism given by the Harvard University has a strike 329, dip 8 and rake 110 degrees, indicating a thrust fault mechanism with a NW-SE trending plane. Based on the shallow aftershock distribution and the known trend of the decollement plane in the subduction zone it is reasonable to accept the NE trending shallow dipping fault plane at a focal depth of about 30 km. The earthquake ruptured a large segment (1200 km x 200 km) along the trench from northern Sumatra to northern Andaman Islands, as evidenced by the distribution of a large number of aftershocks of moderate size. However, the unidirectional northward propagation of the rupture and seismicity is quite surprising, and hardly any aftershock activity was observed south of it along the Australia-Sunda plate boundary. Modeling of the fault plane and rupture propagation using teleseismic broadband data (Yagi, 2004; Yamanaka, 2004; Ji, 2004; Borges et al.,

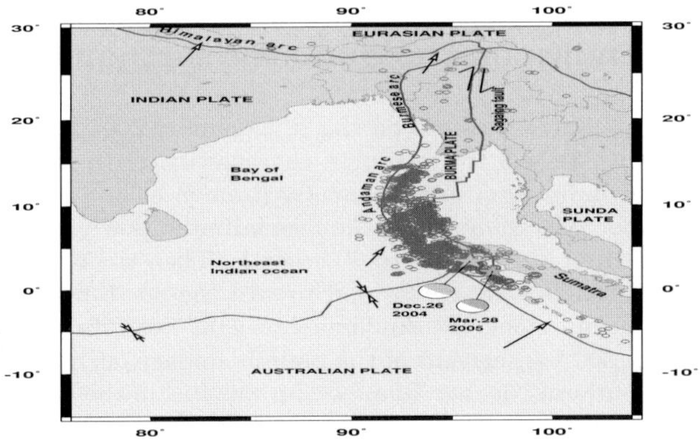

Fig. 4. Location map of the 26 December 2004 and the 28 March 2005 Sumatra earthquakes along with their focal mechanism solutions. The earthquakes occurred near the quadruple junction of four tectonic plates - India, Burma, Australia and Sunda - in one of the most complex tectonic scenarios.

2005) broadly indicates that the rupture propagated mostly in the NNW and then in the NS directions. The enormous amount of energy released suggests that the 26 December 2004 Sumatran earthquake was a mega-thrust that must have slipped suddenly after a lock-up of several decades, causing fault movement of about 10-15 m (Sieh et al., 2005).

4 The 28 March 2005 Nias earthquake

On 28 March 2005, hardly 3 months after the great Sumatran earthquake of 26 December 2004, another great earthquake of magnitude 8.7 occurred just about 250 km southeast of the 2004 event, close to the Nias Islands. Approximately 1300 people were killed by the earthquake, mostly on the island of Nias. The event caused panic in the region, in view of the recent devastation by the massive tsunami triggered by the 2004 Indian Ocean earthquake, but this earthquake generated a relatively tiny tsunami that caused limited damage. On Nias Islands hundreds of buildings were destroyed and the death toll was at least one thousand. The earthquake lasted for about two minutes in total. In the twenty-four hours immediately after the event, there were eight major aftershocks, measuring between 5.5 and 6.0. The estimated focal depth was close to 30 km and the focal mechanism represented reverse faulting very similar to that of the 2004 Sumatra earthquake.

The rupture of the 28 March 2005 Sumatra earthquake was imaged by Walker et al. (2005) by back-projecting teleseismic P waves recorded by the Global Seismic Network and the Japanese Hi-net to the source. The rupture propagation initiated from the hypocenter and proceeded toward the north lasting for 50 s. A second larger rupture started at about 40 s from the hypocentral time and propagated toward southeast. Apparent average rupture speeds in both directions are between 2.9 and 3.3 km/s, which are slightly faster than those found for the main Sumatra event

5 Research studies on the Sumatra earthquakes

Several studies have been carried out on the 26 December 2004 Banda Aceh (Sumatra) earthquake, with regard to its source characteristics, the rupture dynamics, aftershock patterns and crustal displacements. Ammon et al. (2005) demonstrated that the 26 December 2004 Sumatra earthquake initiated slowly, with small slip and a slow rupture speed for the first 40 to 60 seconds. Then the rupture expanded at a speed of about 2.5 kilometers per second toward the north-northwest, extending 1200 to 1300 kilometers along the Andaman trough. Peak displacements reached about 15 m along a 600 km segment of the plate boundary offshore of northwestern Sumatra and the southern Nicobar Islands. Slip was less in the northern 400 to 500 kilometers of the aftershock zone, and at least some slip in that region may have occurred on a time scale beyond the seismic band.

According to Lay et al., (2005) the 2004 Sumatra-Andaman earthquake had the longest known earthquake rupture (Fig. 5). Short-period seismic body waves (0.5 to 0.25 s) show azimuthally varying durations that indicate that the seismic rupture front propagated to about 1200 km north of the epicenter with a rupture velocity of about 2.0 to 3.0 km/s and that short-period radiation was generated for at least 500 s. Array analysis of 1 to 2 s period seismic waves from Hi-net stations in Japan yielded compatible results.

Centroid-moment-tensor (CMT) analysis of the earthquake was carried out using waveform inversion technique considering multiple point sources to represent a propagating slip pulse (Tsai et al., 2005). The final model consists of five sub-events (Fig. 6), with the southernmost source accounting for the majority of the moment release. The presumed fault planes of the southern sources strike northwest, while those in the north strike northeast, consistent with the geometry of the subduction trench. Broadband arrays at teleseismic distances have been used to image the spatio-temporal characteristics of the seismic energy release during the 26 December 2004 Sumatra earthquake at early observation times (Kruger et al., 2005). Using a non-plane-wave array location technique, a rupture length of about1150 km, duration of about 480 seconds and an average rupture velocity of 2.4-2.7 km/s were confirmed.

The source region of the 26 December 2004 Sumatra tsunami was delineated using wave arrival times obtained from coastal tide gage and satellite altimetry records for the Indian Ocean (Fine et al., 2006) which closely matches the seismic source estimated from broadband geophysical data. Yuan et al., 2005 studied the tsunami

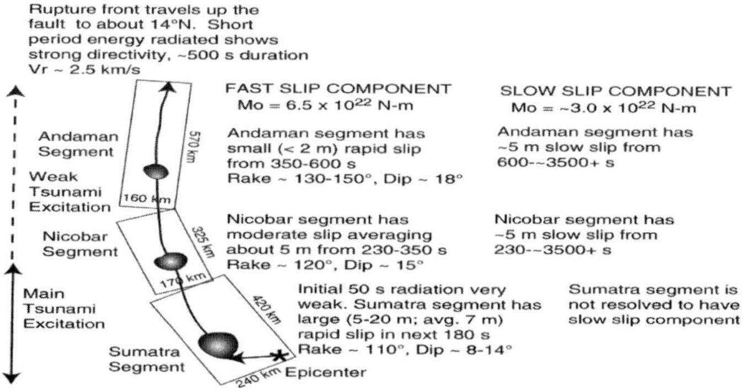

Fig. 5. Summary rupture scenario for the 2004 Sumatra-Andaman earthquake, with three inferred rupture segments (after Lay et al., 2005). The rupture begins at the southeastern edge of the Sumatra segment, with the initial 50 s of rupture characterized by fairly low energy release and slow rupture velocity. The rupture front then expands to the north-northwest at about 2.5 km/s, extending about 1300 km. Short-period radiation tracks the rupture front, with a total duration of about 500 s and clear north-northwest directivity. Large, rapid slip occurs in the Sumatra segment, with some patches having slip as great as 20 m during the first 230 s. The Nicobar segment has weaker slip during the next 2 min, and the Andaman segment fails with little (less than 2m) rapid slip. Slow slip appears to continue in the Nicobar and Andaman segments, with a total duration of about 1 hour. The precise amount of slip and total moment of the slow-slip component are not well resolved, but about 10 m of slip under the Andaman Islands is required to account for the tilt experienced by the islands.

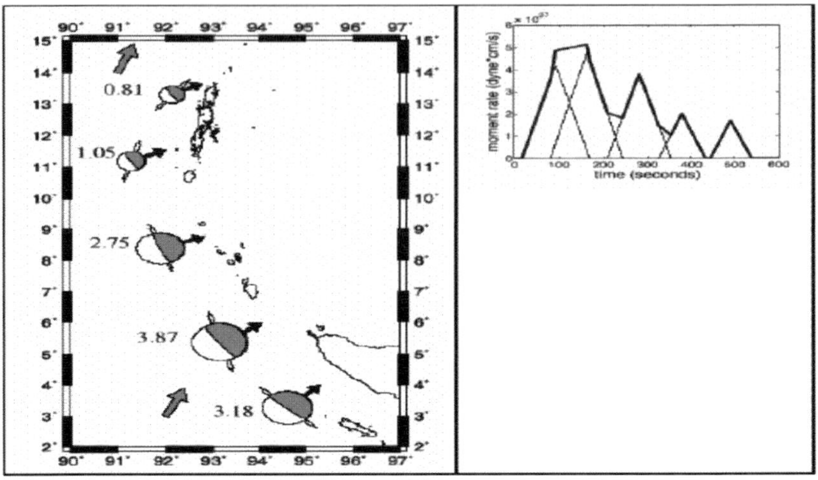

Fig. 6. Sub-event analysis of the 26 December 2004 Sumatra earthquake (after Tsai at al, 2006). (a) Locations and focal mechanisms of the five CMTs of the final model. The areas of the focal mechanisms are proportional to their scalar moments, which are given next to each mechanism in units of 1029 dyne-cm. The needles denote the strike directions of the shallowly dipping nodal planes; the black arrows denote the slip directions projected onto the horizontal plane. The gray arrows show the long-term plate motions of the Indian plate relative to the Eurasian plate (DeMets et al., 1994). (b) Source time function. The thin black lines denote the individual sources. The heavy black line denotes the sum. The zero time corresponds to the hypocentral time of 00:58:53.5.

induced long-period (greater than 1000 s) signals on the horizontal components of the broadband seismometers located on islands in the area. The Gravity Recovery and Climate Experiment (GRACE) satellites observed a 15 microgal gravity change induced by the great December 2004 Sumatra-Andaman earthquake (Han et al., 2006). Coseismic deformation produces sudden changes in the gravity field by vertical displacement of earth's layered density structure and by changing the densities of the crust and mantle. The GRACE observations provided evidence of crustal dilatation resulting from an undersea earthquake.

Several GPS studies have observed the near-field and far field displacements of the 26 December 2004 earthquake (e.g. Fig. 7 [1]). Using far-field GPS sites, about 400-3000 km from the rupture, Banerjee et al. (2005), Catherine et al. (2005), Vigny et al. (2005) and Hashimoto et al. (2006) constrained the fault slip during the December mainshock. A maximum slip of 30 meters was inferred for this earthquake. Banerjee et al. (2005) estimated an average slip along the rupture of 5 meters. Hashimoto et al. (2006) suggested that coseismic slip as large as 14 meters occurred beneath the Nicobar Islands. Gahalaut et al. (2006) improved slip resolution and rupture characteristics using coseismic displacements derived from near-field GPS

[1] Editor's note: Due to conversion to b/w the color code is not visible. Please contact the author for the original color figure.

and estimated coseismic slip of 3.8-7.9 meters under the Andaman Islands and 11-15 meters under the Nicobar Islands. They also estimated coseismic horizontal ground displacement and vertical subsidence along the Andaman-Nicobar Islands of 1.5-6.5 meters and 0.5-2.8 meters, respectively. Both geodetical and seismological slip models agree that the largest slip occurred near the southern end of the rupture zone and diminished northward (Ammon et al., 2005). This is also supported by moment-tensor analysis through long-period waveform modeling by Tsai et al., (2005).

The co-seismic and post-seismic deformation of the 28 March 2005 earthquake was studies by Kreemer et al. (2006) using continuous GPS data. Up to 5 m of co-seismic slip has been reported, with slip results comparable to that of Walker et al. (2005). According to Briggs et al. (2006) the seismic rupture produced major tectonic deformation above a 400 km strip of the Sunda megathrust. Measurements from coral microatolls and GPS stations reveal trench-parallel belts of uplift up to 3m high on the outer-arc islands above the rupture and a 1m deep subsidence trough farther from the trench. Surface deformation reflects more than 11m of fault slip under the islands and a pronounced lessening of slip trenchward.

6 What caused two Great Earthquakes in three months?

Interestingly, both the earthquakes, 26 December 2004 and 28 March 2005, have almost identical focal depths as well as focal mechanism solutions. Normally such large earthquakes have return periods of the order of a century or more. Hence,

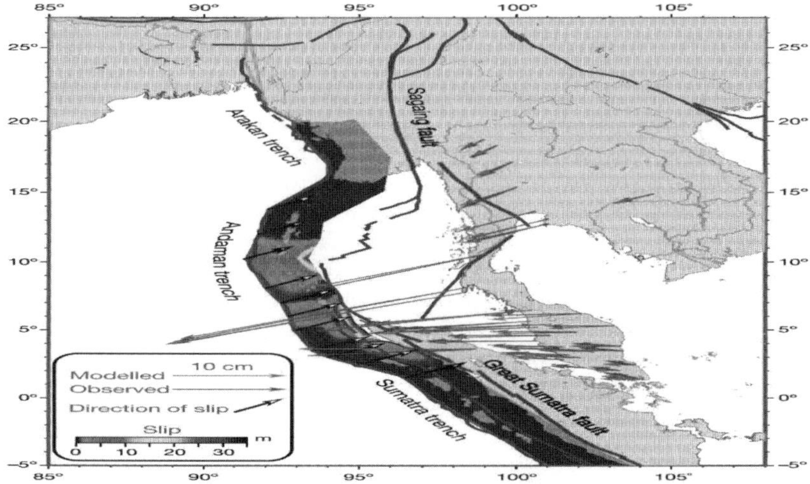

Fig. 7. Best-fit elastic model of the co-seismic displacements using non-homogeneous slip along the rupture plane (after Vigny et al., 2005). Colour code indicates the amount of slip from 0 (dark blue) to 35m (dark red). Note that the area with significant slip is approximately 1,100 km long. Measured displacements (purple vectors) and modelled deformation (green vectors) are also depicted. Thin red lines depict major faults

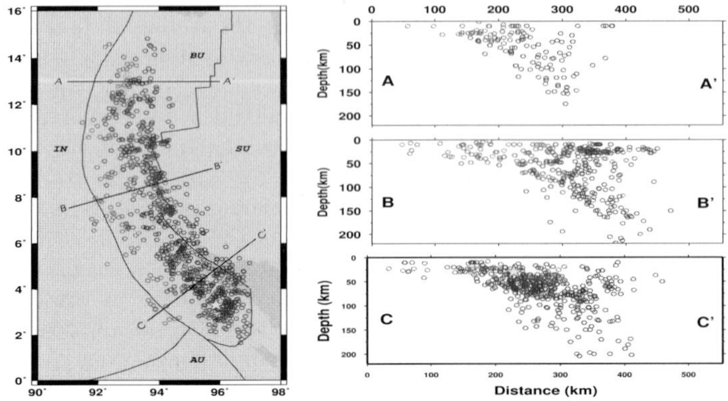

Fig. 8. The epicentral map of the Andaman-Sunda region in (a) plan view and (b) depth sections, indicating the locations of earthquakes prior to the 26 December 2004 Sumatra earthquake (blue circles) and the aftershocks (red circles). It can be seen that the aftershocks have ruptured a fresh zone, which was probably locked up for several decades .

the close spatial proximity of the two great earthquakes occurring within a very short span of 3 months, raises important questions about the mechanism of stress lock-up over large time periods in convergent plate boundary zones. In the present case, a pertinent question is whether the 26 March event was an aftershock of the 26 December event, and if so whether a lock-up of the total energy of the two is feasible. Or could the first earthquake have triggered the second one?

The epicentral locations of these two great earthquakes very close to the quadruple plate junction comprising the India, Burma, Australia and Sunda plates (Fig. 8 [2]) suggests that the difference in styles of India-Burma and Australia-Sunda plate subductions aided by the nascent development of the diffuse Indo-Australian boundary in between could be responsible for the occurrence of these two great earthquakes. As the Australian plate overrides the Indian plate with an anticlockwise rotation that produces the highest velocity of about 1.2 cm/yr near the plate junction. From the above it appears that the India-Burma and Australia Sunda subduction zones are distinct, with different stress lock-up scenarios separated by a barrier or a break along the Australia-India plate boundary (Rao and Chary, 2005). The release of stress along the India-Burma plate boundary resulting in the great earthquake of 26 December 2004, in turn appears to have triggered the release of stress locked up at the Australia-Sunda plate boundary, within a very short span of 3 months, producing the great earthquake of 28 March 2005.

7 A new seismotectonic model

The Sumatra-Andaman-Nicobar arc is undoubtedly among the world's seismically most active regions where subduction consumes the Indo-Australian plate into the

[2] Editor's note: Due to conversion to b/w the color code is not visible. Please contact the Author for the original color figure.

underlying mantle. In the Andaman-Nicobar region, the Indian plate is known to be subducting beneath the Burmese plate while in the Sumatra region the Australian plate is subducting beneath the Sunda plate. As described earlier, the Indian plate subduction beneath the Burmese plate is highly oblique varying from an almost NS shear at the Burmese arc in the North (Rao and Kumar, 1999) to a northeastward subduction in the southern Andaman arc (Kumar et al., 1996). Further down south, the Australian plate has a much smoother and faster northerly subduction beneath the Sunda plate. The scenario is further complicated by the development of a diffuse convergent deformation zone in the northeastern Indian Ocean, between the India and Australia plates in the last few million years (Gordon et al., 1990), owing to a greater resistance of the Indian plate against the Himalayan collision zone in the north, compared to the smoother subduction of the Australian plate beneath the Sunda arc in the Southeast (Stein and Okal, 1978). Also, the styles of subduction in the India-Burma and Australia-Sunda plate boundary zones are quite dissimilar. While the former has a lower plate velocity (about 5 cm/yr) with greater lithospheric coupling, the latter is governed by a higher plate velocity (about 6.5 cm/yr) with a smoother oceanic subduction. Further, the depths of earthquake occurrence on the subducting slabs in these two regions are very distinct, as indicated by the shallower focal depths (about 230 km) in the Burma-Andaman arc as compared to the much deeper seismicity (about 700 km) in the Australia-Sunda plate boundary zone (Fig. 9).

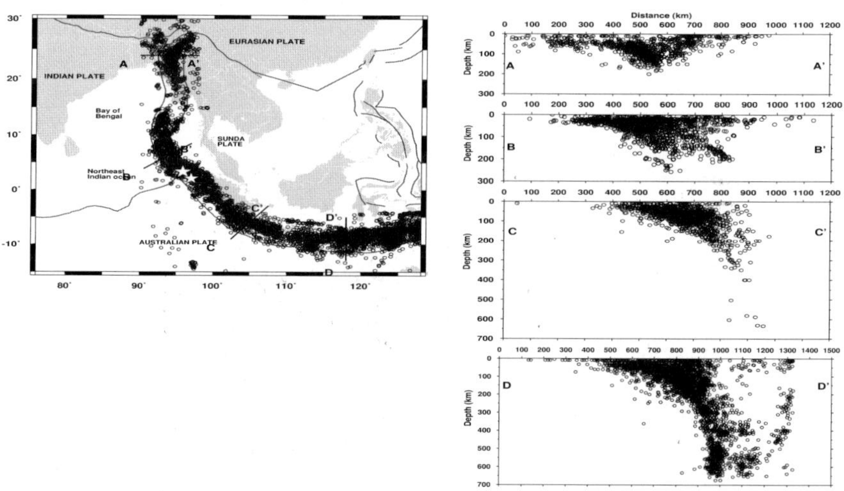

Fig. 9. Seismicity trends and depth sections along the India-Burma (AA' and BB') and Australia-Sunda (CC' and DD') subduction boundaries. Note the variation in the extent of earthquake focal depths, 230 km in the former and 700 km in the latter.

From the above it appears that the gigantic megathrust earthquake of 26 December 2004 was totally confined to the India-Burma segment of the arc, to the north, while the subsequent earthquake of 28 March 2005 was governed by a segment of the Australia-Sunda plate boundary to the south. This is also evident from seismicity data, since the rupture propagation and aftershock distribution of the 2004 earthquake were directed entirely northward and confined to the India-Burma plate boundary, while those of the 2005 earthquake were directed southward along the Australia-Sunda plate boundary. Further, it appears that due to the differential styles of India-Burma and Australia-Sunda plate motions, the once united Indo-Australian lithospheric slab got split in the recent geological past, creating a barrier which prevented lateral transfer of stress across the two zones during the two Great earthquakes of 2004 and 2005 (Fig. 10). Interestingly, this barrier zone is also the region of highest rate of convergence of about 1.2 cm/y between the Australia and India plates, as computed using the NUVEL-1A model of DeMets et al. (1994).

Further, it appears that as the Indian plate plunged violently beneath the Burmese plate during the first earthquake, it exerted a drag force on the adjoining Australian plate, triggering another great earthquake on the already critically stressed segment of the Australia-Sunda plate boundary, with a time lag of about three months. This would however, require evidence of strike-slip motion between the India and Australia plates in the NE-SW direction in the deformation zone. However, absence of any strike-slip earthquakes in this region during the observation

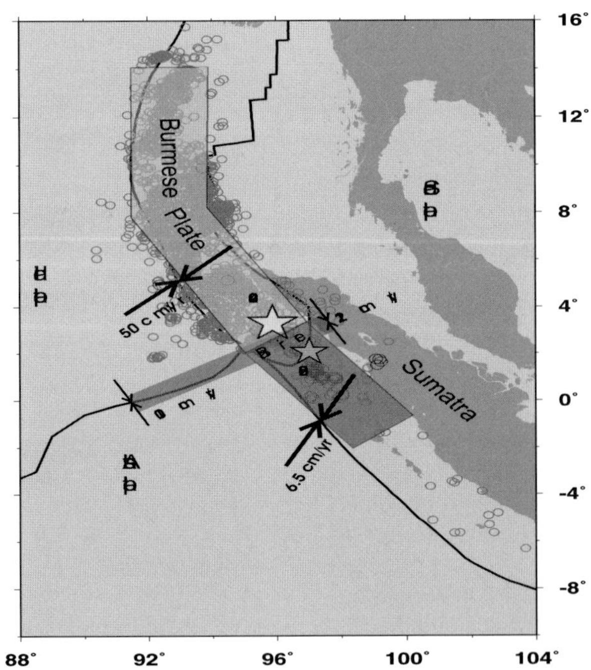

Fig. 10. A schematic of the proposed tectonic model to explain the occurrence of the two large earthquakes of 26 December 2004 and 28 March 2005 near the quadruple plate junction of the India, Burma, Australia and Sunda plates (after Rao and Chary, 2005). The velocities indicated, are computed using the NUVEL1-A plate motion model (DeMets et al., 1994).

period possibly indicates that a slow, aseismic creep, with a strike-slip deformation of the 'wrench fault style' characteristic of this part of the northeastern Indian Ocean region (Neprechnov et al., 1989; Rao and Kumar, 1996) may have been in operation. Interestingly, between the two great episodes of the 26 December 2004 and the 28 March 2005 earthquakes, there was a large cluster of about 150 predominantly strike-slip earthquakes of magnitude 5.0 and above, in the Andaman sea region further northeast during 27-30 January 2005 (Lay et al., 2005). However, it remains an open question whether this most energetic earthquake swarm ever observed globally, has any relationship with the occurrence of the two great Sumatran earthquakes of 2004 and 2005.

8 Plate Junctions and Great Earthquakes

We now focus on another important and interesting issue of why the epicentral locations of the two great Sumatran earthquakes are so close to the quadruple-junction comprising the India, Burma, Australia and Sunda plates (Fig. 4). To investigate the role of multiple plate interaction for generation of great earthquakes, 2D Finite Element Modeling (FEM) method was applied (Rao et al., 2006) using the STAAD.Pro-2004 modeling software. The kinematics of the three plates - India, Burma and Australia - were modelled, assuming homogenous lithospheres. Realistic plate geometry was considered, and horizontal velocities were computed using NUVEL1-A model (DeMets et al., 1994) along each plate boundary. For the India-Burma plate boundary, since no direct velocity estimates exist, constraints from GPS data (Paul et al, 2001; Vigny et al., 2003) were used to partition the India-Eurasia velocity estimates from NUVEL1-A along the India-Burma and Burma-Sunda plate boundaries. Modeling was carried out adopting a discrete mesh with velocity vectors defined at nine nodal points (Fig. 11a). Significantly, the results indicate peaking of the stress field near the plate junction and further southeast along the Sunda arc, close to the epicentral locations of the two great earthquakes (Fig. 11b).

In general, a correlation between great earthquakes and plate junctions appears to be true on a global scale, as is clearly demonstrated by an examination of epicentral locations of the world's 12 largest earthquakes since the year 1900, which are in the magnitude range of 8.5 to 9.5 (Fig. 12). The details of these great earthquakes (USGS, 2006) and their locations with respect to tectonic plates are listed in Table 1. It can be seen that each of these great earthquakes falls very close to multiple plate junctions, with the exception of events 6 and 8 on the Western Aleutian trench. The 1964 Alaskan earthquake (event 2) appears to be a special case of two-plate interaction, but associated with a junction between a subduction zone in the Western Aleutian trench in the north, and a right-lateral strike-slip faulting environment along the San Andreas fault in the east, that seems to act like a triple-junction, aiding stress build-up. On the whole, there appears to be a strong (more than 80 percent) correlation between epicentral locations of the world's largest earthquakes and plate junctions or inflexion points. Finite element modeling of stress field in three dimensions incorporating the plate geometries, geological ages and hence the mechanical properties of the slabs, their thicknesses, depths of penetration of the subducting slabs along different sections, and the differential velocities of plates involved, is likely to shed further light on the current understanding of the mechanism of multiple plate interactions and the resulting large earthquakes.

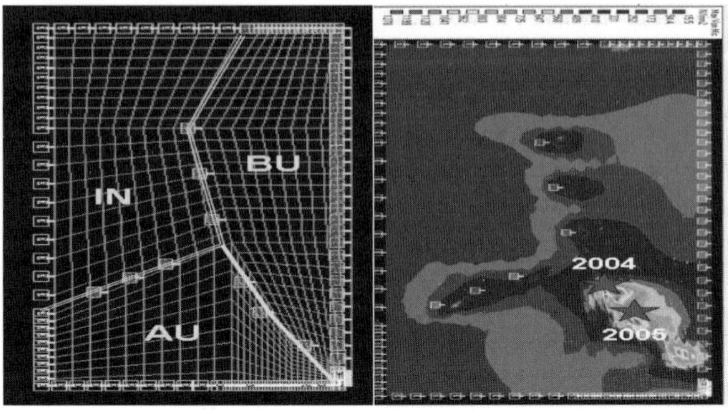

Fig. 11. (a) 2D finite element modeling of stress field around the plate junction comprising the India (IN), Burma (BU) and Australia (AU) plates. Plate motion velocities are computed using NUVEL1-A model and constraints from GPS data. (b) Results of FEM, indicating that the locations of the 2 great Sumatran earthquakes of 26 December 2004 and 28 March 2005 are close to the zones of maximum stress.

9 Tsunamigenic earthquakes in the Indian Ocean

It is strange that the 26 December 2004 Indian Ocean tsunami, the worst ever known globally, should take the world by surprise, since this region has not been devoid of history of tsunami occurrence. A detailed treatise on the past tsunami occurrences in the Indian Ocean region has been brought out by Rastogi and Jaiswal (2006) as indicated in Fig. 13 and listed in table 2. The oldest record of tsunami occurrence corresponds to the November 326 BC earthquake near the Indus delta / Kutch region that set off massive sea waves in the Arabian Sea. The mighty Macedonian fleet of

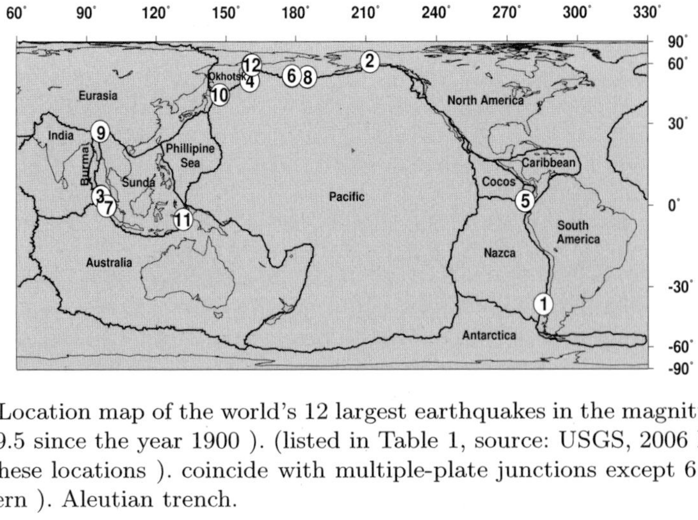

Fig. 12. Location map of the world's 12 largest earthquakes in the magnitude range of 8.5 to 9.5 since the year 1900). (listed in Table 1, source: USGS, 2006 Note that most of these locations). coincide with multiple-plate junctions except 6 and 8 on the Western). Aleutian trench.

Table 1. List of the world's 12 largest earthquakes in the magnitude range of 8.5 to 9.5 since the year 1900, their locations close to plate junctions, an d the plates associated with each earthquake (source: USGS, 2006), (Also, see Fig. 5).

No.	Location	Date UTC	Magnitude	Latitude	Longitude	Associated Plates
1.	Chile	1960 05 22	9.5	38.24 S	73.05 W	NZ, SA, AN
2.	Alaska	1964 03 28	9.2	61.02 N	147.65 W	PA, NA
3.	Sumatra	2004 12 26	9.1	3.30 N	95.78 E	IN, BU, AU, SU
4.	Kamchatka	1952 11 04	9.0	52.76 N	160.06 E	PA, NA, OK
5.	Ecuador	1906 01 31	8.8	1.0 N	81.5 W	NZ, SA, CO,CA
6.	Alaska	1965 02 04	8.7	51.21 N	178.50 E	PA, NA
7.	Sumatra	2005 03 28	8.6	2.065 N	97.010 E	IN, BU, AU, SU
8.	Alaska	1957 03 09	8.6	51.56 N	175.39 W	PA, NA
9.	Assamt	1950 08 15	8.6	28.5 N	96.5 E	IN, EU, BU, SU
10.	Kuril Islands	1963 10 13	8.5	44.9 N	149.6 E	PA, PS, EU, OK
11.	Banda Sea	1938 02 01	8.5	5.05 S	131.62 E	AU, SU, PS, PA
12.	Kamchatka	1923 02 03	8.5	54.0 N	161.0 E	PA, NA, OK

Alexander the Great, returning to Greece by sea after conquest was reportedly destroyed (Lisitzin, 1974). Poompuhar, an ancient flourishing town in Tamilnadu, southern India, was washed away by a tsunami in about 500 AD, a period that matches well with the Krakatoa explosion. A tsunami at Nagapattinam in 900AD apparently destroyed a Budhist monastery and several temples, and killed hundreds of people, according to literature available in the library of Thondaiman kingdom in Puduckottai, Tamilnadu. A tsunami was reportedly observed in the North Indian Ocean on the Iranian coast from a local earthquake between 1 April and 9 May 1008 (Murty et al., 1999). An earthquake occurred during 1524 A.D. off the coast of Dabhol, Maharashtra resulting in a large tsunami that caused considerable alarm to the Portuguese fleet that was assembled in the area (Bendick and Bilham, 1999). A tsunami is also known to have occurred in the Bay of Bengal on April 2, 1762, caused by an earthquake in Bangladesh-Myanmar border region, 257 km SE of Dhaka. The shock caused severe damage at Chittagong and other areas on the eastern seaboard of the Bay of Bengal (Mathur, 1998). The great earthquake of 16 June 1819 in the Kutch region of northwestern India (M 7.8) inundated the town of Sindri and adjoining parts, with the ground sinking apparently by about 5m (Macmurdo, 1821). An earthquake on 11 November 1842 near the northern end of Bay of Bengal caused a tsunami by which waters of the distributaries of the Ganges Delta were agitated. On 19 June 1845, in the Kutch region, the sea rolled up the Kori creek with the Indus overflowing as far westward as the Goongra river, northward to the vicinity of Veyre, and eastward to the Sindree Lake (Nelson, 1846). On 31 October 1847 the small island of Kondul near Little Nicobar was inundated (Heck, 1947; Berninghausen, 1966) by an earthquake whose magnitude could have been greater than 7.5 (Bilham et al. 2005). According to Mihir Guha, former Director General of the India Meteorological Department, a tsunami struck Sunderbans (Bangladesh) in May 1874, killing several hundred thousand people (http://www.freejournal.net) as a result of an earthquake in Bhola district. The earthquake and tsunami effect was felt in several districts, and even Kolkata felt its impact.

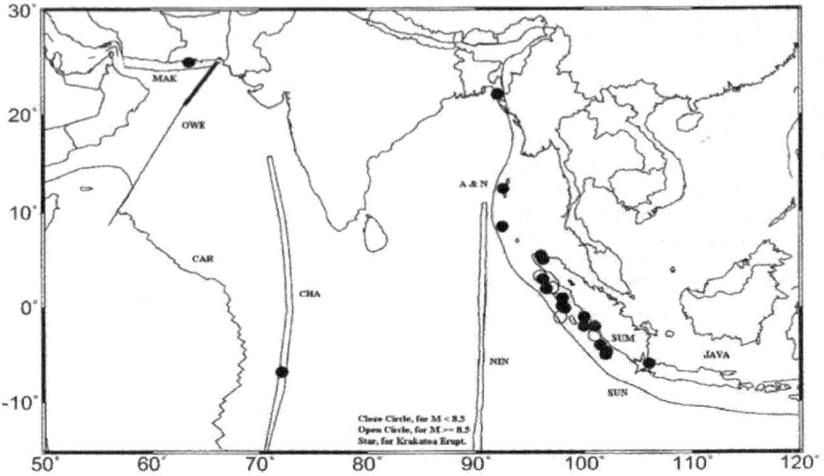

Fig. 13. Locations of tsunamis in the Indian Ocean (after Rastogi and Jaiswal, 2006). MAK - Makran Accretion Zone, OWE - Owen Fracture Zone, CAR - Carlsberg Ridge, CHA - Chagos Archipelago, A-N - Andaman and Nicobar Islands, SUM - Sumatra, NIN - Ninety East Ridge, SUN - Sunda Subduction Zone and JAVA-Java.

Other minor tsunamis of height up to 2m hit the east coast of India in 1842 and 1861 from Sumatra, 1881 from Car Nicobar (Berninghausen, 1966), 1883 from Krakatau, 1884 from Bay of Bengal (Murty et al., 1999), 1907 from Sumatra and 1941 from Andaman. The 1881 tsunami was caused by an M7.9 earthquake in Andaman resulting in 1.2 m high tsunami. The tsunami struck Chennai on the Tamil Nadu coast and then progressed north toward Vishakhapatnam in Andhra Pradesh, the Mahanadi delta in Orissa and Pamban in the Gulf of Mannar. Finally waves less than 0.3 m high were recorded Dublet in West Bengal and Diamond Harbour (Ortiz and Bilham, 2003). The 26 June 1941 Andaman earthquake had a magnitude of 7.7 and triggered a tsunami in the Bay of Bengal (Bilham et al., 2005) with heights of the order of 0.75 to 1.25 m. This tsunami was supposedly witnessed along the eastern coast of India, killing nearly 5,000 people, but was mistaken to be a storm surge, of which no meteorological record could be traced (Murty, 1984).

The deadliest tsunami prior to 2004 in South Asia was in November 1945 (Mathur, 1998), which originated off the Makran coast of Pakistan in the Arabian Sea and caused deaths as far as Mumbai. More than 4000 people are known to have been killed on the Makran Coast by both the earthquake and the tsunami. Significantly, the tsunami is supposed to have reached a height of 17m in parts of Makran ports and caused great damage to the entire coastal region. The tsunami was also recorded at Muscat and Gwadar, Kutch in Gujarat (Pendse, 1945) and as far as Mumbai with a height of 2m.

An earthquake of magnitude 7.7 occurred in 1983, in the Chagos Archipelago of the Indian Ocean, is known to have generated a tsunami of 1.5 m height at Diego Garcia. This is probably a unique occurrence of a tsunami due to an earthquake in a divergent zone, with a normal faulting earthquake mechanism, that too in a

Table 2. List of Tsunamis that affected the Indian Ocean Region (After Rastogi and Jaiswal, 2006), *Note*: Brh=Berninghausen, O&B= Ortiz & Bilham, Mea= Murty et al, Bea= Bilham et al, KKM= K. Krishnamurty, Mac= Macurdo, NG/NO= NGDC/NOAA, Lis=Lisitzin, Mat=Mathur , Nel=Nelson, Old= Oldham, Wikip= Wikipedia

No.	Date	Location	Long	Lat	Mag	MxRup	Source
1	326 BC	Indus/Kutch	-	-	-	-	Lis(1974)
2	About 500AD	Poompuhar	79.52	11.12	-	-	Wikip
3	900AD	Nagapattinam	79.53	10.46	-	-	KKM
4	1008	IranianCoast	60	25	-	-	Mea(1999)
5	1762.04.12	BayofBengal	92	22	-	2(1)	Mat(1998)
6	1819.06.16	Kutch	26.6	71.9	7.8	-	Mac(1821)
7	1842.11.11	N.BayofBengal	90	21.5	-	(3)	Old(1883)
8	1845.06.19	Kutch	23.6	68.37	-	-	Nel(1846)
9	1847.10.31	LittleNicobar Is.	93.667	7.333	7.5 -7.9	-	Brhn(1966), Heck (1947)
10	1868.08.19	Andaman Is.	92.73	11.67	-	4	NG/NO
11	1874	Sunderbans	89	22	-	-	M Guha, Free Journal
12	1881.12.31	W. Car Nicobar	92.43	8.52	7.9	1.2	Brh(1966), O&B(2003)
13	Jan.1882	SriLanka	81.14 E	8.34	-	-	Brh(1966)
14	1883.08.27	Krakatau(Volcanoe)	105.25	-6.06	-	2	Brh(1966)
15	1884	W. BayofBengal	-	-	-	-	Mea(1999)
16	1935.05.31	AndamanNicobar	-	-	7.5	(1)	NG/NO
17	1935.11.25	AndamanNicobar	94	5.5	6.5	-	NG/NO
18	1941.06.26	Andaman Is.	92.5	12.1	7.7	1.25	Bea(2005)
19	1945.11.27	MakranCoast	63.5	25.2	8.0	17	Mea(1999)
20	1983.11.30	Chagosridge	72.11	-6.85	7.7	1.5(2)	NG/NO
21	2004.12.26	Sumatra	95.95	3.31	9.3	30	NG/NO

diffuse deformation zone, rather than on a well defined plate boundary. This event probably adds to our list of potential tsunami producing zones around the Indian Ocean.

10 Seismic characterization of potential tsunamigenic source zones

A detailed investigation of earthquakes and their source mechanisms surrounding the Indian sub-continent has brought out the potential tsunamigenic zones for future consideration (Fig. 14). The major plate boundaries of the Indian plate region are the Himalayan front in the north separating the India and Eurasia plates, the Burma-Andaman arc in the east separating the India and Burma plates, the diffuse Indian Ocean deformation zone to the south separating the India and Australia plates, the mid-oceanic ridge in the southwest separating the India and the Africa plates, and finally the Heart-Chaman fault on the west separating the India and Arabia plates. Additionally we have the Australia-Sunda plate boundary near Sumatra in the southwest and the Africa-Arabia plate boundary near Makaran coast in the west that need to be considered. Importantly, an earthquake has to occur in the oceanic or coastal region and requires considerable magnitude, at least 6.5 to 7.0, to be able to produce a tsunami. Also, it is found that a subduction zone or a convergent plate boundary is the most conducive for production of a tsunamigenic earthquake. On the basis of these criteria, the following regions are identified as capable of generating a tsunamigenic earthquake in future (Fig. 14):

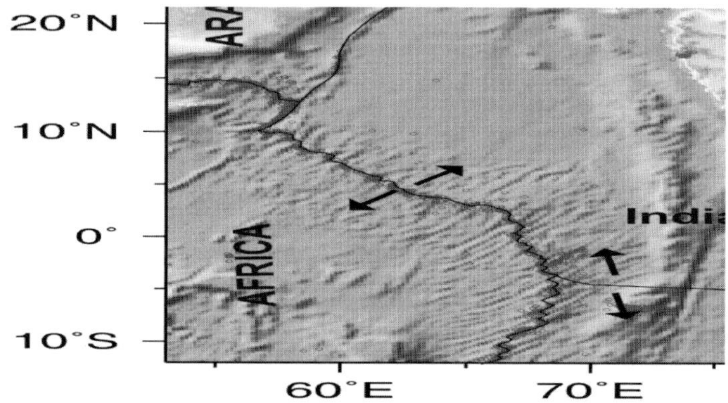

Fig. 14. Identification of potential tsunami zones around the Indian plate region through seismic source characterization. The encircled zones represent the most potential zones namely (I) Andaman-Sumatra zone (II) Makaran subduction zone and (III) Indian Ocean deformation zone and (IV) Offshore Pondicherry.

10.1 Burma-Andaman-Sumatra arc region

The Andaman-Sumatra region is a well known subduction zone that has already produced the most devastating tsunamis for the Indian region. The significant earthquakes in this region occurred earlier in 1797 (M8.4), 1833 (M9.0), 1881 (M7.9), 1907 (M7.8), 1935 (M7.7), 1941 (M7.7), 2000 (M7.9) and 2002, prior to the great devastating earthquakes of 2004 (M9.1) and 2005 (M8.6).

Interestingly, the Burmese arc region further north, does not appear potentially very tsunamigenic as discussed below. While most of it is on land, the southern part of the Burmese arc encompassing the oceanic region in the Bay of Bengal, is suggested to be a region with a low tsunami potential. It has been demonstrated that this region ceases to be an active subduction zone (Rao and Kumar, 1999), with thrust fault earthquakes of lower magnitudes occurring only below a depth of 90 km (Fig. 15 [3]) due to slab detachment in response to gravitational loading on an overturned lithospheric slab with an arc-parallel motion (Rao and Kalpna, 2005).

[3] Editor's note: Due to conversion of color figures to b/w the color code is not visible. Authors may be contacted for the original figures.

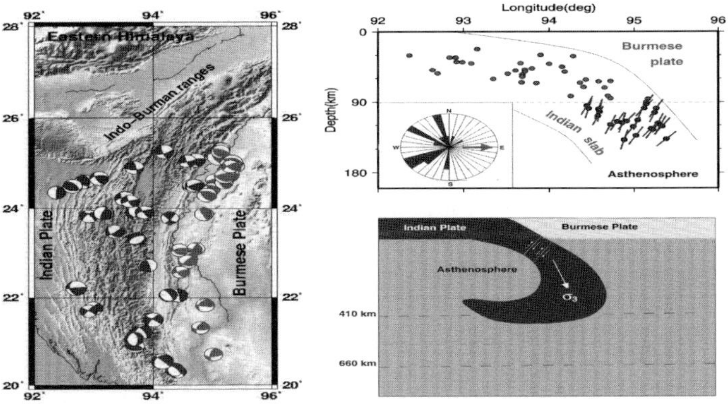

Fig. 15. (left) Map view of Burmese arc, the eastern margin of the Indian plate, indicating focal mechanism solutions in the shallow (red) and the deeper (blue) ranges. (top) Earthquakes plotted in an EW depth section across the Burmese arc indicating the dip angles of the predominantly reverse fault earthquake mechanism solutions along with their azimuthal variation (inset). (Bottom) Evidence for an overturned slab at 410 km discontinuity indicating the process of slab detachment at the base of the lithosphere, in a direction away from that of the slab dip, which explains the reverse fault mechanism earthquakes (after Rao and Kalpna, 2005).

10.2 Makaran subduction zone

The Makran Subduction Zone (MSZ) in the Northern Arabian Sea (Figs. 16 & 17) has been formed by the northward under-thrusting of the Arican plate with respect to the Arabian plate further north at a very shallow angle of about 20 degrees. The east-west trending MSZ is more than 800 km-long. Large earthquakes along the Makran Subduction Zone (MSZ) have generated destructive tsunamis in the past (Berninghausen, 1966). The most significant one was the 28 November 1945 earthquake of M7.8 with a focal depth was 25 km. Although the historic record is incomplete, it is believed that tsunamis from this region had significant impact on several countries bordering the Northern Arabian Sea and the Indian Ocean. The tsunami generated along the MSZ on November 28, 1945 was responsible for great loss of life and destruction along the coasts of Pakistan, Iran, India and Oman (Qureshi, 2006; Mokhtari and Farahbod, 2005; Pararas-Carayannis, 2006). The potential for future tsunami generation along the Makran coast of Pakistan have been dealt with in great detail by Pararas-Carayannis, 2006, based on a thorough review of recent geophysical surveys and seismic data.

10.3 Indian Ocean deformation zone

This region south of the Indian peninsula forms the India-Australia plate boundary, which is believed to be nascent and diffuse. The region depicts both convergence in the east and divergence in the west as the Australian plate overrides the Indian plate in an anti-clockwise direction (Fig. 14). This region has already experienced an M7.7 earthquake on the Chagos Archipelago in 1983, generating a tsunami with 1.5 m

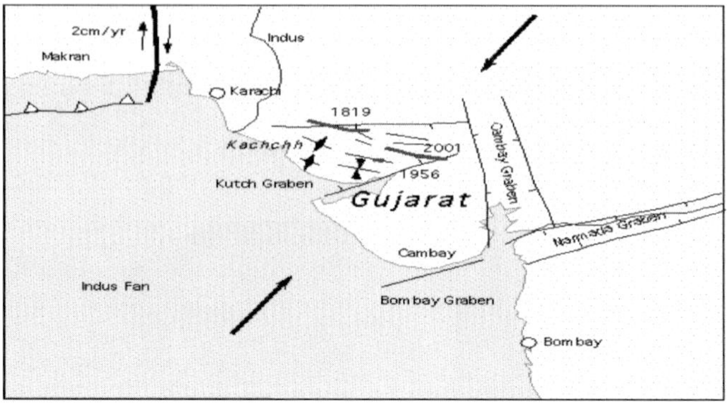

Fig. 16. Tectonics of the Makaran subduction zone indicating the Kutch, Bombay, Cambay and Namacia graben regions (after Pararas-Carayannis, 2006).

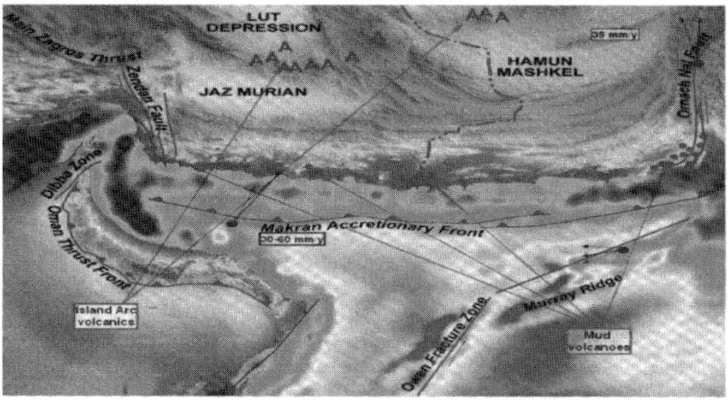

Fig. 17. The Makran accretionary prism and the zone of tectonic subduction in the northern Arabian Sea (after Pararas-Carayannis, 2006).

wave height at Diego Garcia (Rastogi and Jaiswal, 2006). Normally earthquakes with reverse fault mechanism that occur along subduction zones are the most potential earthquakes that comprise both, the energy levels required for deforming the ocean floor, and the vertical directivity required for displacing huge amounts of the ocean waters that can generate a tsunami. However, often it is possible that normal fault earthquakes can generate the requisite water displacement to generate a tsunami as in the case of the 1983 earthquake, marking this region as potentially tsunamigenic in future. Another potential tsunamigenic source zone in the Indian Ocean is to the east where a nascent, diffuse convergent plate margin is in the offing during the last few million years (Gordon et al., 1990). Although, considering the very low plate velocities of 0.5 to 1 .0 cm/y, the probability of a large earthquake in this region appears slim, it would be worth taking this into account in view of its proximity to Sri Lanka and the southern tip of India. Interestingly, further south of this region, an earthquake of magnitude 7.9 did occur on 18 June 2000.

10.4 Pondicherry Offshore region

An earthquake of magnitude M 5.5 occurred on 25 September 2001 off Pondicherry, on the east coast of India, which is a considerable event for the southern Indian peninsular region, considering the exisiting levels of observed seismicity. Located at 11.95 N and 80.23 E, the epicentre falls over the continental slope, about 40 km off Pondicherry at 1900 m water depth with a focal depth of 10 km (Fig. 18). The epicentre of the earthquake falls over a reported fault zone interpreted as an offshore extension of a lineament (Murty et al., 2002). The focal mechanism solution of the earthquake suggests thrust faulting with a small strike-slip component on a preferred fault plane striking east-west. Although the likelihood of a much larger earthquake capable of producing a tsunami appears low, this is one of the potential regions that require a careful consideration since a tsunami, if generated would be very local and hence damaging.

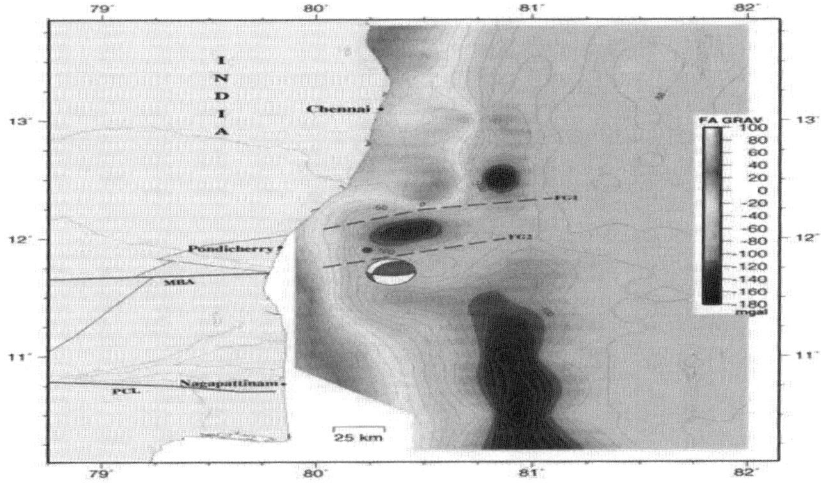

Fig. 18. The Pondicherry offshore region along the east coast of India, along with the focal mechanism solution of the 25 September 2001 earthquake of magnitude 5.5. Also shown is the free-air gravity anomaly and the inferred faults extending into the sea (after Murty et al., 2002).

11 Tsunami hazard in the Indian Ocean and approaches for mitigation

The 26 December 2004 Tsunami took the world by surprise with its manifestation in the most unexpected part of the globe. This apart, the absence of a proper Tsunami warning system in the Indian Ocean made it almost impossible to forecast its arrival. While precise prediction of earthquakes is almost impossible anywhere in the world, the prediction of arrival times of a tsunami at different places is very much within reach. Particularly, in the case of the Indian sub-continent with the most potential tsunamigenic earthquakes occurring in the Sumatra region, a clear 2 hours gap is available for proper assessment and authentic warning to be issued, unlike in the Pacific Ocean where often the time gap is only of the order of a few minutes.

The occurrence of such a gigantic earthquake on 26 December 2004 has caused considerable concern among scientists with regard to the future seismic hazard potential of the region. Bilham (2005) elaborated on the concept of seismic gaps all along the convergent Indian plate margins that draw our attention to several great earthquakes that are over due. McCloskey et al., 2005 used the slip distribution to calculate the stress perturbation tensor subsequent to the 2004 Sumatran earthquake. Results showed a stress increase of up to 5 bars in the 50 km of the Sunda trench next to the rupture zone (Fig. 19). Another great earthquake of magnitude 8.6 did occur around the Nias islands about 3 months later. The unusual proportions of these 2 great earthquakes occurring quickly in the same region eludes the scope of any reliable expectation of things to come in the near future, as aftershocks of large magnitude continue to occur in large numbers even two years after the worst ever tsunamigenic earthquake struck the world.

One of the prime drawbacks in the Indian Ocean region is the lack of regional cooperation for joint seismological station networks, with real time data acquisition, quick detection of earthquakes, identification of their size, and estimation of the

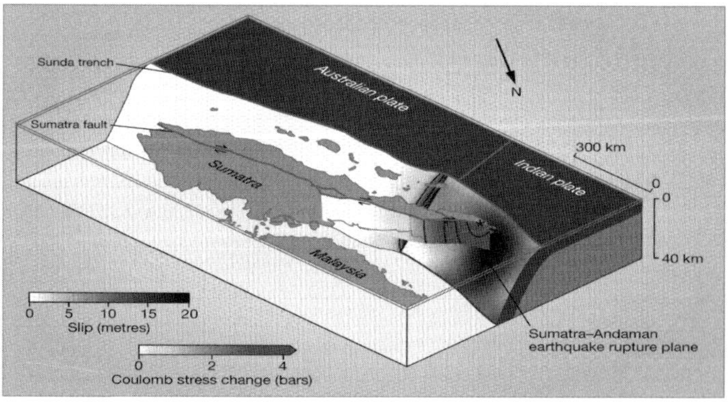

Fig. 19. The Sumatran subduction zone with the three-dimensional stresses projected on to structural geometry and geography of the region (after McCloskey et al., 2005). Grey-scale values on the rupture plane represent the amount of slip in meters experienced on the southernmost 450 km of the Sumatra-Andaman earthquake. Colour-scale values represent the co-seismic stress changes on the Sunda-trench subduction zone and the Sumatra fault. Stress contours (in black) show 2-bar intervals, starting from a maximum of 8 bars on both faults. Essential features of the calculated stresses are robust to changes in the slip distribution in recent long-period inversions, which show continuation of slip to the north for a total rupture length of about 1,200 km. Black asterisk indicates location of the devastated Indonesian city of Banda Aceh.

source parameters. These are essential for development of a Regional Earthquake Alert and Tsunami Warning System.

The following steps need to be adopted for future tsunami hazard mitigation in the Indian Ocean region:

11.1 Installation of broadband seismological networks in the Indian Ocean

The most crucial requirement for a real time tsunami warning system in the Indian Ocean is the deployment of a seismograph station network linked through satellite connection for real time detection, analysis and parameter estimation of large impending earthquakes capable of generating a tsunami. It is therefore the need of the hour to strike regional cooperation amongst the effected countries in the Indian Ocean for setting up of such a joint network.

11.2 Real time earthquake data retrieval through satellite communication

Majority of tsunamis are caused by large earthquakes that occur in off coast areas. Hence it is most pertinent to retrieve earthquake data in real time, through satellite communications to be further processed automatically for generating a possible earthquake alert.

11.3 Automatic analysis for earthquake location and magnitude estimation

Standard approaches exist for automatic detection of an event and subsequent processing using standard software, as used by several global agencies like the United States Geological Survey (USGS) and the European Mediterranean Seismological Center (EMSC). These programs are capable of providing earthquake source parameters in real time and to alert the concerned agencies within a few minutes. The information comprising magnitude, location, distance, focal depth and focal mechanism can also be assessed to decide the possibility of a tsunami generation due to the earthquake.

11.4 Installation of tide gauge systems

In addition to the seismological network, it is desirable to have a tide gauge network to compliment the efforts in tsunami detection subsequent to the earthquake. The tide gauge data which will be available online through VSAT would enable confirmation of tsunami generation conclusively.

11.5 Earthquake alert and Tsunami warning system

Based on the above inputs, it would be possible to activate an earthquake alert and tsunami warning to all the concerned agencies. The latter, however, would be done after a very careful analysis based on the obtained source parameters and also the computed models of tsunami propagation at the data base.

11.6 Acquisition of coastal bathymetry and geomorphological data

Currently the coastal bathymetry and geomorphological data in the Indian Ocean region is very meager and major efforts need to be launched by the countries in the Indian Ocean region in this direction. Detailed modeling studies with the above inputs would provide a clear picture of tsunami propagation and inundation at different places along the coastlines for different scenario earthquakes.

11.7 Sensitization of people

One of the most crucial factors in tsunami related deaths has been lack of proper awareness, the classic example being the 2004 Indian Ocean tsunami where 200,000 people died even though the lag time at several places was up to 2.5 hours. Apart from a good communication network the need to sensitize people on aspects of basic awareness as well as precautionary steps, cannot be over-emphasized.

References

[1] Ammon et al. (2006) Rupture Process of the 2004 Sumatra-Andaman Earthquake. Science 308:1133-1139
[2] Banerjee, P, Politz FF, Burgman R (2005) The size and duration of the Sumatra-Andaman earthquake from far-field static offsets. Science 308:1769-1772
[3] Bendick R, Bilham R (1999) A search for buckling of the SW Indian coast related to Himalayan Collision. In: Macfarlane A, Sorkhabi RB, Quade, J (eds) Himalaya and Tibet: Mountain Roots to Mountain Tops. Geol Soc Amer Special paper 328:313-322
[4] Berninghausen, WH (1966) Tsunamis and Seismic Seiches reported from regions adjacent to the Indian Ocean. Bull Seism Soc Am 56: 69-74
[5] Bijwaard, H, Spakman W, Engdahl ER (1998) Closing the gap between regional and global travel time tomography. J Geophys Res 103:30055-30078
[6] Bilham, R, Engdahl R, Feld N, Satyabala SP (2005) Partial and complete rupture of the Indo-Andaman plate boundary 1847-2004. Seism Res Lett 76:299-311
[7] Bilham, R (2006) Dangerous tectonics, fragile buildings, and tough decisions. Science 311: 1873-1875
[8] Borges, JF, et al, (2005) http://www.emsccsem.org/Doc/Sumatra-ruputre-process-2004.pdf
[9] Briggs, et al. (2006) Deformation and slip along the Sunda megathrust in the Great 2005 Nias-Simeulue earthquake. Science 311:1897-1901
[10] Catherine, JK, Gahalaut VK, Sahu VK (2005) Constraints on rupture of the December 26, 2004, Sumatra earthquake from far-field GPS observations. Earth Planet Sci Lett 237:673-679
[11] Chamot-Rooke, N, Jestin F, deVoogd B (1993) Intraplate shortening in the central Indian Ocean determined from a 2100 km long north-south deep seismic reflection profile. Geology 21:1043-1046

[12] Chandra, U (1984) Tectonic segmentation of the Burmese-Indonesian arc. Tectonophysics 105:279-290

[13] Curray, JR, Moore D, Lawyer L, Emmel F, Raitt R, Henry M, Kieckhefer R (1979) Tectonics of the Andaman sea and Burma. In: Watkins JS, Montadert L, Dickerson P (eds) Geological and Geophysical Investigations of Continental margins. AAPG Memoir 29:189-198

[14] DeMets, C, Gordon RG, Argus DF Stein S (1990) Current plate motions. Geophys J Int 101:425-478

[15] DeMets, C, Gordon RG, Vogt P (1994) Location of the Africa-Australia-India triple junction and motion between the Australian and Indian plates: Results from an aeromagnetic investigation of the central Indian and Carlsberg ridges. Geophys J Int 119:893-930

[16] Eittreim, S, Ewing J (1972) Mid-plate tectonics in the Indian ocean. J Geophys Res 77: 6413-6421

[17] Fine, IV, Rabinovich AB, Thomson RE (2006) The dual source region for the 2004 Sumatra tsunami. Geophys Res Lett 32:L16602 doi: 10.1029/2005GL023521

[18] Fitch, TJ (1972) Plate convergence, transcurrent faults and internal deformation adjacent to SE Asia and Western Pacific. J Geophys Res 77:4432-4460

[19] Gahalaut, VK, Nagarajan B, Catherine JK Kumar S (2006) Constraints on 2004 Sumatra-Andaman earthquake rupture from GPS measurements in Andaman-Nicobar Islands. Earth and Planetary Science Letters 242:365-374

[20] Geller, CA, Weissel JK, Anderson RN (1983) Heat transfer and intraplate deformation in the central Indian ocean. J Geophys Res 88:1018-1032

[21] Gordon, RG, DeMets C, Argus DF (1990) Kinematic constraints on distributed lithospheric deformation in the equatorial Indian ocean from present motion between the Australian and Indian plates. Tectonics 9:409-422

[22] Han et al. (2006) Crustal Dilatation Observed by GRACE after the 2004 Sumatra-Andaman Earthquake. Science 313:658-662

[23] Hashimoto, MN, Choosakul M, Hashizume S, Takemoto H, Takiguchi Y, Fukuda, Frjimori K (2006) Crustal deformations associated with the great Sumatra-Andaman earthquake deduced from continuous GPS observation, Earth Planets and Space 58:127-139

[24] Heck, NH (1947) List of seismic sea waves. Bull Seism Soc Am 37:269-286

[25] Ji, C (2004) http://www.gps.caltech.edu/jichen /Earthquake/ 2004/aceh/aceh.html

[26] Kreemer, C, Blewitt G, Maerten F (2006) Co- and postseismic deformation of the 28 March 2005 Nias Mw 8.7 earthquake from continuous GPS data. Geophys Res Lett 33:L07307, doi:10.1029/2005GL025566

[27] Kruger, F, Ohnberger M (2005) Spatio-temporal source characteristics of the 26 December 2004 Sumatra earthquake as imaged by teleseismic broadband arrays. Geophys Res Lett 32: L24312, doi:10.1029/2005GL023939

[28] Kumar, MR, Rao NP (1995) Significant trends related to the slab seismicity and tectonics in Burmeese arc region from Harvard CMT solutions. Phys Earth and Planet Inter 90:75-80

[29] Kumar, MR, Rao NP Chalam SV (1996) A Seismotectonic study of the Burma and Andaman arc regions using Centroid Moment Tensor data. Tectonophysics 253:155-165

[30] Lay et al. (2005) The Great Sumatra-Andaman earthquake of 26 December 2004. Science 308:1127-1133
[31] Le Dain, AY, Tapponier P, Molnar P (1984) Active faulting and tectonics of Burma and surrounding regions. J Geophys Res 89:453-472
[32] Levchenko, OV (1989) Tectonic aspects of intraplate seismicity in the northeastern Indian ocean. Tectonophysics 170:125-139
[33] Levchenko, OV, Ostrovsky AA (1992) Seismic seafloor observations: a study of anomalous intraplate seismicity in the northeastern Indian ocean. Phys Earth Planet Inter 74:173-182
[34] Lisitzin, E (1974) Sea Level Changes. Elsevier, Oceanographic Series New York
[35] Macmurdo, Captain (1821) Account of the earthquake which occurred in India in June 1819. Edinburgh Phil J 4:106-109
[36] Mathur, SM (1998) Physical Geology of India. National Book Trust of India, New Delhi
[37] Maung, H (1987) Transcurrent movements in the Burma-Andaman sea region Geology 15:911-912
[38] McCloskey et al. (2005) Earthquake risk from co-seismic stress. Nature 434:291
[39] Minster, JB, Jordon TH, Molnar P, Haines E (1974) Numerical modelling of instantaneous plate tectonics. Geophys J R Astron Soc 36:541-576
[40] Minster, JB, Jordon TH (1978) Present-day plate motions. J Geophys Res 83: 5331-5354
[41] Mokhtari, M, Farahbod AM (2005) Tsunami Occurrence in the Makran Region. Tsunami Seminar Tehran 26 February 2005
[42] Mukhopadhyay, M, Das Gupta S (1988) Deep structure and tectonics of the Burmese arc : constraints from earthquake and gravity data. Tectonophysics 149:299-322
[43] Murty, TS, Bapat A, Prasad V (1999) Tsunamis on the coastlines of India. Science of Tsunami Hazards 17:167-172
[44] Murty, TS (1984) Storm surges- meteorological ocean tides. Bull. Fisheries Research Board of Canada Ottawa
[45] Murty, GPS, Subrahmanyam AS, Murthy KSR, Sarma KYLNS (2002) Evidence of fault reactivation off Pondicherry coast from marine geophysical data. Current Science 83:1446-1449
[46] Nelson, Captain (1846) Notice of an earthquake and a probable subsidence of the land in the district of Cutch, near the mouth of Koree, or the eastern branch of the Indus in June 1845. Geol Soc London Quart J 2:103
[47] Neprechnov, YP, Levchenko OV, Merklin LR, Sedov VV (1988) The structure and tectonics of the intraplate deformation area in the Indian Ocean. Tectonophysics 156:89-106
[48] Ni, JF, Speziale MG, Bevis M, Holt WE, Wallace TC, Seager WR (1989) Accretionary tectonics of Burma and the three dimensional geometry of the Burma subduction. Geology 17:68-71
[49] Ortiz, M, Bilham R (2003) Source area and rupture parameters of the 31 December 1881 Mw = 7.9 Car Nicobar earthquake estimated from tsunamis recorded in the Bay of Bengal. J Geophys Res 108 B4 2215 doi: 10.1029/ 2002JB001941
[50] Pararas-Carayannis (2006) The potential of tsunami generation along the Makran Subduction Zone in the northern Arabian Sea. Case Study: The Earthquake And Tsunami Of November 28, 1945. Science of Tsunami Hazards 24:358

[51] Paul, J, Burgmann R, Gaur VK, Bilham R, Larson KM, Ananda MB, Jade S, Mukul M, Anupama TS, Satyal G, Kumar D (2001) The motion and active deformation of India. Geophys Res Lett 28:647-650
[52] Pendse, CG (1945) The Makran earthquake of the 28th November 1945. India Met. Dept. Scientific Notes 10:141-145
[53] Qureshi, RM (2006) Vulnerability of Pakistan coast to tsunami. Possible applications/Role of nuclear techniques RCARO workshop (20 - 24 February, 2006) Daejon Korea
[54] Rao, NP, Kumar MR (1996) Deformation tectonics of the diffuse Indo-Australian plate boundary using centroid moment tensor data. Current Science 70:238-242
[55] Rao, NP (1999) Active tectonics of the margins and stable parts of the Indian shield. Ph.D. Thesis, Tokyo Univ.
[56] Rao, NP, Kumar MR (1999) Evidences for cessation of Indian plate subduction in the Burmese arc region. Geophys Res Lett 26:3149-3152
[57] Rao, NP, Kalpna (2005) Deformation of the subducted Indian lithospheric slab in the Burmese arc. Geophys Res Lett 32 L05301, doi: 10.1029/ 2004GL022034
[58] Rao, NP, Chary AH (2005) What caused the Great Sumatran Earthquakes of 26 December 2004 and 28 March 2005? Current Science 89:449-452
[59] Rao, NP, Catherine JK, Kumar RP (2006) Plate junctions junctions and great earthquakes: The Sumatran tectonic puzzle. Geophys Res Lett (submitted)
[60] Rastogi, BK, Jaiswal RK (2006) A catalog of tsunamis in the Indian Ocean. Science of Tsunami Hazards 25:128-143
[61] Satyabala, SP, (1998) Subduction in the Indo-Burma region : Is it still active? Geophys Res Lett 25:3189-3192
[62] Sieh, K (2005) What happened and what's next?. Nature 434:573-574
[63] STAAD Pro (2004) A Structural Analysis and Design Software. Research Engineers International
[64] Stein, S, Okal E (1978) Seismicity and tectonics of the Ninety east ridge area Evidence for internal deformation of the Indian plate. J Geophys Res 83:2233-2245
[65] Stein, S, Gordon RG (1984) Statistical tests of additional plate boundaries from plate motion inversions. Earth Planet Sci Lett 69:401-412
[66] Stein, S, Okal E (2005) Speed and size of the Sumatra earthquake. Nature 434: 581-582
[67] Sykes, LR, (1970) Seismicity of the Indian ocean and possible nascent island arc between Ceylon and Australia. J Geophys Res 75:5041-5055
[68] Tsai et al. (2005) Multiple CMT source analysis of the 2004 Sumatra earthquake. Geophys Res Lett 32:L17304 doi:10.1029/2005GL023813
[69] USGS (2006): http://earthquake.usgs.gov/regional/world/10-largest-world.php
[70] Verma, RK, Mukhopadhyay M, Ahluwalia MS (1976) Earthquake mechanisms and tectonic features of northern Burma. Tectonophysics 32:387-399
[71] Vigny, C, Socquet A, Rangin C, Chamot-Rooke N. Pubellier M, Bouin M, Bertrand G, Becker M (2003) Present-day crustal deformation around Sagging fault, Myanmar. J Geophys Res 108:ETG 6-1-6-10
[72] Vigny, C, Simons WJF, Abu S, Ronnachai Bamphenyu, Chalermchon Satirapod, Nithiwatthn Choosakul, Subarya C, Socquet A, Omar K, Abidin

HZ, Ambrosius BAC (2005) Insight into the 2004 Sumatra-Andaman earthquake from GPS measurements in southeast Asia. Nature 436:201-206 doi:10.1038/nature03937

[73] Walker, KT, Ishii M, Shearer PM (2005) Rupture details of the 28 March 2005 Sumatra Mw 8.6 earthquake imaged with teleseismic P waves. Geophys Res Lett 32:L24303, doi:10.1029/2005GL024395

[74] Weissel, JK, Anderson RN, Geller CA (1980) Deformation of the Indo-Australian plate. Nature 287:284-291

[75] Wiens, DA (1986) Historical seismicity near Chagos: A complex deformation zone in the equatorial Indian ocean. Earth Planet Sci Lett 76:350-360

[76] Yagi, Y (2004) http://iisee.kenken.go.jp/staff/yagi/eq/ Sumatra2004/Sumatra2004.html

[77] Yamanaka, Y (2004) http://www.eri.u-tokyo.ac.jp/sanchu/ Seismo-Note/2004/EIC161e.html

[78] Yuan, X, Kind R, Pedersen HA (2005) Seismic monitoring of the Indian Ocean tsunami. Geophys Res Lett 32 :L15308, doi:10.1029/2005GL023464

Index

Printing: Krips bv, Meppel
Binding: Stürtz, Würzburg